THIN SHELL
CONCRETE
STRUCTURES

THIN SHELL CONCRETE STRUCTURES

DAVID P. BILLINGTON
Professor of Civil Engineering
Princeton University

First Edition, 1965
Second Edition, 1982

McGRAW-HILL BOOK COMPANY
New York   St. Louis   San Francisco   Auckland
Bogotá   Hamburg   Johannesburg   London   Madrid
Mexico   Montreal   New Delhi   Panama   Paris
São Paulo   Singapore   Sydney   Tokyo   Toronto

Library of Congress Cataloging in Publication Data
Billington, David P.
Thin shell concrete structures.
"1st edition, 1965".
Includes bibliographical references and index.
1. Shells (Engineering)   2. Reinforced concrete
construction.   I. Title.
TA660.S5B52 1982   624.1′83462   81-5073
                                           AACR2

ISBN 0-07-005279-4

1 2 3 4 5 6 7 8 9 0   KPKP   8 9 8 7 6 5 4 3 2 1

The editors for this book were Joan Zseleczky and Iris
Cohen, the designer was Elliot Epstein, and the production
supervisor was Sally Fliess. It was set in Baskerville by
Bi-Comp, Incorporated.

Printed and bound by The Kingsport Press.

CONTENTS

4 ANALYSIS AND DESIGN OF DOMES AND SHELL WALLS
139

5 ANALYSIS OF CIRCULAR CYLINDRICAL SHELLS
184

6 DESIGN OF CYLINDRICAL SHELL ROOFS
218

7 MEMBRANE THEORY FOR TRANSLATION SHELLS OF DOUBLE CURVATURE
258

8 BEHAVIOR AND DESIGN OF FOLDED PLATES
291

9 STABILITY AND SAFETY

311

10 ROOF DESIGN

330

PREFACE TO THE SECOND EDITION

In the 17 years since the first edition, at least five new developments affecting thin shell concrete structures make a second edition desirable. First, the pioneering designers have mostly either died or retired from design so that it is now possible to present a coherent *historical perspective* on the first 60 years of concrete shell design. More importantly, this new interest in history carries with it the emphasis on long-term full-scale structural behavior as a prerequisite to future design. Studies of the recent past permit a fuller view of design and emphasize to future designers the great extent to which successful structures were built without detailed mathematical theories but with a sound engineering foundation in structures, structural behavior, and construction practice. Because of the recent realization by the profession of the importance in the record of built shells, I have rewritten the first chapter to include a historical perspective and to emphasize the simplified but realistic overall behavior of the various shell systems. I have replaced the general mathematical theory by more specific developments in the chapters on analysis (2, 3, 5, and 7) and added an appendix summarizing general theory.

The second new development has been the widespread use of large-scale *computer programs* for structural analysis. This has reduced reliance on classical methods, provided insight into previously unsolved problems, and held out the hope for automatic optimized design. Although classical methods are no longer practical for such systems as barrel shells, they are still useful as guides to understanding behavior, and thus I have kept some of the developments in Chapters 5 and 6 as before. Still, in Chapter 6, a new section on computer analysis of barrels provides an introduction to current practice. With wind-loaded shell walls and gravity-loaded hyperbolic paraboloids, however, the computer has been more than a convenience. For the former, it provides a means of comparing bending theory with membrane theory and for providing a much deeper insight into how complex loads are carried by various different forms. Design by membrane theory provided essentially a correct basis for design in the hands of experienced shell designers; but bending theory gotten from computer-based numerical solutions gives the inexperienced engineer the opportunity to explore, through harmonic analysis, how forms like hyperboloids carry loadings that vary substantially over the shell surface. A large part of Chapter 2 deals with such questions of behavior and draws heavily on computer-based solutions. For hyperbolic paraboloids, the bending theory by classical means was never able to clarify the behavior of commonly built systems, and here the computer was essential not just in providing insight but also in removing a number of misconceptions about structural response. Thus, Chapter 7 describes in some detail the recent research which has clarified behavior.

Finally, the computer has held out for some the hope that it can be used to find forms, to replace a central part of the design. Until now such ideas, while suggestive, are hardly definitive and even have the potential for leading inexperienced designers along the wrong path. Pioneers like Maillart, Freyssinet, Torroja, Nervi, Finsterwalder, and Candela all got forms by another route—one which, in the words of the most theoretically inclined of that group (Torroja), was neither purely rational nor purely visual (*The Structures of Eduardo Torroja,* New York, 1958, p. 7)

> . . . but rather both together. The imagination alone could not have reached such a design unaided by reason, nor could a process of deduction, advancing by successive cycles of refinement, have been so logical and determinate as to lead inevitably to it. . . .

In short, the best designs have always been personal choices by designers who are well grounded in structural analysis but at the same time have strong aesthetic sensitivities. The contemporary example of Heinz Isler serves to show how a well-trained engineer can develop a variety of forms in economical shells without using the computer. This revision, therefore, does not centrally focus on the computer, although its insights are incorporated where they have design significance. All engineers in the late twentieth century need to know the computer well; all designers need to keep from relying on it for their basic structural experience.

One of the best modern examples of this problem with the computer arose in the third new development in thin shells since the first edition of this book: the rapid evolution of very high *natural draft cooling tower thin shells*. The greatest impetus to their study came in November 1965 with the collapse of three out of eight 375-ft-high, 5-in.-thick towers at Ferrybridge, England, in a moderate (by U.S. standards) windstorm. Research began immediately, and the story of that research interpretation and the subsequent failure of another tower in Scotland designed on the basis of that interpretation makes up a part of the new Chapter 9. Also, the Ferrybridge experience stimulated long-overdue studies of full-scale wind loadings on such high towers. The results of these studies are given in Chapter 2. Because the discussion on shell walls gives much more on basic loadings and structural behavior, I have put it in Chapter 2 (where it previously was in Chapter 3) and put the discussion of domes in Chapter 3. This exchange makes sense on theoretical grounds as well since cylinders, which form the main part of the discussion on shell walls, are the simplest of all shell geometries. Moreover, the axisymmetrical bending theory for cylinders has always been used as the design basis for domes and should come first. Thanks to Professor Gaylord, editor of the *Structural Engineering Handbook,* I now recognize that the same tables given for cylinders can be used for domes—another argument for putting cylinders first.

The fourth new development has come in *hyperbolic paraboloids*. These visually appealing shells stimulated designers in the 1950s to explore their uses in a variety of different forms. Unfortunately, the membrane theory was the only practical basis for analysis. When used by engineers not already experienced in the full-scale structural behavior of thin shells, this simple theory gave at times misleading results. A small number of such shells failed and a larger number experienced difficulties. It became apparent that the behavior of all such shells could not be represented by the membrane theory and that this general geometry needed to be studied form by form. Here the numerical results from computer-based analyses helped define the problems. Even more important was the close observation of existing hyperbolic paraboloids and the realization that they could be understood by simplified ideas. For example, for the groined vault, the bending study gives results very close to those from the membrane theory, whereas for the gabled shell, the membrane theory gives highly misleading results even though experienced designers would not be misled. Therefore, in Chapter 1, hyperbolic paraboloids are introduced not as one general form but as a series of different types, each behaving differently. In Chapter 7, I have added a detailed review of the gabled shell to show its behavior and to suggest how successful designing has been done. Finally, Chapter 9 includes discussion on the failure of a gabled shell. On the basis of the experience of the last 17 years, these impressive forms appear to hold out great promise for successful design once the profession recognizes the origins of past difficulties and sees how they can be overcome.

The past difficulties with cooling towers, hyperbolic paraboloids, and some other isolated shell examples have led to the fifth and last major development over the past 17 years, a clearer understanding of the *safety of concrete shells*. Much full-scale experience and related research has led to clearer pictures of the buckling capacity of concrete domes, cylinders, and hyperboloids; of the role of imperfections in hyperboloid behavior; of cracking and yielding in barrel shells and cooling towers; and of creep and large deflections in flat, thin-shell roofs. All of these questions are taken up in a new chapter on safety.

I would like to have added a chapter on aesthetics, but my ideas on the subject are not yet in a form suitable for this revision. A new generation of designers has taken over from the pioneers. While the works of these designers are not yet in a form suitable for summary, this group does show the future promise for thin shell concrete structures. Some have assumed during the past few years that thin shells are going out of fashion and that they will not be built in great numbers in the future. Generally, three reasons exist for this pessimism: first, that their analysis and design are too complex and expensive for most projects; second, that their forms are too expensive to build in high-cost labor economies; and third, that they visually have lost their appeal to the general public and to the architect. While we could easily argue against each of these reasons, the most stunning refutation of them all stands in the works of a few contemporary designers like Heinz Isler and John Christiansen, whose recent designs clearly overcome any general objections to thin shells. In Switzerland, a country where the cost of building is similar to that in the United States, Heinz Isler has designed over 250 thin shell projects that have not only been built but have been extraordinarily thin and economically competitive. Christiansen's design for the immense 1975 Kingdome, built at less per seat cost than that of the Astrodome, shows thin shell concrete structures to have the potential for even the greatest of spans. These and other designers, often working under the same difficulties as their predecessors Maillart, Nervi, Torroja, and Candela, will like the earlier designers show how new forms of great efficiency, economy, and elegance can arise in the hands of designers who focus on structural behavior, construction practice, and aesthetic forms. To help clarify the works of Isler and Christiansen, I have added a final chapter on roof design which is intended to be suggestive of the future.

In addition to the acknowledgments made in the first edition, I am greatly in the debt of a number of engineers whose ideas and works have helped shape this second edition. As with the first edition, Anton Tedesko has provided documents and counsel; in addition, for this edition he has taught me much about the thin shell history which appears in the rewritten first chapter.

Over the past 20 years, Alexander Scordelis and I have carried on a continuous dialogue on thin shells, and his influence permeates this edition including the simplified introductions of Chapter 1 and the reworking of sections on cylindrical shells and folded plates. John Abel and I have worked together on numerous shell problems over the past decade, especially on cooling towers. It is his finite-element analyses that underlie the major revision of Chapter 2 on shell walls and much of the new Chapter 9 on safety. To the Swiss shell designer Heinz Isler, I owe much of Chapter 10; he has shared his radically new ideas with me and has personally taken me to see many works. In the same way, John Christiansen has provided substantial help in the understanding of American practice and especially of his Kingdome design.

William Schnobrich's research, especially on hyperbolic paraboloids, has provided me with new insight as has the long design experience of Milo Ketchum. The late George Boos, chairman of the Folded Plate Committee, was both a good friend and the model of a practitioner seeking a clearer understanding of shell behavior. Adolph Bouma of the Netherlands has helped me through both his cylindrical shell research and his design ideas. Also the ideas of Stephen Medwadowski on design as well as those of Mario Salvadori have helped me think more deeply about history and aesthetics in structures.

The new sections on cooling towers have come from research sponsored largely by Research Cottrell, Inc., which has generously allowed me to publish data on its designs. I am especially indebted to Dennis Carleton-Jones, Jerry Morrow, and Ed Pavlini of Research Cottrell along with William Scriven and Luc Langeroc of Hamon Sobelco for their support. Also, grants from the National Science Foundation, with Robert Mark and with John Abel, supported work drawn on for this revised edition. Robert Mark's model tests on shells have contributed to my understanding of shell behavior and the wind-loading work of Norman Sollenberger and Robert Scanlan has influenced the rewriting of Chapter 2. I am particularly indebted to Professor Sollenberger for his continual counsel on all professional matters.

For much of the numerical work in this present edition I am indebted to former students at Princeton, especially Arthur Hedgren and Peter Darvall for their cylindrical shell studies, Mustapha Pultar and George Riera for work on folded plates, David Elms on hyperboloids, and Richard Larrabee and Ray Steinmetz on cooling towers.

Susan Gallagher did many calculations and sketches and Christine Wiita-Dworkin also helped with calculations. Peter Cole's research on cooling towers and his development of computer programs together with John Abel were invaluable to the rewriting of Chapter 2.

Tony Wang and Peter Lee helped me in theoretical questions essential for the new Chapter 9, and Harry Harris collaborated in a study of buckling tests which strengthened that same chapter. Ahmet Cakmak, as civil engineering chairman, gave me much needed support; and the ideas of my architect colleagues at Princeton, especially William Shellanan, Heath Ficklider, and Robert Geddes, have helped shape my thoughts on design.

A special tribute is due to Julian Dumitrescu who has read much of the manuscript critically and who has collaborated with me on numerous shell studies during the writing of this new edition. His design background and his deep insight into structural mechanics have been of inestimable value.

Finally, the shift in emphasis of this second edition, away from general mathematical theories and toward a more historical perspective, is due largely to the historical studies I have pursued since 1965. These have centered on the life and works of Robert Maillart, whose emphasis on simplified analyses, on well-built and competitive structures, and on the aesthetic potentials for concrete forms are reflected in the rewriting of Chapter 1 and in the new Chapter 10. Support for this research into the history and aesthetics of structures came from the National Science Foundation, National Endowment for the Humanities, and the National Endowment for the Arts. In addition, several long-term grants from the National Endowment for the Humanities, augmented by grants from the Ford, the Rockefeller, the Sloan, and the Mellon Foundations, to Robert Mark and me helped shape this perspective.

During the five years in which this revision took place, I have been graced with three excellent secretaries, Betty Mate, Thelma Kieth, and Anne Chase, each of whom has typed substantial parts of the manuscript with great skill and good humor. I am also grateful to my son, David, for preparing the complete index for this revision.

Once again my family has supported this work, and especially my wife, Phyllis, has given just the type of help I most needed.

David P. Billington

PREFACE TO THE FIRST EDITION

This book has been written to link together the mathematical theory of thin shells with the practical problems of reinforcing typical concrete structures. The focus is on large-scale structures, with considerable emphasis placed on developing a single method of analysis for all types of thin shell systems.

The basic mathematical derivations essential to studies of thin shells have been developed directly from elementary principles of statics, geometry, and material properties. This full mathematical treatment is complemented by a discussion of the physical behavior of various thin shell systems and concluded by a description of the detailed reinforcement of typical thin shell concrete structures. Thus, one may follow the general formulations of the first chapter through to their reduction to specific thin shells in succeeding chapters and on to the analysis and dimensioning of a complete system.

The book seeks both to base structural design on general mathematical analyses and to show the practical implementation of these analyses in the final dimensioning of large-scale structures. This dual emphasis was developed during my experiences as a structural designer with Roberts and Schaefer Company and has been reflected in the graduate course on thin shell structures given at Princeton University since 1959.

The book is designed both for university courses on thin shell structures and for reference by the practicing designer. Whereas the student will focus more on the general theory, the designer can work directly from the wide variety of specific thin shell systems which are analyzed and dimensioned. Much of this material has also proved useful in a graduate course for architectural students at Princeton.

For my early interest in concrete structures I am deeply indebted to the late Professor Gustav Magnel, an inspiring teacher in whose laboratory I had the privilege of working some years ago. I am equally indebted to Mr. Otto Grunewald, of Roberts and Schaefer Company, whose unique analytic skill with highly complex thin shell concrete structures shaped my approach to structural analysis, and to Dr. Anton Tedesko, also of Roberts and Schaefer Company, whose emphasis upon the integration of practical design with mathematical formulations has strongly influenced this book.

I owe much to Professor Norman J. Sollenberger, who gave valuable support and constant encouragement to this effort, and to my other present colleagues at Princeton as well as to my former associates at Roberts and Schaefer Company.

I am also grateful to the many graduate and undergraduate students at Princeton who have labored through the growth of this book. Some have contributed to the text and numerical work. I especially thank David Elms, Roderick Gibbons, Arthur Hedgren, and Mustafa Pultar, who have done many of the computations, and whose puzzled expressions have often led to useful revisions.

I express my appreciation to Princeton University for financial support in preparing this work, to the Princeton Engineering Association for a generous summer grant, and to Mrs. Robert McQuadc for her excellence in deciphering and in typing the manuscript.

Finally, I acknowledge the understanding and encouragement of my wife, to whom, being more responsible for this completed task than even she realizes, this book is dedicated.

David P. Billington

THIN SHELL
CONCRETE
STRUCTURES

1

HISTORICAL PERSPECTIVE ON THIN SHELL CONCRETE STRUCTURES

1-1 DESIGN AND HISTORY

This text is for those interested in the design of thin shell concrete structures. Such structures should be designed by engineers who have an understanding of how thin shells can be analyzed to show safe behavior under loads, can be planned for economical construction, and can be shaped for attractive appearance. Thus design properly includes attention to analysis, construction, and appearance with the primary goal of creating forms which are permanent, economical, and handsome. The two types of knowledge that inform all of the best structural designs are scientific and historical. Most of this text presents the scientific knowledge essential to design. Equally important, but vastly more difficult to define precisely, is the historical basis for design, which this chapter seeks to describe.

Scientific knowledge as defined in this text means knowing how to define structural behavior, expressed in displacements and forces, to the end of discovering how a given form carries loads safely with a minimum of materials, i.e., discovering how to design an efficient structure. Chapters 2 through 9 center on calculating displacements and forces from which come safe shell thickness and reinforcing layouts.

But science, by which we here mean primarily classical physics, seeks only to understand how given forms behave, whereas engineering design must first choose a form and it is for that activity that historical knowledge is required. For engineering structures, historical knowledge means two kinds of study. First, it means studying what has been accomplished in the past to the end of choosing forms that reflect successful experience and avoiding those that have been found defective. Second, historical knowledge means studying how successful designers in the past proceeded in the light of limited scientific knowledge, cost restrictions, and aesthetic ideals. The first study is on the evolution of form within the profession, while the second is on the development of individual style by designers. This historical perspective will therefore organize itself around the two themes of form and personality. The major shell forms, in approximate chronological order of development, are cylindrical shell walls as fluid containers, domes, barrel shells, folded plates, hyperbolic paraboloids, elliptic paraboloids, and nongeometrical shells. In describing the history of each form, this historical perspective will focus on a few pioneers.

In what follows, each major shell form will be discussed historically by reference to individual engineers and to elementary mathematical ideas. Engineers were intent upon developing new forms that would use minimum materials and would behave safely. To understand the problems that earlier engineers faced, it is essential to have a brief introduction to thin shell analysis put in an elementary mathematical form.

Safe behavior implies a scientific analysis of how the shell carries loads. We can identify three types of such an analysis. First is a very simplified one which seeks to give an overall picture and which is essential for preliminary design. This analysis usually emphasizes equilibrium only, but seeks also to include some ideas about ultimate load capacity, buckling, vibration, foundation settlements, and volume change effects such as creep, shrinkage, and temperature. Such an analysis is at times deceptively simple because proper use of it depends upon the experience of the designer. Sometimes a very simple picture does not properly represent the behavior and can give misleading results, as we shall see with hyperbolic paraboloids.

The second type of scientific analysis might be called a more detailed one from which enough information comes to check concrete dimensions and to lay out reinforcing steel. This analysis usually includes not only equilibrium but also compatibility and seeks to get stresses and displacements. Its goal is to take a predetermined form and proportion it efficiently to use as little material as possible. Often this serves as the final analysis for actual designs.

The third type of analysis, the more rigorous type, is frequently developed in research programs and finds significant use in giving insight into the limitations of the other two, less rigorous types of analyses. Sometimes, on large-scale projects, this rigorous analysis finds use, but for sizes and forms that are not unusual it is frequently too expensive or too complicated. At the same time, carefully selected results of such a rigorous analysis are invaluable for designers.

Each of these types of analyses gives results that depend heavily upon the assumptions made about the actual structure: its geometry, its material properties, and its boundary conditions. The boundaries in turn require consideration of the complete structural system, including arches, beams, columns, and foundations onto which the thin shell is built. The assumptions about geometry and materials depend strongly on construction conditions, including tolerances (imperfections) and concrete strength at decentering. Sometimes research-type analyses will give insight into these complex questions, but more frequently full-scale observations on similar shells will provide the most helpful basis for reasonable assumptions.

Finally, the aspect of design which is the most difficult to discuss in general terms is appearance. There are two general problems: The first is that many well-known thin shells were designed primarily for appearance and have consequently either not performed well under load or proved uneconomical to build. The second is that many engineers do not consider appearance to be part of engineering design. That is to say, they consider safe behavior and economical construction to be the engineer's province and appearance to be solely the architect's domain. These two general problems form the content of the last chapter in this book.

1-2 SIMPLIFIED ANALYSES

The leading idea in any simplified shell analysis is to relate the behavior to some more elementary forms such as rings, cantilevers, beams, and arches. Usually a thin shell concrete structure can be thought of as a combination of such forms, and the validity of any equivalent structure must be somehow justified: by physical testing, by more rigorous analysis, or by observations of successful full-scale behavior.

The advantage of a simplified analysis is that the designer can spend more time thinking about assumptions, about appropriate forms, and about construction all at the crucial stage of preliminary design. Time need not be focused primarily on detailed analysis. In the following sections, we present qualitative descriptions of the behavior of each of the major shell forms treated in this book. Later chapters will give the more detailed analysis for each type and will discuss limitations on the simplified pictures presented here.

The simplified analyses are all based upon considering the thin shell to be defined by its middle surface, shown in Fig. 1-1 as halfway between the inner and outer edges of the shell. Furthermore, the internal forces used in shell analysis are defined as stress resultants (forces per unit length of middle surface) such as

$$N_\theta \, dy = dy \int_{-h/2}^{h/2} \sigma_\theta \, dz \tag{1-1}$$

and stress couples (bending moments per unit length of middle surface)

$$M_y r \, d\theta = r \, d\theta \int_{-h/2}^{h/2} \sigma_y z \left(1 - \frac{z}{r}\right) dz \tag{1-2}$$

so that the units of N_θ are kips per foot and of M_y are foot-kips per foot. Positive N_θ is tension, and positive M_y is where σ_y is tension (positive) in the positive z direction.

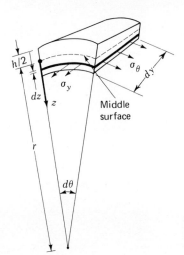

Fig. 1-1 Definitions for thin shells.

The term z/r in Eq. (1-2) arises because the cross section is trapezoidal. Where it appears with unity, the term z/r can be neglected in expressions like (1-2). Stresses are found simply by dividing N_θ by h or dividing M_y by $h^2/6$. When N'_θ appears, the prime denotes values based upon the assumption that all $M = 0$; this assumption leads to the membrane theory of shells, which is discussed in section 1-12.

1-3   CYLINDRICAL SHELL WALLS

Cylindrical shells as water tanks and gas holders were among the first types of structures built of reinforced concrete.[1] G. A. Wayss described the membrane theory in-plane ring tensions for these in 1887, along with a number of applications.[2] Robert Maillart made the first studies which considered bending as well as membrane stresses in fixed-base cylinders.[3] These cylinders were for the gas holders which he constructed for the city of St. Gallen in 1902. Not only were these the largest circular concrete tanks then built, but their calculations, done by graphic statics, took into account nonlinear variations in wall thicknesses. The mathematical theory as given in Chap. 2 was applied first in its modern form by H. Reissner in 1908[4] and given by von Emperger in 1910.[5] Cylindrical shells loaded axisymmetrically were therefore well defined theoretically before World War I. The ideas first used by Maillart and formulated by Reissner are so basic to thin shell theory that they need to be clarified here by the introduction of some elementary mathematical ideas.

Figure 1-2a shows a ground storage reservoir whose structure is a circular cylinder loaded by an internal pressure that increases linearly with depth. Figure 1-2b shows the horizontal ring tension N_θ in the wall for two types of base restraint: free and fixed; Fig. 1-2c shows the vertical bending moment M_y acting on horizontal sections for the fixed case, there being no moment for the case of no base restraint (free). For the free case, all the horizontal pressure is carried by the horizontal rings which expand into tension according to the formula (which Wayss had already presented in 1887)

$$N'_\theta = \gamma(H - y)r \tag{1-3}$$

Where the stress resultant N'_θ represents the force per unit length in the y direction,

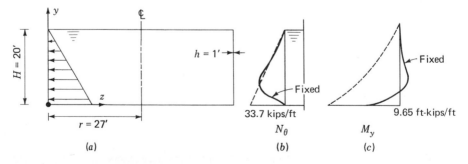

Fig. 1-2   Circular cylinder with internal pressure.

γ is the density of the fluid, $H - y$ is the distance down from the cylinder top, and r is the radius. For the fixed case, the base cannot expand or rotate and hence $N_\theta = 0$ there. The solid lines in Fig. 1-2b and c show the fixed case (the type of solution that Maillart got graphically and Reissner analytically), whereas the dashed lined in Fig. 1-2c shows M_y for the case where $N_\theta = 0$ throughout. This last case occurs when the shell-ring stiffness $k_r = Eh/r^2$ is very small compared to the shell bending stiffness $D = Eh^3/12(1 - \nu^2)$, i.e., when the ratio

$$\frac{k_r}{D} = \frac{12(1 - \nu^2)}{h^2 r^2} = 4\beta^4 \tag{1-4}$$

is small. This ratio indicates the extent to which a cylindrical shell carries loads by in-plane axial forces such as N_θ or by out-of-plane flexural forces such as M_y and Q_y, the horizontal shear in the radial direction. As the quantity β decreases, the rings carry less and the cantilevers more; the cylinder acts less like a shell and more like a retaining wall. The shell behavior depends upon β, and we can summarize the way it does as

1. The smaller the r and hence the larger the curvature, defined as l/r, the more the load is carried as a shell, i.e., carried by the in-plane stress resultant N_θ.
2. The smaller the h, the more the load is carried as a shell.
3. When the shell edge is restrained, the region close to that edge exhibits bending which damps out away from that edge.

In general, therefore, shell behavior depends upon curvature, thickness, and boundaries. The simplified ideas presented here for the cylindrical tank suggest how these three factors influence the design, but a more detailed analysis is essential to show the limitations to these simplifications. For example, the curves of Fig. 1-2 for the fixed-base tank are based upon the simplifying assumption that H will be large enough so that the influence of the fixed base will not be felt near the tank top. Because of this assumption, reducing β will not lead to $N_\theta \to 0$ without a more detailed analysis. For use in design, these questions require the fuller explanation found in Chap. 2. These ideas were well developed before World War I for axisymmetrically loaded cylindrical shell walls.

1-4   DOMES

For domes the situation was different. Johann Schwedler had presented the membrane theory in 1866,[6] and Wayss had given it in his 1887 text on reinforced concrete. But when von Emperger's *Handbuch* summarized dome theory in 1909, only that same membrane theory appeared.[7] This theory still seemed satisfactory to designers because, unlike at the base of a fixed cylinder in water tanks, there were no obvious errors in the membrane theory. Even today it provides a reasonable basis for preliminary design as the following simplified discussion will show.

The dome in Fig. 1-3a carries vertical axisymmetrical loads much like a series of pie-shaped arches along meridian lines. If the resultant vertical load is R, then at any

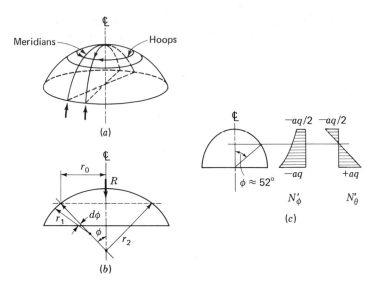

Fig. 1-3 Definitions and behavior of domes.

level defined by its slope ϕ (Fig. 1-3b) the meridional stress resultants ($r_0 = r_2 \sin \phi$)

$$N'_\phi = -\frac{R}{2\pi r_0 \sin \phi} \tag{1-5}$$

A second equilibrium equation, for forces perpendicular to the surface, is merely an extension of Eq. (1-3) to account for the double curvature

$$\frac{N'_\theta}{r_2} + \frac{N'_\phi}{r_1} + p_z = 0 \tag{1-6}$$

where r_2 is the principal radius of rotation, r_1 the principal radius of meridional curvature, and p_z the local radial pressure. When $r_1 = \infty$ and $\phi = 90°$, then $\sin \phi = 1$, $r_0 = r_2$, and Eq. (1-6) reduces to (1-3) if $p_z = -\gamma(H - y)$.

Unlike arches, these meridional strips can carry loads without bending even if their axial curve is different from the beam bending moment diagram for the loading. This is so thanks to the stress resultants N'_θ which act as stiff rings to prevent meridional bending.

Thus unlike arches, the form of the dome is not critical to the reduction in bending, thanks to the rings; but as with arches, the edge supports of a dome are important. Consider a dome for which only vertical supports exist. The arch-type forces, called meridional stress resultants at the edge (Fig. 1-3a) being in the plane of the dome surface, require at the edge both a vertical and a horizontal reaction. A relatively stiff edge ring can provide this horizontal reaction $H = N'_\alpha \cos \alpha$ by stretching under a ring tensile force (where ϕ at the edge is α)

$$T = Hr_0 = -N'_\alpha r_2 \cos \alpha \sin \alpha \tag{1-7}$$

In the dome itself, the simplified theory (the membrane theory) gives values for N'_ϕ and N'_θ shown in Fig. 1-3c. For loads distributed uniformly over a spherical shell surface, N'_θ will be compression as long as α is less than about 52°. Integrating N'_θ along any meridian from $\phi = 0$ to $\phi = \alpha$, we would find that the total compression force C equals the ring tension T of Eq. (1-7).

The dome-edge-ring system is thus in equilibrium horizontally. Vertically, of course, the reaction $V = N'_\alpha \sin \alpha$ when integrated around the base ring must equal the total vertical weight R of the dome loading as given in Eq. (1-5).

To summarize dome behavior,

1. The vertical gravity loads are carried by meridional arch-type forces.
2. There will be negligible bending along the meridians so long as the rings are intact.
3. The dome can be supported solely by vertical reactions so long as a stiff ring is provided to take the horizontal component of the edge meridional force.

These ideas were satisfactory so long as spans were small and thicknesses were relatively large as they were before World War I.[7] By 1915, a complete mathematical theory including bending was developed largely by Ernst Meissner (1883–1939) at the Federal Technical Institute in Zurich,[8] but there was little incentive to use it because of its complexity and because of the successful building up to that time based upon membrane theory.

The impetus for bringing mathematical theory into practical use came in 1922 when the Zeiss optical firm needed to build a hemispherical dome in Jena, East Germany, to test the new planetarium unit to be installed in the Deutsches Museum in Munich. Walter Bauersfeld (1879–1959) of the Zeiss firm made the design for this first thin shell dome in collaboration with Franz Dischinger of the firm of Dyckerhoff and Widmann A. G. who had already built many reinforced concrete structures for Zeiss.[9] Following the successful completion of that 1 1/5-in.-thick dome with a diameter of about 52 ft, Bauersfeld together with the builders took out patents and decided to pursue more broadly the development of thin shell structures.

Their next opportunity arose again in Jena where another firm agreed to cover its new factory building in a flat 6-cm-thick dome with a 40-m-plan diameter and a rise of only 7.87 m. Here, however, because of the flatness, the sloping shell edge needed a stiff ring; whereas in the hemispherical dome, $T = 0$ and no such ring had been needed. This connection between thin shell and thick ring introduced discontinuities in the shell behavior leading to significant bending (as described in Chap. 4). Bauersfeld understood that problem qualitatively and made simplified calculations upon which to base his design. The shell was completed in 1924. He then asked J. W. Geckeler to look into the theory more fully. Geckeler's studies published in 1926 clarified the mathematical theory and put it into a form easily used by designers. By recognizing that the bending was confined to a zone near the edge ring, Geckeler simplified the mathematics to such an extent that the procedure developed for cylindrical shells could be used for domes.[10]

Geckeler's basic idea was to separate the analysis into two parts: in the first, the shell stresses are found entirely from equilibrium equations (the membrane theory); in the second, bending stresses are studied only in those parts of the shell near discontinuities, i.e., usually just at edges. His argument (which we present in Chap. 3) consisted first of giving the full equations developed by Meissner; second of arguing physically on the basis of the principle of St. Venant that bending effects are confined to a region near the edges, thereby simplifying the full equations radically; and third of showing that the simplified equations give results that are negligibly different from those of the full theory and about the same as test results.

To see the idea behind Geckeler's development, we can consider the effect of boundary conditions (Fig. 1-4) on both the arch and the dome structures. If we give the arch a horizontal push at one side, this force must be held in equilibrium by an equal and opposite force acting at the other support. The only way that the force can be transmitted through the system is up through the arch, over the crown, and down the other side. The effect clearly is to produce high bending moments throughout the entire arch system, with a maximum moment at the crown. We see, therefore, that the arch is not only restricted in its most efficient form by the nature of the loading, but it is also very sensitive to foundation displacements or edge forces.

Let us give a corresponding horizontal force to the dome. Such a force may be considered as a uniform horizontal thrust applied all around the circular edge of the dome. In cross section it would appear that this horizontal force would create bending moments throughout the dome similar to those created in the arch. This is, however, not the case. As the horizontal force tends to bend the shell, and thus to be carried up a meridian, the rings again come into play and cause a rapid damping of the bending so that at a relatively short distance from the edge the bending effect is no longer observable. Thus edge forces in equilibrium applied to an arch propagate throughout

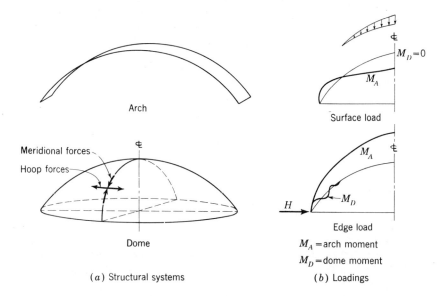

Fig. 1-4 Comparison of dome and arch.

the entire structural system and create large bending moments, whereas similar forces acting on a dome create bending moments in a very narrow region near the edge and have generally no effect throughout a large portion of the structure.

This edge effect in the dome is similar to that which we have already observed for the vertical cylinder with restrained base. Indeed the extent of the bending in a dome can be well estimated by using the same analysis as for a cylinder where the ratio β has the same meaning, except r is now r_2 the radius of curvature of the dome edge, defined as $r_2 = r_0/\sin \alpha$.

Geckeler's ideas only work for shells of revolution, but they make the solution so simple that any trained engineer could follow the analysis. Although the stimulus for his work came from concrete structures, Geckeler saw the field of application largely in terms of steam boiler shells (Dampfkesseln). Two years later he included the results in his 168-page chapter on elastostatics in the German *Handbook for Physics.* [11]

Geckeler's work was taken up by Franz Dischinger (1887–1953), who began with Dyckerhoff and Widmann in 1913. Dischinger had worked directly with Bauersfeld on the Jena dome in 1922, and by 1932 he had directed the design and construction of the major reinforced concrete thin shell domes built up to then. His major writings on domes was a 221-page section in the twelfth volume of the third edition of von Emperger's *Handbuch.* Appearing in 1928, this treatise presented for the first time both a systematic analysis procedure and a variety of completed examples of concrete shells. [12]

Unlike Geckeler's treatise, Dischinger's was focused fully on concrete thin shells and included the analysis for edge rings. He gave an extended discussion of membrane theory solutions for wind loadings and gravity loadings all for various types of domes such as spherical, conical, elliptical, and cycloidal ones. He also included certain mathematically defined variations in shell thickness. He presented Geckeler's method for bending and gave the worked example of the flat dome which he had built in 1924. Some approximate methods for nonrotational domes appeared along with a brief discussion of barrel shells with only the membrane theory presented. The analysis section closed with a full membrane theory study of polygonal domes.

What dischinger's section on applications shows is the radical difference between the design ideas before 1922 and those following the Zeiss-Dyckerhoff and Widmann collaboration at Jena. Dischinger himself identified the primary technical difference by comparing three domes: one over St. Peter's Basilica in Rome (1590), one over the Centennial Hall in Breslau (1913), and one of several over a market hall in Leipzig (1929) under construction as he wrote. [13] The following table summarizes the main differences:

Date	Place	Span, meters	Rise, meters	Weight, metric tons
1590	Rome	42	21	10,000
1913	Breslau	67	16	6,340
1929	Leipzig	76	17.5	2,160

The Pantheon, completed in 124 A.D., was the widest spanning masonry dome (it is concrete) until its 43.5-m span was surpassed in 1913 by the Breslau dome. Yet, in spite of reinforced concrete, that modern work weighed 63 percent of the weight of the stone Renaissance dome over St. Peter's. The major change occurred after 1922, the Leipzig domes each weighing one third the 1913 Breslau structure.

The primary difference between the Breslau and Leipzig domes is that the former is conceived of as a series of arches and rings supporting a roof, whereas the latter is a self-supporting thin shell roof stiffened at its polygonal folds by thin ribs. The difference showed the impressive efficiency possible with thin shell concrete structures. Dischinger had observed in his first article on shells that concrete domes designed as thin shells could be built in spans just as great as steel domes.[14] His prophecy has proved fully correct over the half century following, as seen by the steel Superdome (1975) and the concrete Kingdome (1975); these two largest domes in the world are of essentially the same span.

It was the practical construction problem of building a very light dome to test the Zeiss planetarium that had stimulated both Dischinger and Geckeler to find a suitable mathematical theory for domes. One significant result of the 1922 experiment was the creation of a patented system which became known as the Zeiss-Dywidag (Dyckerhoff and Widmann A. G.) method. Mostly the patents were for construction techniques, such as the network of reinforcement upon which concrete was sprayed. Still, in the development of thin shells the primary result was that a firm of designer-builders with long experience in reinforced concrete gained confidence in the use of very thin roof structures, and part of their confidence stemmed from the satisfactory development of a simple mathematical theory.

1-5 BARREL-SHELL ROOFS

But Dischinger recognized quickly that roofs with circular plans were rare and that the much broader application required adaptation of their new ideas to shells covering rectangular spaces.[15]

Dischinger first tried in 1923 to design a dome on a nearly square plan, but the calculations proved too difficult, and he then turned to a simpler geometry.[16] This meant shells of single curvature—barrel shells—and Bauersfeld was not convinced that the thin shell ideas would carry over from the much stiffer double curvatures of domes. He made calculations by membrane theory and then by small-scale model tests which proved Dischinger to be on the right track.

Based on the initial success, the Zeiss Company had Dyckerhoff and Widmann build in 1924 a barrel shell for one of their factory buildings in Jena, and the following year the builders erected a larger-span barrel-shell roof for themselves.

Then, during the summer of 1925, Dyckerhoff and Widmann tested a 10-m-long, 4-m-wide, 1.5-cm-thick elliptical barrel-shell model and determined that it could carry 500 kg/m² (102 psf) of surface load with negligible deflection, proving that the structure behaved as a shell, i.e., that the major stresses were in-plane.[17] At last in 1926, they built the first barrel shell commercially; it was called the Dywidag-Halle. Built at the Gesolei Fair in Düsseldorf, it had a span of 23 m, the width of each barrel was nearly 12 m, and the shell thickness was 5 cm.[18]

The success of those early barrels was due in large measure to a young engineer soon to become recognized as perhaps the most gifted and versatile concrete structural designer of his generation, Ulrich Finsterwalder (b. 1897). His father, a distinguished professor of mathematics in Munich, had broad interests leading to major contributions to the development of photogrammetry and to the theory of glacial movements. [19]

While serving on the Western front, young Ulrich was captured and during the 2 years' imprisonment by the French proceeded to study mathematics. Following release, he entered the Institute of Technology in Munich, from where he graduated as a civil engineer in 1923. Already in 1922 he had heard through his friend J. W. Geckeler of the new ideas coming from Jena by Bauersfeld and Dyckerhoff and Widmann. Geckeler told him about the barrel-shell idea with which Bauersfeld was grappling, and Finsterwalder became intrigued. One of his professors, A. Föppl, had already studied barrel shells, had tested a model, and knew of its extraordinary strength; but Föppl had no theory for it. That quandary was just what appealed to Finsterwalder, who proceeded to develop a theory for his final school project. [20]

This was the membrane theory of barrel shells which both Geckeler and Dischinger presented in their 1928 treatises [21] and which Bauersfeld had independently developed as well. Thus when Finsterwalder graduated, it was only natural that he should go right to work for Dyckerhoff and Widmann, where he remained for his entire career.

The mathematical idea developed by Bauersfeld and Finsterwalder can be most simply seen by recognizing that under surface loads (dead-load g, for example), a pipe shell (Fig. 1-5) behaves like a beam spanning between transverse diaphragms where the shear and bending stresses are plotted (See Fig. 1-5c). The troublesome

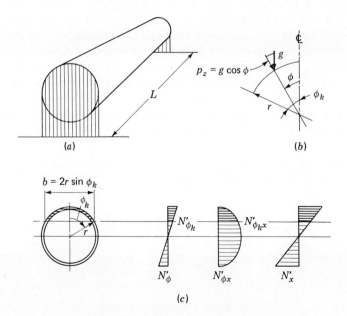

Fig. 1-5 Definitions and behavior of barrel shells.

consequence of these results is that for a barrel shell (the pipe section cut at a level where $b = 2r \sin \phi_k$) this theory requires reactions equal to N'_{ϕ_k} and $N'_{\phi_k x}$ along the lower edge lines of the barrel.

Finsterwalder's earliest solution to this problem was to change the cross section from a circle to an ellipse (Fig. 1-6) with $a = r \sin \phi_k$ and $b = r(1 - \cos \phi_k)$. At the bottom edge the tangent is vertical, $\phi = 90°$, and $N'_{\phi_k} = 0$. The edge shear $S \neq 0$, so he placed a straight member to be connected along the free edge designed to carry the shear. This straight edge became a tension tie with the shell N'_x being, according to the membrane theory, all in compression. As long as the span between transverse supports was not too great, this simplified idea worked well enough, but the practical use of barrels was overly inhibited by this restricted theory. Unlike domes, the membrane theory was not a reasonable simplified theory because it did not satisfy equilibrium, i.e., the overall bending stresses (N'_x/h) were all compression. Furthermore the elliptical section required very steep side slopes on the barrel and hence created construction difficulties. Another idea had to be found, and Finsterwalder was not long in discovering it. By 1930 he had developed a new basis for design. [22]

His new idea rested upon a clear understanding both of the ideal behavior of barrel shells and of a practical method of developing a shell theory to their solution. The ideal behavior envisioned by Finsterwalder was simply that of a beam with a thin curved cross section.

Figure 1-7b shows a rectangular area covered with a slice of a cylinder. It is a long-barrel slice because the ratio of span length L to radius curvature r is large, i.e., $L/r \geq 2.5$. As such, the system with longitudinal edge beams acts as a beam of curved cross section spanning between transverse arch supports. To get some idea of the advantages Finsterwalder saw in this type of structure, it is useful to compare it to the beam and slab system.

Consider the framing plan shown in Fig. 1-7a, in which the loads are essentially carried transversely by the slab and longitudinally by the beams. We may improve this system by building the slab and the beam monolithically, thus obtaining some T-beam action and thereby increasing considerably the stiffness of the beams. Since the beam is relatively shallow compared to its span, we know that the simple flexural theory is valid and that the magnitudes of the stresses are directly proportional to the depth and vary as a straight line from top to bottom. Even in the slab the loads are essentially carried transversely by bending moments, the sizes of which are functions of the transverse span.

If we now imagine that the slab is given a curvature and that the longitudinal beams are reduced in cross section, as shown in Fig. 1-7b, we have a typical simply

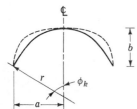

Fig. 1-6 Barrel shell cross sections.

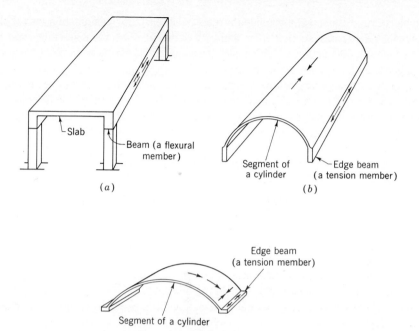

Fig. 1-7 Comparison of barrel shells and slabs.

supported barrel shell spanning the same distance as the beams in Fig. 1-7a. The structural action, however, is significantly different.

Let us first consider the transverse slab, which is now curved. We might think that this would act as an arch, but it does not because the slender edge beam is unable to sustain horizontal forces. Therefore, although the curved slab will try to act as an arch, the thrust contained at the springing lines will be very small. The principal structural action of this system is found to be longitudinal. In the previous case the beam had to carry the entire load with perhaps some additional flange help from the slab, but in this case the entire system acts as a beam with curved cross section to carry the load.

The principal action, therefore, of the barrel shell with small flexible edge beams is longitudinal bending, but the bending stresses are within the plane of the shell itself. Thus we may liken the entire shell-and-edge-beam system to a beam with the compression stresses near the crown and with the tension stresses concentrated in the edge beams on either side. In fact, one approximation for a certain class of barrel shells which makes use of the simple flexural theory is based entirely upon this physical picture of the system.

However, the simple flexural theory, which leads to a straight-line stress distribution, requires that all points within the cross section of a member deflect exactly the same amount, which is the case in a solid rectangular beam. In a T beam with wide flanges, the extremities of the flanges may not deflect the same as the web, but the simple flexural theory still provides a reasonable basis for design. In a barrel shell, however, the cross section may undergo substantial lateral distortion, and it is principally this distortion which causes the longitudinal stresses to depart from the straight-line distribution of the beam theory.

Figure 1-8 shows six different examples of the longitudinal stress resultants N_x where the ratio of L/r varies. In addition to these N_x values, there can be transverse bending moments M_ϕ owing to the lateral distortion of the cross section. This distortion is substantially reduced either by intermediate ties or diaphragms on a single barrel or by connection of a series of barrels together in the transverse direction. Then the designer can use a simplified analysis that considers the shell as a beam with longitudinal stresses determined by beam theory as plotted in Fig. 1-8. This simplified approach can be extended to the transverse bending by an arch analysis for which tables are given in Chap. 6.

It is roughly true that the longer the span in comparison with the transverse-chord width, the more the entire cross section behaves as a beam. The shorter the span compared with the chord width, the more the structure behaves as an arch with a supporting sloped deep beam near the edge (see Fig. 1-7c).

Finsterwalder, having clearly in mind the behavior, set out to develop a method of analysis from which to find the shell stresses and the required reinforcement. Since he had begun with the membrane theory, Finsterwalder saw the problem much as Geckeler had seen the dome—in two steps: first membrane and then bending. For the shell without edge beams, he first solved the membrane theory and found the N'_ϕ and $N'_{x\phi}$ at the free longitudinal edges. In step two he applied $T_L = -N'_{\phi k}$ and $S_L = -N'_{\phi k x}$ at those edges and determined the shell stresses by a bending theory. The true shell behavior was a combination of results from the two steps.

The major difficulty was the bending theory because now the formulation led to an 8th-order partial differential equation. Finsterwalder neglected some terms and then converted the resulting equation to an ordinary differential equation by expanding the load in a Fourier series in the direction of the span. This follows the Lévy-type solution for plates and forces the transverse boundaries to provide simple supports;

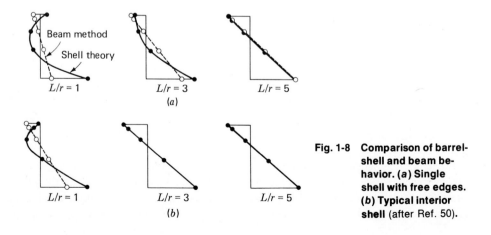

Fig. 1-8 Comparison of barrel-shell and beam behavior. (a) Single shell with free edges. (b) Typical interior shell (after Ref. 50).

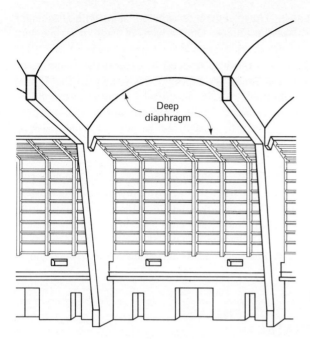

Fig. 1-9   **Wholesale market hall in Frankfurt, 1928** (after Ref. 25).

i.e., displacements in the vertical plane (v and w) must be zero as also must be forces perpendicular to that plane (N_x and M_x). This restriction implies a thin deep diaphragm at the supports which became a strong visual feature of many barrel shells built until World War II (Fig. 1-9). He then attacked the resulting 8th-order ordinary differential equation and developed solutions for individual edge line loads, such as S_L. This approach appears in Chap. 5 in the same form as Finsterwalder first did it in the late 1920s. He also developed a method of introducing edge beams and multiple shells; Hubert Rüsch further developed these questions in his 1931 dissertation,[23] and the essential mathematical basis for barrel shells was in a tractable if not too accessible form.

Partly it was the complicated mathematics of this geometrically simple form that kept other engineers from designing such shells. Partly also the inexperience of builders held back development. Freyssinet working as a designer-builder in France did construct some remarkable shells in the 1920s, but they had little influence on other engineers, especially in the United States.[24] It was almost entirely through the work of Dyckerhoff and Widmann that thin shell concrete structures first came to America, and they came therefore through the two primary forms of the Z-D (Zeiss-Dywidag) system, the dome and the barrel. The person most responsible for these early shells was Anton Tedesko (b. 1903).

Graduated in 1926 from the Institute of Technology in Vienna, Tedesko had spent several years in the United States before beginning to work with Dyckerhoff and Widmann in 1930. This was just the time when a collection of extraordinary engineers all working together were developing the mathematical and the constructional ideas for concrete thin shells. These engineers included, along with Dischinger and Finsterwalder, Rüsch and Wilhelm Flügge. On the basis of the experience with very large domes, such as at Leipzig, and very large barrels, such as the 1928 Market Hall in Frankfurt (Fig. 1-9),[25] the firm decided to expand their operations abroad, and Tedesko because of his American experience was sent to the United States in early 1932. There he affiliated with an older friend from Vienna, John E. Kalinka, then working for the Roberts and Schaefer Company in Chicago, a firm of designer-builders experienced in concrete structures largely for coal handling facilities and for concrete arch bridges done by the earlier firm of Bush, Roberts, and Schaefer.

The first American brochure for thin shells appeared in January 1932 announcing Roberts and Schaefer Company as the agent for the Z-D system and showing many of the German works by then completed.[26] The development in the United States from 1932, largely through barrel shells, culminated 20 years later in an unusual publication by the American Society of Civil Engineers: Manual 31 in which Finsterwalder's basic approach was put in tabular form and made accessible to a wide circle of American engineers.

1-6 INTRODUCTION OF THIN SHELLS IN THE UNITED STATES

From 1932 to 1934 Tedesko worked in Chicago closely with Roberts and Schaefer while still an employee of Dyckerhoff and Widmann, promoting the Z-D system. The Hayden Planetarium was its first American application. Tedesko had introduced the engineers Weiskopf and Pickworth to the Z-D work on European planetariums and got them to invite Roberts and Schaefer to serve as consultants for the design and construction of a concrete thin shell to be bid in competition with a Gustavino tile dome alternative.[27] The concrete design, being cheaper, was chosen and constructed by shotcrete as were the early domes in Germany.

The second American thin shell roof by the Z-D system was a barrel shell for the Brook Hill Farm dairy building at the Century of Progress World's Fair of 1933 in Chicago.[28] There were five 36-ft-span 14-ft-wide multibarrels covering a 36 × 70 ft interior space. This time Roberts and Schaefer were the engineers with Tedesko as principal designer. At the time of demolition, a series of carefully controlled load tests were carried out by Roberts and Schaefer assisted by the Portland Cement Association and closely observed by engineering staff members of the University of Illinois.[29] These tests characterized the close relationship developed between that association and Tedesko all to the end of promoting concrete shells.

Up to 1932 there had been almost nothing written in the United States on concrete thin shells. The first article to gain significant attention was written by Tedesko in 1931 but not published until 1932 and then anonymously. Probably this article led to the Brook Hill Farm design, and certainly it made many American engineers aware for the first time of the new ideas coming from Europe.[30]

The going was, however, slow. The depression had arrived, and most American engineers were uninformed about shell theory and construction practice. The first

Fig. 1-10 Ice hockey arena, Hershey, 1936. (Courtesy Anton Tedesko.)

major thin shell in the United States came in 1936 (Fig. 1-10), with the decision by the Hershey Chocolate Company to build an ice hockey arena. Not only was this 232-ft-span 340-ft-long roof by a considerable margin the largest thin shell concrete structure on the continent, but it was built by the Hershey Company itself as "one of several community institutions built by the Hershey interests for the purpose of welding their 2200 citizen employees into a big happy family."[31] Since the arena holds over 7000 people, its welding capacity extended well beyond the chocolate works, and once opened on December 19, 1936, after only 8 months for complete construction, it proved a great success both for the spectators and for the engineering profession. This immense barrel, still in excellent condition after over 40 years of service, was designed by Tedesko and he personally supervised the construction.[32] This was the first short-barrel shell, i.e., where the ratio $r/l > 0.6$, it being 3.4 for the Hershey geometry.

In the short barrel, the 3½-in. shell carries the load essentially as an arch by N_ϕ membrane stress resultants down to near the edge where the shell, over about its lower 25 ft of arc length, behaves like a deep beam carrying most of the entire roof load to the arches (Fig. 1-10). This design would serve as a prototype for a large number of arenas and hangars designed over the next two decades by Roberts and Schaefer Company. The great size, the extreme thinness, and the impressively short completion time all served to demonstrate that concrete shells could be built in the United States. Hershey was the major turning point of thin shells in North America. Following that shell, a large number of shell roofs were built until the outbreak of World War II, all designated as Z-D shells. In fact, they were designed following published analyses and built by well-known methods. The proprietary aspect was primarily growing construction experience and access to engineers who were used to the detailed calculations necessary for shell analysis.

Roberts and Schaefer had announced in 1932 that their "rights include the possession of the intricate engineering formulae essential to the preparation of the design and construction of such structures under this system."[33] Of course even by 1932 such formulas had been openly published, but the force of Roberts and Schaefer's claim lay much more in the reluctance of most American engineers to study such mathematical works. This reluctance was expressed by the only two American engineers to discuss in 1936 a paper by the Swiss Herman Schorer that was the first paper on shell theory published in an American technical journal. I. K. Silverman noted that "Even with the simplifications introduced into this paper, the application of the theory requires considerable computations . . ." and F. W. Seidensticker (who worked for Roberts and Schaefer) observed that "knowing from experience that many readers are reluctant to follow through a series of difficult mathematical derivations, the writer fears that, in many cases, only the 'synopsis' and 'conclusions' of this excellent paper will be read."[34] By contrast, each of the three European discussors (Wilhelm Flügge, Tedesko, and Finsterwalder) emphasized how clear Schorer had made the problem in that he employed substantial mathematical simplifications.

Meanwhile, Roberts and Schaefer designed numerous industrial and civic shells as long barrels,[35] short barrels,[36] and domes[37] up to 1941, when the war effort changed the focus to military warehouses and hangars. The military use was stressed by Kalinka in an article emphasizing the hazard-proof nature of concrete shells and showing a dramatic photo of a bombed shell roof church in Finland where the shell remained standing in spite of a gaping hole where the bomb crashed through the shell.[38]

Following the war, the long-barrel, warehouse-type shells underwent a basic change connected to the growing American experience and the gradual confidence in physical model testing over mathematical theories. In late December 1940, the U.S. Army Quartermaster Corps took bids on three types of structural designs for a large warehouse in Columbus, Ohio: a steel and wood design, a steel design, and a concrete shell one. The concrete shell was the least expensive and was thus built in 1941 to the design of Roberts and Schaefer Company.[39] The project included three buildings, each 182 ft wide and 1562 ft long, covered by four barrels, each about 45 ft wide, with a span of 38 ft between transverse diaphragms (Fig. 1-11). Under Tedesko's direction, the diaphragms were made to project above the roof as heavy frames following the requirements of the mathematical theory pioneered by Finsterwalder. These frames visually dominated the roof and complicated the roofing and flashing. They were put above the roof to ease the sliding of the movable forming below and thus to speed construction.

Tedesko was not satisfied with that solution and in 1950 directed tests on the property of Roberts and Schaefer Company of a large-scale concrete model barrel

Fig. 1-11 U.S. Army warehouse, Columbus, 1941. (Courtesy Anton Tedesko.)

shell where the shell was merely thickened into rib bands over the columns to avoid structural and roofing discontinuities. From measurements, Tedesko and his colleagues concluded that such ribless barrels could be safely and efficiently built in spite of there being no satisfactory mathematical theory (the original idea was developed in correspondence with Dischinger by then a professor in Berlin).[40] These rib bands plus columns had earlier been analyzed as frames to take shell forces, and the 1950 test was made to verify that analysis.

The opportunity to design a major ribless shell arose in the mid-1950s with the large warehouse for the U.S. Air Force at Olmsted base near Middletown, Pennsylvania. The cast-in-place ribless shell roof project, consisting of sixteen units each of 200 ft by 200 ft, was bid competitively against a precast shell alternative and proved to be significantly less expensive.[41] Compared to the similar warehouses at Columbus, the postwar design is visually quite different owing to the removal of the heavy, mathematically suggested, transverse frames (see Fig. 6-4).

A second major change occurred in shell design just following World War II, this one primarily in short-barrel shells. With air transportation—including the military—rapidly increasing, a large number of hangars appeared between 1945 and 1952. Here the conditions were different because the shell behaved mainly as a thin arch while the ribs served both as stiffeners against buckling and as load-carrying arches. By shaping the roof more as a catenary, Tedesko and his colleagues were able to

Fig. 1-12   U.S. Navy hangar, San Diego, 1941. (Courtesy Anton Tedesko.)

Fig. 1-13   U.S. Air Force hanger, Limestone, Maine. (Courtesy Anton Tedesko.)

achieve lighter ribs. A typical prewar design (Fig. 1-12)[42] compared to a typical post-war one (Fig. 1-13)[43] shows how much lighter the ribs became even though they still projected sharply above the shell to increase their stiffness and to ease the movement of the form centering.

Meanwhile, following the war, Charles S. Whitney, after forming a partnership with Othmar Ammann, began to design shell structures in competition with Roberts and Schaefer Company; Ammann and Whitney's shells began to appear in 1948.[44] Whitney had written extensively about concrete arches as early as 1925 and even contributed in that year a discussion on domes which showed an understanding of the dome-ring interaction stresses, though he then believed that "an exact analysis of them does not seem possible."[45]

Whitney's entrance into thin shell design provided competition previously nonexistant to Roberts and Schaefer and led to designs based upon somewhat different ideas about structural behavior.

Just after the war, the American Society of Civil Engineers formed a subcommittee of the Committee on Masonry and Reinforced Concrete to study thin shells. The committee was chaired by Whitney, and included H. H. Bleich, M. G. Salvadori, and Schorer. The result of the committee's work was a 177-page manual prepared largely by committee member Alfred L. Parme, of the Portland Cement Association. [46]

This remarkable document provided a simplified presentation of Finsterwalder's original ideas, tables that were easy to use, full theoretical derivations, and a comprehensive discussion of barrel-shell behavior. It was the central document in opening up thin shell design in the United States, making it accessible to a wide circle of engineers rather than only to those with long experience. Part of its power derived from the long successful record of structures already built to designs by Robert and Schaefer, and part derived from the relative ease with which designers could now calculate stresses. With that manual, which is the basis of Chap. 6, thin shell design entered a new phase in the United States. Not just the manual but also the increased publicity on shells coming from outside the country caused a buildup in demand for such structures from owners and from architects.

1-7 FOLDED PLATES

Folded-plate roofs followed the development of barrel shells by only several years. In 1929 Craemer described a roof of 24 m span of the cross-sectional form shown in Fig. 1-14a. Craemer also showed examples of folded plates used for coal silos. His work and that of others were summarized in 1932 by W. Petry at the first Congress of the International Association for Bridge and Structural Engineering in Paris. [47] Petry showed the early development of silos by comparing three works by the firm of Wayss and Freytag. The first in 1915 (Fig. 1-15a) had heavy beams between the hoppers and a stiff ribbed frame supporting the wall. The second in 1921 (Fig. 1-15b) shows substantial improvement; the heavy hopper beams are replaced by a thickening of the intersecting walls and the interior wall ribs are gone. But the most efficient form came in 1925 (Fig. 1-15c) with the elimination of the hopper beams entirely and the full integration of walls and hoppers; a true folded-plate system had emerged wherein the loads were carried by the plates through bending to the folds and then by the plates as deep beams through in-plane behavior.

Possibly the first designer to see the possibility for fully integrating the walls and hoppers to form a folded-plate system was Robert Maillart, whose 1915 Swiss patent

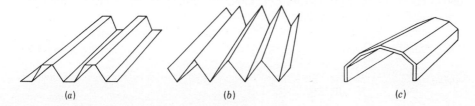

(a)   (b)   (c)

Fig. 1-14 Typical folded-plate cross sections.

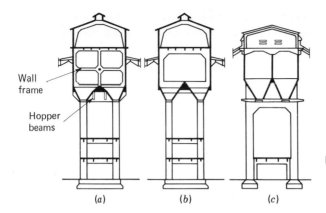

Wall frame

Hopper beams

Fig. 1-15 Development of folded-plate silos. (From W. Petry, 1932.)

(a)   (b)   (c)

shows nearly the same form as the Wayss and Freytag 1925 design.[48] Maillart's walls were hollow to permit air to circulate and to allow for a stiffer structure with less material.

Throughout the 1930s, engineers studied folded plates to develop reliable methods of analysis, but it was not until after the war that the use of folded plates became widespread for roofs. A summary of the methods of analysis appeared in 1963 with a recommendation that the method referred to in Chap. 8 be used.[49] Since then, more general computer methods have become common.[50]

Many roof structures are made by the connection of a series of plates folded in such a way that the resulting structure is similar to a barrel shell (Fig. 1-14c) or to a multiple-barrel shell (Fig. 1-14a and b). The overall behavior is similar, the main difference being that transverse bending moments are greater because of the lack of curvature. Transversely, the barrel carries as an arch, at least over the central part of the shell, whereas the folded plate carries as a continuous plate on flexible supports.

1-8 THE HYPERBOLIC PARABOLOID

Following domes and barrels, the next thin shell form to engage the profession was the hyperbolic paraboloid. Here the pioneering was by French, Italian, and Spanish designers and began with the recognition of those special geometric properties which make the surface easy to analyze and appealing to the designer's visualization of construction. Both factors rely upon the straight-line generators (Fig. 1-16) whose geometry is defined by

$$z = \frac{y^2}{h_2} - \frac{x^2}{h_1} \tag{1-8}$$

where

$$\frac{\partial^2 z}{\partial x^2} = -\frac{2}{h_1} \quad \text{and} \quad \frac{\partial^2 z}{\partial y^2} = \frac{2}{h_2}.$$

Hence, h_1 and h_2 are equal to twice the principal radii of curvature of the surface at the origin $x = y = 0$ and are good approximations to twice the radii of curvature along the coordinate lines over the entire surface where that surface is flat, i.e., where $\partial z/\partial x$ and $\partial z/\partial y \ll 1$. The surface can be generated either by parabolas as in Eq. (1-8) or by straight lines (see Fig. 1-16). A simpler form of Eq. (1-8) arises when the surface is defined by coordinates parallel to the straight-line generators (Fig. 1-17). Then

$$z = \frac{c_0}{a_0 b_0} uv = kuv \tag{1-9}$$

where k is the twist of the surface defined as the change in the slope

$$k = \frac{\partial}{\partial u} \frac{\partial z}{\partial v} = \frac{\partial}{\partial v} \frac{\partial z}{\partial u} = \frac{1}{r_{uv}} = \frac{1}{r_{vu}}$$

Probably Ferdinand Aimond gave the earliest presentation of this geometry for roof shells in 1933.[51] There he developed the simplest equilibrium equation by showing that vertical loads, assumed to be uniformly distributed over the horizontal projection of the surface (Fig. 1-18), can be carried entirely by shear forces. We can see this by rewriting Eq. (1-6) to include the twist $r_{xy} = r_{yx}$:

$$\frac{N'_x}{r_x} + \frac{N'_y}{r_y} + \frac{N'_{xy}}{r_{xy}} + \frac{N'_{yx}}{r_{xy}} + p_z = 0 \tag{1-10}$$

In the case where x and y are straight-line coordinates, $r_x = r_y = \infty$ and since $N'_{xy} = N'_{yx}$, Eq. (1-10) reduces simply to

$$N'_{xy} = \frac{p_z}{2} r_{xy} = \frac{p_z}{2k} \tag{1-11}$$

Aimond recognized that this analysis neglects the influence of boundary conditions, and for that reason he provided for stiff edge beams. His brief article concluded

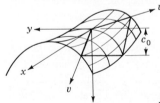

Fig. 1-16 Hyperbolic paraboloids with curved boundaries.

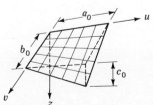

Fig. 1-17 Hyperbolic paraboloids with straight boundaries.

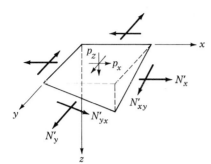

Fig. 1-18 Membrane theory stress resultants in hyperbolic paraboloids.

with the statement that "the very simple calculations can be done in a very intuitive geometrical form, which makes their checking easier. Their construction in concrete or metal, is made easy by their mode of generation, which only uses straight lines."

Three years later, Aimond published an extensive 111-page mathematical treatise on hyperbolic paraboloids using the membrane theory.[52] Shortly thereafter, M. L. Hahn described the design and construction of a large garage designed by Aimond.[53] The analyses developed by Aimond remained the basis for design for over three decades. Just as with domes and barrels, early designers relied on the membrane theory; but unlike the other two shell forms, these straight-line shells seemed not to need the corrections of a bending theory.

The major impetus to design came from Felix Candela (b. 1910), the Spanish architect trained in Madrid who left for Mexico in 1939 following the Civil War. After working in various offices, he designed his first shell, a barrel vault, in 1949 and his first hyperbolic paraboloid in 1951. By 1954, when he was invited to speak at the First American Conference on Thin Shells, he was an international figure.[54] The 1963 book, *Candela, the Shell Builder,* confirmed his position as one of the half-dozen best-known structural designers in reinforced concrete.[55] In that same year, he wrote a spirited attack on a report of the concrete shell committee of the American Concrete Institute. He had been a member of the committee, but he declined to sign their report because of its overemphasis, in his view, on analysis for stresses and rules for reinforcement.

The problem that Candela saw was that mathematical formulations got in the way of design imagination, and what he emphasized was the simplicity of calculation while at the same time a recognition of the unlimited potential for form. His flair for lecturing, his lively writing style, and the stunning photographic beauty of his Mexican structures all stimulated engineers and architects in the United States.

After Candela, the principal force in popularizing these shells was the Portland Cement Association and especially Alfred Parme. Following Candela's 1955 paper in the *Journal of the American Concrete Institute,*[56] Parme began to develop simple ways of visualizing behavior, which he published in 1958.[57] The main idea was that these structures behave as two systems of arches, one in compression and one in tension (Fig. 1-19).

In surfaces with straight edges, these arches carry loads to those edges where

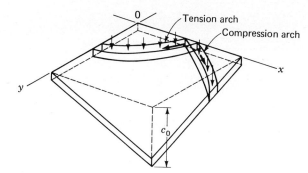

Fig. 1-19 Simplified explanation of hyperbolic paraboloid behavior.

they are resolved into edge shears. These shears require edge members which are thereby loaded axially. This simple model formed the basis for numerous designs, but some troubles arose just because of one virtue emphasized by Candela, i.e., the variety of forms possible. Some of these forms did behave as the membrane theory predicted, but some did not, and before a reliable bending theory existed there was no way by calculations to show which was which. There had been many attempts to develop a bending theory, but none had succeeded in providing a solution for the variety of boundary conditions found in practice. Only with the development of computer-based numerical methods was the solution found by calculation. Ironically, the uses of these forms began to diminish at the same time as the numerical methods began to yield useful results. This situation arose partly because the simplified methods used widely up until about 1970 could not provide a reasonable basis for design in all cases, and partly because some shells exhibited structural defects. In general, where the designers were engineers with substantial experience in thin shell concrete structures, such defects did not arise. Gradually, it became clear that hyperbolic paraboloids needed to be thought of not as one type of form but rather as several different types, of which at least four had been well developed by the 1970s: the double cantilever, the inverted umbrella, the gabled vault, and the groined vault. Each behaves differently enough to merit separate discussion. The first two behave essentially as cantilever beams, while the latter two as stiffened arches.

Cantilever-Type Hyperbolic Paraboloidal Roofs

Figure 1-20 shows two common types of roofs, each with straight-line boundaries and each taken from a saddle surface (Fig. 1-16). The membrane theory assumes that vertical loads are primarily carried by the intersecting parabolic arches, part by the compression arches and part by the tension ones. The arch forces are brought to the straight-line edges where their components perpendicular to those edges cancel and the components parallel to the edges add to give shear forces along the edge beams. These edge beams, in turn, carry the shears by axial tension or compression.

In the double cantilever (Fig. 1-20a) these edge beams carry compression thrusts to the supports; in the inverted umbrella (Fig. 1-20b) the interior slanted beams also

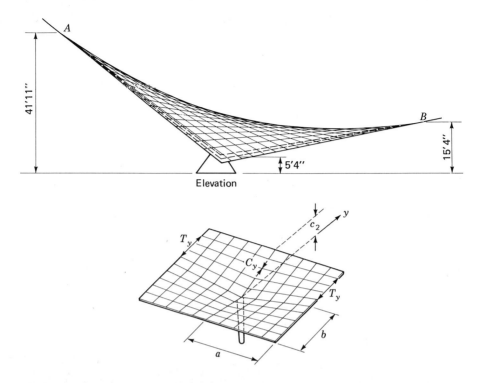

Fig. 1-20 Hyperbolic paraboloid forms.

carry compressions to the support, while the exterior flat beams carry tensions for which prestressing is often supplied.

These simplified descriptions of behavior are not reliable, as research-type analyses have shown. The problem is that the laterally flexible edge beams tend to deform and the compression arches tend to carry more load and to carry it more directly to the supports and less toward the edge beams. It is best therefore to consider each system separately and, as we have done with other systems, to relate each to a more elementary form.

The double cantilever is best thought of as a cantilever with the transverse crown carrying longitudinal tension and the straight line edge beam carrying longitudinal compression such that together the total cantilever moment is resisted. This means that much of the shell near the supports will be in longitudinal compression as well as in shear. Of particular importance is the flat region near the cantilever tip where the moment arm between shell crown and edge beam can be very small. Such overall bending must always be fully resisted, and near the tip there must be enough bending resistance to prevent local failure. Moreover, the straight edge beams tend to add weight onto the shell. Where live loads can be important, the possible unbalanced load is often resisted by vertical columns along the edge beams.

The inverted umbrella can also be seen as a cantilever in which two opposite straight exterior edge beams carry the tension and the slanting shell and ridge beam carry compression. As with the cantilever saddle, the overall equilibrium for unsymmetrical loads needs careful consideration. All bending from such loads must be resisted by the column. Where the structure includes a series of such umbrellas, lateral connections will prevent the single unit from having to act alone for unsymmetrical loadings (Fig. 1-21). The inverted umbrellas carry as cantilevers in two directions, unlike the cantilever saddles. To compute T_y the tension in a pair of edge beams, the overall moment due to a load q uniformly distributed over the horizontal plane (where $c_1 = c_2 = c$ and the u and v axes are replaced by x and y):

$$M_y = \frac{q2ab^2}{2} \tag{a}$$

where for simplicity we have dropped the subscripts for a and b, and

$$T_y = \frac{M_y}{c} = \frac{qab^2}{c} \tag{b}$$

whereas the compression in the slanting ridge beam will be

$$C_y = \frac{T_y}{\cos \theta_y} \tag{c}$$

and the vertical component of C_y will be

$$V = T_y \tan \theta_y = \frac{qab^2}{c} \frac{c}{b} = qab \tag{d}$$

Fig. 1-21 Berenplaat Filter Building roof, 1965.

or one-quarter the total vertical load as it must be. This simple analysis neglects the shell entirely; in fact, that part of the shell near the horizontal edges will be in tension and that part near the sloping ridge in compression. Of course the shell also acts as the web between the tension and compression flanges and therefore must carry shear stresses even though such are usually very small. Furthermore, at the flat exterior corners there is a danger that the overall bending might not be resisted if too little steel is present. These questions of overall cantilever-type behavior are central to design, but local behavior is also significant. For example, if large edge beams are used, they will load the shell more and in addition cause bending where they join the shell.

Gabled Hyperbolic Paraboloidal Roofs

The third commonly used hyperbolic paraboloid is the four-quadrant gable (Fig. 1-22). According to the membrane theory, the interior flat ridge beams carry compressions which are equilibrated at the midspan, and the exterior slant beams carry compressions to the supports. However, detailed analyses by numerical methods show that the horizontal ridge beam carries very little compression at midspan and furthermore has the effect primarily of adding dead weight to the shell. This beam thus can introduce significant bending and vertical deflections in the flat parts of the shell. Moreover, the placement of this beam can be important since the shell along its middle surface introduces compression into that horizontal member. When the beam projects below the shell, this shell compression introduces a positive bending moment in the beam which adds to the dead-load moment at midspan and can result in dangerous vertical deflections.

Another way to visualize behavior is to consider the overall shell as two flat intersecting diagonal arches in which the crown compression stresses are well estimated by finding the arch thrust per foot $H = ql^2/8c = q(2\sqrt{2}a)^2/8c = qa^2/c$ for the case where $a = b$ and computing the resulting stress

$$\sigma = \frac{H}{h} = \frac{qa^2}{ch}$$

Where $a = b$ and each diagonal arch carries all the load in its quadrant.

To the extent to which the load is concrete dead weight, the stress is independent of shell thickness; but it is directly related to the shell rise or more correctly to the shell curvature

$$\frac{1}{r} = \frac{d^2z/ds^2}{[1 + (dz/ds)^2]^{3/2}}$$

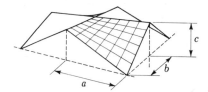

Fig. 1-22 Gabled hyperbolic paraboloid.

and along the diagonal $z = cs^2/2a^2$; then at the crown where $s = 0$ and $dz/ds = 0$,

$$\frac{1}{r} = \frac{d^2z}{ds^2} = \frac{c}{a^2} \qquad (f)$$

and hence the arch stress there is a typical ring stress as in Eq. (1-3) where the load q is carried by

$$\sigma = \frac{N'_\theta}{h} = \frac{qr}{h} = \frac{qa^2}{ch} \qquad (g)$$

What is significant is that the membrane theory of thin shells when applied to these gabled hypars (Sec. 7-4) results in stresses that are half those found in (g). This important difference arises because the membrane theory results depend upon the hanging arches carrying half the load to the gables and the ridge beams; this in turn takes no account of the deformations, and since the shell tends to move inward along the gable boundaries (Fig. 7-14), the tension thrust is relaxed and most of the load goes as compression in the diagonal direction directly to the corner supports. This discussion, in summary, shows that a proper design of gabled hypars depends upon

1. A correct overall picture of behavior as diagonal compression arches
2. A sufficient curvature at midspan to keep stresses low
3. A stiffening of the flat central part by reinforcement and properly placed ribs that are no heavier than necessary
4. Sufficient stiffness at the supports to prevent horizontal displacements

In such shells a major question will be creep and large midspan deflections all contributing to buckling. The summary just given leads to design considerations aimed at avoiding such failures.

1-9 GROINED VAULTS

Figure 1-23 shows a simple groined vault made up of two intersecting segments of paraboloids. To get an overall picture of the behavior, it is simplest to assume a vertical gravity load to be uniformly distributed over the horizontal projec-

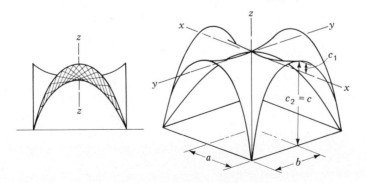

Fig. 1-23 Groined vault hyperbolic paraboloid.

tion of the surface. For a flat roof, that is not an unreasonable assumption for dead loads, and it is normally a reasonable one for snow loads. Then the parabolic arch strips of each cylinder carry those vertical loads q by pure axial forces whose horizontal component in one direction is constant and has the value in the x direction of

$$H_x = \frac{qa^2}{2c} \qquad (a)$$

and whose vertical component varies as

$$V_x = qx \qquad (b)$$

At the groin these forces join with similar ones from the arch strips in the y direction to load the diagonal arch formed by the groin valley with forces in the diagonal direction. These are, where for simplicity we take a square plan with $a = b$, an in-plane shear

$$H_s ds = H_x dy \cos \alpha + H_y dx \sin \alpha \qquad (c)$$

where

$$\frac{dy}{ds} = \sin \alpha \qquad \text{and} \qquad \frac{dx}{ds} = \cos \alpha$$

so that

$$H_s = (H_x + H_y) \sin \alpha \cos \alpha \qquad (d)$$

where $H_s = H_x = H_y$ when $a = b$. The vertical force V_x can be expressed as qx and V_y as qy so that

$$V_s ds = qx\, dy + qy\, dx \qquad (e)$$

$$V_s = V_x \sin \alpha + V_y \cos \alpha \qquad (f)$$

The sum of these forces along the diagonal of length $s = d = a/\cos \alpha$ gives for the corner reactions ($\cos \alpha = \sqrt{2}/2$),

$$H_0 = \int_0^d H_s\, ds = \frac{qa^2}{2c} d = \frac{qa^3}{\sqrt{2}c} \qquad (g)$$

and

$$V_0 = \int_0^d V_s\, ds = qa^2 \qquad (h)$$

This latter value is simply the total load on one-quarter of the square plan as it must be, and the former expression is the diagonal horizontal thrust. Resolving that into the x and y directions would give

$$H_{x0} = H_0 \cos \alpha = \frac{qa^3}{2c} \qquad (i)$$

The groin carries the horizontal and vertical components of the shell loading and it can do so as a three-hinged arch. This is so because, at the crown, the valley loses all its depth and only the shell thickness resists bending; or, in other words, the groin will act about like a hinge there. At the supports, these shells are often supported by hinges to avoid bending there.

Next we explore the bending in the groin by writing the simple expressions for bending about the crown hinge from where for any point S along the s axis (the horizontal projection of the diagonal groin)

$$M_S = -\int_0^S V_s s \, ds + \int_0^S H_s z \, ds + V_S S - H_S z \tag{j}$$

To get simple expressions for V_s and H_s, we can write (d) and (f) in terms of expressions like (a) and (b) to get

$$V_s = 2q \frac{ab}{d^2} s \tag{k}$$

$$H_s = \left(\frac{qa^2}{2c} + \frac{qb^2}{2c}\right) \sin \alpha \cos \alpha = \frac{qd^2}{2c} \sin \alpha \cos \alpha = \frac{qab}{2c} \tag{l}$$

where $d^2 = a^2 + b^2$ and $b/d = \sin \alpha$, etc.

Next we find the forces at any section S by

$$V_S = \int_0^S V_s \, ds = \frac{qab}{d^2} S^2 \tag{m}$$

$$H_S = \int_0^S H_s \, ds = \frac{qab}{2c} S \tag{n}$$

Finally, with $z = cs^2/d^2$, the M_S are found by putting (k), (l), (m), and (n) into (j) to give

$$M_S = -\frac{2qab}{d^2} \frac{S^3}{3} + \frac{qab}{2d^2} \frac{S^3}{3} + \frac{qab}{d^2} S^3 - \frac{qab}{2d^2} S^3$$

$$= \frac{qab}{d^2} S^3 \left(-\frac{2}{3} + \frac{1}{6} + 1 - \frac{1}{2}\right) = 0 \tag{o}$$

which demonstrates that for a parabolic groined vault with vertical loads distributed uniformly over the horizontal projection of the surface, there will be no bending in the groin when the groin is taken as three-hinged and only static equilibrium is considered. By extension, this very simplified discussion suggests that in flat groined vaults one would not normally expect much bending along the groins under gravity loads. On the other hand, where the analysis considers gravity dead loads and where the vault is relatively steep, some bending will arise. An example published by Scordelis shows that for a square-plan vault with a rise of 37.5 ft and with $a = b = 50$ ft, the groin moments due to dead and live loads are still small.[58] His results and those of others show that the loads do deviate somewhat from such simple paths as found here and that the directions of the principal stresses show a tendency near the groin to turn toward the corners, thus indicating that some of the shell next to the groin participates in carrying the forces to the corner supports. Finite-element studies of these groined vaults show that this simplified analysis does give a reasonable basis for studying form, and, unlike for the gabled vault, the studies in equilibrium seem to give reasonable estimates of forces in both shell and groin, where reasonable means reasonably close or, if not, at least values that provide for a reasonably conservative design.

It is crucial to emphasize, however, that this simplified analysis does not include deformations and it does not include any study of boundary conditions other than equilibrium. Thus for any full-scale design, the questions of creep, cracking, and support movement are still to be studied. These questions become more important as the scale increases. The equilibrium analysis does not change for any increase in scale and if relied upon too heavily for large structures, great difficulties can arise.

1-10 THE ELLIPTIC PARABOLOID

There has continuously been the desire to support shells on points and yet take advantage of the efficiency of dome behavior. Parme presented the membrane theory of these shells in 1958,[59] but there have not been many published examples of completed shells of this form. The form is nevertheless of interest because of the light it sheds upon shell behavior, especially compared to domelike shells based upon nongeometric forms.

Figure 1-24 shows an elliptical paraboloid whose form has the equation

$$z = \frac{c_1 x^2}{a^2} + \frac{c_2 y^2}{b^2} \tag{a}$$

The curvatures along the x and y axes, because one slope is always zero, are

$$\frac{1}{r_x} = \frac{\partial^2 z}{\partial s^2} = \frac{2c_1}{a^2} \tag{b}$$

along the y axis and

$$\frac{1}{r_y} = \frac{\partial^2 z}{\partial y^2} = \frac{2c_2}{b^2} \tag{c}$$

along the x axis. If the shell is flat, i.e., if $\partial z/\partial x \ll 1$ and $\partial z/\partial y \ll 1$, then (b) and (c) can be taken as constant over the surface so that arcs of circles with radii r_x and r_y approximate the parabolas.

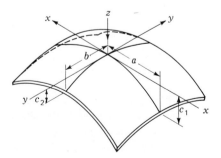

Fig. 1-24 Elliptic paraboloid.

The stress in the shell can be estimated by taking about the x axis the overall static moment as in (a) of Sec. 1-8 to get for the case where $a = b$, $M_x = qa^3$. To get the compression in the shell, we again assume the tension taken by horizontal ties between adjacent supports. Then the lever arm of the overall moment will be $z_0 = c_2 + c_1 - z_t$, where z_t is the distance from the crown of the shell to be centroid of the section at $y = 0$. This centroid is about $c_1/4$ for a flat surface ($\partial z/\partial x \ll 1$ and $\partial z/\partial y \ll 1$) taken as a circular curve, and thus where $c_1 = c_2$,

$$z_0 = \frac{7}{4} c_2 \qquad (d)$$

so that the total compression will be

$$C = \frac{M_x}{z_0} = \frac{4}{7} \frac{qa^3}{c_2} \qquad (e)$$

The area of the section will be simply

$$A \approx 2ah \qquad (f)$$

so that the average stress resultant

$$N'_{y,\text{av}} = \frac{Ch}{A} = \frac{2}{7} \frac{qa^2}{c_2} \qquad (g)$$

This compares with values from a more detailed analysis (see Chap. 7) which give $N'_y = 0.25\, qa^2/c_2$ at $x = y = 0$ and $N'_y = 0.50\, qa^2/Kc_2 \approx 0.47\, qa^2/c_2$ at $x = \pm a$ and $y = 0$ for the case where $c_1/a = 0.2$ and $a = b$ where K is a parameter of curvature. Thus (g) gives a reasonable estimate of the stresses over the central part of the shell, but is about one-third low in this case near the edges.

Actually, the edge stresses and the center stress can be estimated by considering the shell to be arch strips. The edge, where $x = \pm a$ and $y = 0$, acts as an arch strip spanning between adjacent supports as in (a) of Sec. 1-9, so that

$$H_y = N'_y = \frac{qb^2}{2c_2} = 0.5 \frac{qa^2}{c_2} \qquad (h)$$

which is the same as the more detailed analysis, apart from the slight influence of the slope in the x direction (K in the equations of Chap. 7).

At the center of the shell, it acts as two diagonal arch strips spanning between diagonally opposite supports so that, for $a = b$ and $c_1 = c_2$,

$$H = N' = \frac{q(\sqrt{2}a)^2}{2 \times 2(c_1 + c_2)} = 0.25 \frac{qa^2}{c_2} \qquad (i)$$

in both diagonal directions. This isostatic point, therefore, gives also $N'_x = N'_y = N'$ which is the same as from the more detailed theory.

This last analysis makes clear that the shell carries the loads through arch strips directed toward the supports. However, the arch strips between the edge ones and the diagonal central ones do not carry their forces directly to the support, but rather produce in-plane shear at the edges thus requiring an edge arch beam.

1-11 SHELLS DESIGNED BY PHYSICAL ANALOGIES

When Jurgen Joedicke published his shell book surveying progress up to 1962,[60] the various surfaces of rotation and translation dominated. Designers had widely used cylinders, domes, barrels, folded plates, hyperbolic paraboloids, and elliptic paraboloids. Joedicke did include a few other forms, but it then seemed that thin shell concrete structures were those whose surfaces could be defined by well-known mathematical formulas.

The two decades since then marked two trends: first that fewer thin shell concrete roof structures appeared, and second that a few engineers, dissatisfied with the shape restrictions imposed by the mathematically defined forms, began to explore other ways of designing. Foremost among these engineers was the Swiss Heinz Isler, who by 1980 had designed more thin shell roofs structures than any other contemporary practitioner. Joedicke had recognized Isler's potential as early as 1962, but it took many more years before the international engineering profession grasped what Isler was trying to achieve.

At the second Congress of the International Federation of Prestressing in Amsterdam in 1955, Isler presented his earliest ideas, which included pneumatic forms obtained by inflating a rectangular-plan rubber membrane clamped along its four horizontal edges.[61] The upward shape was then in pure tension under upward pressure so that downward pressure would put the same shape (in concrete) into pure compression. The straight horizontal edges being in tension would be pre-

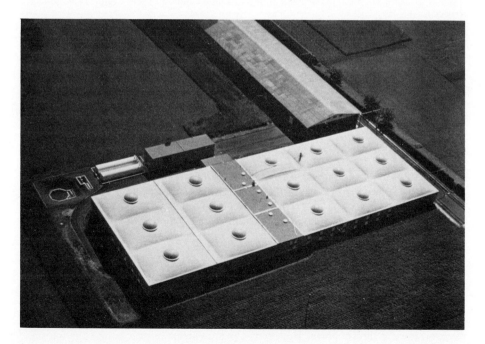

Fig. 1-25 Pneumatically formed shell. (Courtesy Heinz Isler.)

Fig. 1-26   Sicli Building, Geneva, 1969. (Courtesy Heinz Isler.)

Table 1-1

Type of shell	Examples	Method of definition
Sculptural	Simulation of shapes in nature, freely sculpted shapes, and geometric or structural shapes used decoratively rather than structurally	Arbitrary
Geometric	Cylinders, spheres, cones, hyperbolic paraboloids, elliptic paraboloids, segments of these, and others	Mathematical formulas
Structural	Membranes under tension, pneumatic membranes, flowing viscous fluid, and hanging reversed membranes	Physical analogies

stressed, and the entire rectangular-plan roof would be supported by columns at the four corners (Fig. 1-25).

Then in 1959 at the first Congress of the International Association for Shell Structures, Isler created a minor sensation with his new ideas including the pneumatic forms and a second major type: the hanging membrane reversed.[62] Here he simply draped a cloth from any types of supports, coated it with a quick-setting plastic, and when it hardened, turned the shell form over to create a domelike shape. Since the hanging membrane was in pure tension, the reversed membrane under dead load would be in pure compression. In this way Isler has been able to design shells on unusual plan forms without recourse to edge stiffener beams or arches (Fig. 1-26).

In summarizing his approach in 1980, Isler identified three types of shells, as shown in the Table 1-1.[63]

Because Isler's work is so suggestive for the future of thin shell concrete structures, the last chapter will take up his ideas in some detail and at the same time relate them to the tradition of thin shell design that includes Maillart, Freyssinet, Dischinger, Finsterwalder, Nervi, Torroja, Candela, and several more recent designers in the United States.

1-12 BASIS FOR THE MEMBRANE THEORY

A general theory of thin shells is overly complex for the needs of the designer, especially since general shell analysis is commonly done by numerical methods which do not necessarily use such general theories. Two special theories are, however, useful as a basis both for some simplified analyses and for some more detailed analyses. They are the membrane theory and the shallow shell theory (see Chap. 5).

The membrane theory rests on the main assumption that the shell carries loads solely by in-plane stresses and therefore all bending moments and out-of-plane shearing forces are taken as zero.

From this main idea, there follow three significant consequences:

1. The state of stress (membrane stress resultants) in the shell is completely determined by equations of equilibrium, i.e., the shell is statically determinate.
2. The boundary conditions must provide for those shell edge forces which are computed from the equations of equilibrium.
3. The boundary conditions must also permit those shell edge displacements (translations and rotations) which are computed from the forces found by the membrane theory.

Thus the membrane theory needs three equations of equilibrium plus expressions for displacements in terms of the membrane stress resultants.

The formal derivation of the equations for this theory will be given for each system in subsequent chapters.

1-13 GENERAL METHOD OF ANALYSIS

The following general method of analysis is used frequently throughout this book:

1. The loading is considered to be resisted entirely by the membrane stress resultants.
2. The forces and displacements at the shell boundaries computed from the membrane stress resultants will, in general, not be compatible with the known boundary conditions.
3. Forces and displacements (edge effects) must be applied to the shell boundaries to remove the incompatibilities resulting from the membrane stress resultants.
4. The sizes of the edge effects necessary to remove the incompatibilities caused by the membrane stress resultants are found by solving compatibility or equilibrium equations for the shell boundaries.

The above method exactly duplicates the force method of analysis frequently used for statically indeterminate frameworks where

1. The *primary system* is obtained by reducing the general theory to the membrane theory which, in fact, reduces to three equations with three unknowns, and hence to a statically determinate system.
2. The *errors* correspond to the incompatible edge effects.
3. The *corrections* correspond to unit edge effects derived from a bending theory.
4. *Compatibility* is obtained by determining the size of the corrections required to remove the errors in the membrane theory.

To gain some insight into the physical significance of this type of analysis, we may again make an analogy between a thin shell system and a planar system. The truss system of Fig. 1-27a is statically indeterminate externally with regard to its reactions and statically indeterminate internally because of its rigid joints.

We may remove the redundant reactions and thus operate on the primary system of Fig. 1-27b which is highly statically indeterminate because of the bending resistance at the joints. Solution of the system, without regard to the redundant external reactions but retaining the internal bending stiffness of the system, is analogous to the particular integral for a thin shell system. If we neglect the rigid joints and operate on the primary system of Fig. 1-27c, we have a much simpler problem which is analogous to the membrane theory solution for thin shell systems where all bending resistance is neglected. Just as the design of many truss systems may be satisfactorily based on such an approximation, so also may many systems of thin shells.

Finally, the effects of the redundant reactions must be included. These reactions are analogous to the boundary conditions which are determined by the solution to the homogeneous equation. These conditions normally require that the bending resistance of the thin shell be included even though in a truss system we can often continue to base our analysis on the moment-free truss.

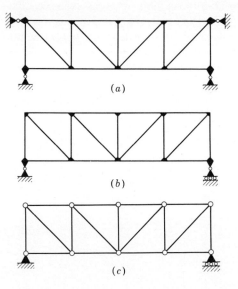

(a)

(b)

(c)

Fig. 1-27 Truss analogy for thin shell behavior.

1-14 CLASSIFICATIONS OF SHELL SYSTEMS

The most general classification of thin shells is by gaussian curvature, illustrated in Fig. 1-28. Shells of positive gaussian curvature, sometimes called *synclastic* shells, are formed by two families of curves both with the same direction. Spherical domes and elliptic paraboloids are examples. Shells of zero gaussian curvature or singly curved shells are formed by one family of curves only; some examples are cylinders and cones. Shells of negative gaussian curvature, sometimes called *anti-clastic,* are formed by two families of curves each in opposite directions. Hyperbolic paraboloids and hyperbolas of revolution are examples.

Mathematically the gaussian curvature of a surface is defined as the product of the principal curvatures:

$$K = \frac{1}{r_x}\frac{1}{r_y}$$

where r_x and r_y are the principal radii of curvature.

One important characteristic difference among these shell types is the propagation of edge effects into the shell.

For shells of positive curvature, the edge effects tend to damp rapidly and are usually confined to a narrow zone at the edge. Thus in these shells the membrane theory will often be valid throughout the entire shell except just at the boundaries. This rapid damping has been clearly demonstrated for spherical domes.

For shells of zero curvature the edge effects are damped but tend to extend further into the shell than for shells of positive curvature.

Finally, for shells of negative curvature the damping is markedly less than for the others. Thus the boundary effects tend to become significant over large portions of the shell.

Another useful classification divides shells into rotational and translational systems. Domes and tanks are normally surfaces of rotation, whereas cylindrical barrels, elliptic paraboloids, and hyperbolic paraboloids are surfaces of translation. This classification is helpful in visualizing the analysis and the construction, but it has little general merit with regard to structural action. For example, the behavior of the shallow spherical domes normally used for roofs is about the same as that of shallow

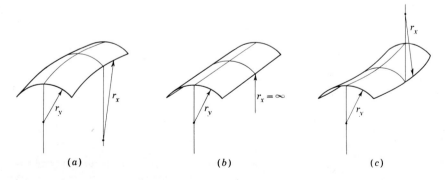

(a)   (b)   (c)

Fig. 1-28 Definitions of curvature.

elliptic paraboloids. In fact, it has been pointed out that for commonly used roof dimensions the maximum difference between the surfaces of rotation and of translation will be considerably less than the thickness of the shell itself (less than 1 in.). Furthermore, as described in Ref. 64, a small constructional deviation in shell form can result in large differences in geometry between the designed and the constructed shell.

1-15 DESIGN AND ANALYSIS

The main justification for thin shell concrete structures lies in their thinness, the achievement of which in turn depends primarily upon the minimization of bending moments and the prevention of buckling. The relationship between the overall design, the setting of form, and the structural analysis is thus an iterative one in which the initial choice of form is the crucial question. That choice should be made such that the moments are as small as possible, and thus often a simplified theory will give a reasonable estimate of behavior. The goal should be that the overall design permits the analysis to be simple. It is far better to alter the overall form so that a simple analysis becomes reliable, rather than to seek safety through complicated analyses of questionable forms. It is far more important to study the loading assumptions, the boundary conditions, and the overall design by simplified means.

The immense CNIT three-point supported shell built in Paris by N. Esquillon (Fig. 1-29) shows an example where the original form was chosen to achieve a simple picture of structural behavior, in this case as three arches.

1-16 DESIGN AND CONSTRUCTION

The thinness and curvature which make shells efficient by reducing materials create problems of construction. It is, therefore, not surprising that most of the best-known pioneers in thin curved concrete structures had long experience in construction as well as in design: Freyssinet, Maillart, Nervi, Finsterwalder, Esquillon, and Candela. In the United States, the early shells built to designs associated with Anton Tedesko were successful largely because of the close cooperation between his firm—Roberts and Schaefer Company—and the builders. Indeed, although Roberts and Schaefer Company did not build the thin shells that they designed, they were a building firm, and they worked closely in the 1930s with Dyckerhoff and Widmann—Finsterwalder's firm of designer-builders.

The problems that these pioneers had to consider can be summarized under two main topics: (1) scaffolding and form (2) reinforcement and concrete. In each subsequent chapter there will be further discussion of specific systems. Here the goal is only to state some general ideas that have wide application.

One major question regarding scaffolding and forming is that of reuse as a means of reducing per-square-foot costs. This reuse can occur either by precasting or by casting in place. Precasting can lead to the reduction in scaffolding altogether, but it also leads to an increase in field connections. As some well-known examples show, precasting is often not an economical solution. Casting in place on a scaffolding which then is moved has been considered to be more economical than precasting in many

Fig. 1-29 CNIT thin shell, Paris.

cases. This is so in particular where the shell is a long barrel and the scaffold can move along a straight track. On the other hand, for domes of rotation, the scaffold can be built for only several pie-shaped segments and reused by rotating to a new position. This method, however, changes design because the partial shell must carry its dead load as an arch without the stiffening effect of the hoops.

Another design question, that of sufficient concrete strength, arises when the scaffolding is moved and a section of shell must then carry its loads early in its life. The young concrete in roof shells can be subjected to higher creep and thus greater deformations than older concrete.

The reinforcement and the concrete influence construction largely in two ways. The simpler the reinforcing pattern, the easier to build, even though simplicity may sometimes lead to more reinforcement than a pattern carefully detailed to reflect the calculated stresses. The concrete problem mainly arises where the design requirement for a stiff mix to achieve small creep and shrinkage conflicts with the construction desire for a workable mix to reduce honeycombing and to permit rapid placement.

1-17 DESIGN AND APPEARANCE

The fine visual potential inherent in thin shells has been exploited by many designers. Apart from the striking appearance of their shells, the excellence of these designers' solutions depended upon their concern for minimizing materials and costs.

When these two fundamental constraints are not recognized as essential to the success of their works, then designers can easily fall into one of several traps in design.

The first trap is to become strongly attracted to a shell geometry without first understanding its structural behavior. For example, because of the well-publicized elegance of hyperbolic paraboloids, a number of such shells got built to designs that were made by inexperienced designers and the results were often not good.

The second trap is to believe that any shell form can be built partly because shells are so efficient and partly because technology is so advanced. A series of well-publicized examples of such designs using shells outside the experience of successful forms have shown how such ideas lead to excessive costs, to excess materials, and at times to questionable structural performance. Examples of these are the Sidney Opera House and the Montreal Olympic Stadium.

The third trap and the most basic one is to imagine that fine appearance is something separate from fine engineering. The primary reason why the best-known shell designers have made such striking works is that they conceived of design as the integration of efficiency, economy, and elegance. Their works are not elegant just because they are efficient and economical; there are too many examples of such works which are not visually striking. Rather, the best designers work within the disciplines of efficiency and economy and at the same time let their aesthetic sense direct their choices to ones of visual appeal. They have understood that engineering design at its best includes aesthetics as a central concern, and they have succeeded best because they have been not just analysts but designers, not just applied scientists but also artists.

REFERENCES

1. In fact, reinforced concrete originated in thin shells designed by Lambot for boats and by Monier for water tanks before 1870.
2. Wayss, G. A., *Das System Monier,* Berlin, 1887, pp. 33–34 and 117–126.
3. Wuezkowski, R., "Flüssigkeitsbehälter," in F. von Emperger (ed.), *Handb. für Eisenbetonbau,* vol. 5, Berlin, 1910, p. 329; and for the calculations, pp. 483–490.
4. Reissner, H., "Ueber die Spannungsverteilung in Zylindrischen Behälterwanden," *Beton und Eisen,* vol. 7, no. 6, Apr. 22, 1908, pp. 150–155.
5. Wuezkowski, *op. cit.,* pp. 325–355.
6. Hertwig, A., *Johann Wilhelm Schwedler,* Verlag von Wilhelm Ernst and Sohn, Berlin, 1930.
7. Kohnke, R., "Die Kuppelgewölbe," in F. von Emperger (ed.), *Handb. für Eisenbetonbau,* vol. 4, Berlin, 1909, pp. 546–569.
8. Geckeler, J. W., "Die Love-Meissnersche Theorie rotations symmetrischer Schalen," *Handb. der Phys.,* vol. 6, Berlin, 1928, pp. 238–246.
9. Bauersfeld, W., "Development of the Zeiss-Dywidag Process," lecture given on Dec. 12, 1942, in Berlin; published in J. Joedicke, *Shell Architecture,* Reinhold Publishing Corporation, New York, 1963, pp. 281–283.
10. Geckeler, J., "Uber die Festigkeit achsensymmetrischer Schalen, insbesondere von Dampfkesseln," *Z. des Vereines dtsch. Ing.,* vol. 70, 1926, pp. 163–168. This is essentially an abstract of Geckeler's treatise, published separately as J. Geckeler, *Uber die Festigheit achsensymmetrischer Schalen,* Forschungsarbeiten des Vereines deutscher Ingenieure 276, Berlin, 1926.
11. Geckeler, J. W., "Elastostatik," *Handb. der Phys.,* vol. 6, chap. 3, 1928, pp. 141–208.
12. Dischinger, F., "Schalen and Rippenkuppeln," in F. von Emperger (ed.), *Handb. für Eisenbetonbau,* vol. 12, 3d ed., Berlin, 1928, pp. 151–371.

13. *Ibid.*, p. 326. For St. Peter's and the Pantheon see B. Fletcher, *A History of Architecture,* 14th ed., Charles Scribner's Sons, New York, 1948, pp. 157, 642, 647.
14. Dischinger, F., "Fortschritte im Bau von Massivkuppeln," *Der Bauingenieur,* Heft 10, 1925, p. 362.
15. Bauersfeld, *op. cit.,* p. 283.
16. Dischinger, F., "Eisenbetonschalendacher Zeiss-Dywidag zur Ueberdachung weitgespannter Räume," *First International Congress for Concrete and Reinforced Concrete,* vol. 1, Liège, Belgium, 1930, pp. 262–291.
17. Dischinger, *op. cit.,* 1928, pp. 257–259.
18. Dischinger, F., and U. Finsterwalder, "Die Dywidag-Halle auf der Gesolei," *Der Bauingenieur,* no. 48, 1926, pp. 929–931.
19. Rüsch, H., "Ulrich Finsterwalder zu seinem funfzigsten Dienstjubilaum, Sein Lebensweg als Mensch und Ingenieur," *Festsch. Ulrich Finsterwalder,* 50 Jahre Für Dywidag, Dyckerhoff and Widmann, Karlsruhe, 1973, pp. 9–18.
20. *Ibid.*, p. 10.
21. Geckeler, *op. cit.,* 1928, pp. 260–265; and Dischinger, *op. cit.,* 1938, pp. 260–263.
22. Finsterwalder, U., "Die Querversteiften Zylindrischen Schalengewölbe mit kreissegmentformigem Querschnitt." *Ing.-Arch.,* vol. 6, 1933, pp. 43–65. The work was submitted for a prize competition to the Prussian Academy of Construction on Jan. 15, 1930, where it won second prize.
23. Rüsch, H., "Theorie der querversteiften Zylinderschalen für schmale, unsymmetrische kreisseg-mente," dissertation, Universitätsverlag von Robert Noske in Borma-Leipzig, E. Germany, 1931.
24. Joedicke, J., *Shell Architecture,* Reinhold Publishing Corporation, New York, 1963, pp. 11 and 14.
25. Dischinger, F., and U. Finsterwalder, "Die Grossmarkthalle Frankfurt a.M.," *Z. des Vereines dtsch. Ing.,* vol. 73, no. 33, Aug. 17, 1929, pp. 1145–1148.
26. "Zeiss-Dywidag," Roberts and Schaefer Co., Bull. No. 138, January 1932.
27. Bertin, R. L., "Construction Features of the Zeiss Dywidag Dome for the Hayden Planetarium Building," *J. ACI,* vol. 31, May–June 1935, p. 449.
28. *Eng. News-Rec.,* June 14, 1934, p. 775.
29. "Thin Concrete Shell Roof Tested under Large Unsymmetrical Load," *Eng. News-Rec.,* Nov. 7, 1935.
30. "Thin Concrete Shells for Domes and Barrel-Vault-Roofs," *Eng. News-Rec.,* Apr. 19, 1932, p. 537. Before this article there had appeared three publications by A. Florison one in *Eng. News-Rec.* of Feb. 21, 1929; one in *West. Constr. News* of Apr. 10, 1929; and one in the *Trans. ASCE,* 1930, p. 1173 (a discussion). There had also appeared an article in *Archit. Forum,* July 1931, and *Constr. Methods,* April 1932.
31. Witmer, D. Paul, "Sports Palace for Chocolate Town," *Archit. Concr.,* vol. 3, no. 1, 1937, p. 3.
32. Tedesko, Anton, "Z-D Shell Roof at Hershey," *Archit. Concr.,* vol. 3, no. 1, pp. 7–11; and Anton Tedesko, "Large Concrete Shell Roof Covers Ice Arena," *Eng. News-Rec.,* Apr. 8, 1937.
33. "Zeiss-Dywidag," *op. cit.,* 1932, p. 1.
34. Discussion to Herman Schorer, "Line Load Action on Thin Cylindrical Shells," *Trans. ASCE,* vol. 101, 1936, p. 767.
35. Tedesko, A., "Tire Factory at Natchez," *Eng. News-Rec.,* Oct. 26, 1939.
36. Tedesko, A., "Thin Concrete Shell Roof for Ice Skating Arena," *Eng. News-Rec.,* Feb. 16, 1939.
37. Tedesko, A., "Point Supported Dome of Thin Shell Type," *Eng. News-Rec.,* Dec. 7, 1939; and Eric C. Molke, "Elliptical Concrete Domes for Sewage Filters," *Eng. News-Rec.,* Nov. 9, 1939.
38. Kalinka, J. E., "Monolithic Concrete Construction for Hangars," *Mil. Eng.,* January–February 1940.
39. "Large Concrete Warehouses Built with Moving Falsework," *Eng. News-Rec.,* Apr. 24, 1941. The bidding prices appeared in a promotional drawing made by Tedesko on Jan. 16, 1941.
40. Tedesko, A., "Multiple Ribless Shells," *J. Struct. Div.,* ASCE, vol. 87, October 1961.
41. "Cast in Place Concrete Shell Structures Bid below Price of Precast Structures," *Civ. Eng.,* September 1957.
42. Tedesko, A., "Wide-Span Hangars for the U.S. Navy," *Civ. Eng.,* vol. 11, no. 12, December 1941, pp. 697–700.
43. Allen, J. E., "Construction of Long-Span Concrete Arch Hangars at Limestone Airforce Base," *J. ACI,* vol. 46, February 1950, pp. 405–414.

44. Gray, N., "Form on Traveler Speeds Arch Construction," *Eng. News-Rec.*, vol. 141, 1948, pp. 108–110; and Boyd G. Anderson, "Thin Shell Arch Hangars Span 257 Feet," *Eng. News-Rec.*, vol. 141, 1948, pp. 55–57.
45. Whitney, C. S., "Discussion," *Trans. ASCE*, vol. 88, 1925, pp. 116–119; and in the same volume also by Whitney, "Design of Symmetrical Concrete Arches," pp. 931–1029, and "Discussion," pp. 1241–1243.
46. "Design of Cylindrical Concrete Shell Roofs," *Man. Eng. Pract., ASCE*, no. 31, 1952.
47. Petry, W., "Plates and Shells in Reinforced Concrete," *Proceedings*, First Congr. Int. Assoc. Bridge and Struct. Eng., held at Paris in 1932, published in Zurich, 1932, pp. 267–302.
48. Maillart, R., "Silomagazin" (Storage Silos), Swiss Pat. 72011, Oct. 9, 1915.
49. Phase I Report of the Task Committee on Folded Plate Construction, *J. Struct. Div.*, ASCE, vol. 89, December 1963.
50. Scordelis, A. C., "Analysis of Cylindrical Shells and Folded Plates," *Concr. Thin Shells*, ACI, Publ. SP-28, 1971, pp. 207–236.
51. Aimond, F., "Thin Shell Roofs in the form of a Hyperbolic Paraboloid," *Le Genie Civ.*, vol. 52, no. 8, Feb. 25, 1933.
52. Aimond, F., "Treatise on Statics of Parabolic-Hyperboloidal Shells Not Stiff in Bending," *Publ.*, Int. Assoc. Bridge and Struct. Eng., vol. 4, 1936.
53. Hahn, M. L., "Application des voiles minces en paraboloide hyperbolique à la construction d'un garage en Beton Armé," *Trav.*, May 1937, pp. 207–215.
54. *Proceedings of a Conference on Thin Concrete Shells*, MIT, Cambridge, June 21–23, 1954. Candela's two papers are on pp. 5–11 and 91–98.
55. Faber, C., *Candela, the Shell Builder*, Reinhold Publishing Corporation, New York, 1963.
56. Candela, F., "Structural Applications of Hyperbolic Paraboloidal Shells," *J. ACI*, January 1955, pp. 397–415.
57. Parme, A. L., "Shells of Double Curvature," *Trans. ASCE*, vol. 123, 1958, p. 989.
58. Scordelis, Alexander C., "Analysis and Design of HP Groined Vaults," *Proceedings*, IASS World Congr. Space Struct., Montreal, 1976, pp. 561–568.
59. Parme, A. L., *op. cit.*, 1958.
60. Joedicke, J., *Shell Architecture*, Rheinhold Publishing Corporation, New York, 1963.
61. Isler, H., Discussion in Sess. IIIB, *Second Congress of the Fédération Internationale de la Précontrainte*, Amsterdam, 1955, pp. 736–742.
62. Isler, H., "New Shapes for Shells," *Bull. IASS*, no. 8, paper C-3, 1960.
63. Isler, H., "New Shapes for Shells—Twenty Years Later," *Heinz Isler as Structural Artist*, The Art Museum, Princeton University, N.J., 1980.
64. Tottenham, H., "Approximate Solutions to Shell Problems," *Proceedings of the Second Symposium on Concrete Shell Roof Construction*, Teknisk Ukeblad, Oslo, 1958.

2
ANALYSIS OF SHELL WALLS

2-1 INTRODUCTION

Thin shell walls have been most commonly used as pressure vessels or as dams. However, they can be used as bearing walls for tower structures. The structural systems treated in this chapter are shells which have the form of surfaces of revolution: circular cylinders, cones, and hyperboloids. The first two systems are normally liquid containers; the hyperboloid will be analyzed as a cooling tower. The membrane theory will be presented for all three types. The circular cylinder is the simplest and the most frequently utilized shell wall system and will be discussed in detail. The bending theory will be presented for vertical cylindrical walls monolithically connected to circular plates both as floor slabs and as roof slabs. The analysis of domed roof slabs connected to vertical cylinders will be given in Chap. 4.

Shell walls under wind load have become a significant design problem with the wide use of tall cooling towers. Therefore, this chapter concludes with an introduction into the bending analysis of shell walls for nonaxisymmetrical loading. The emphasis here is on the shell behavior, on the differences between results from membrane and bending theories, and on the influence of form on structural response. A discussion of cooling tower design comes in Chap. 4.

2-2 STRESS RESULTANTS FROM THE MEMBRANE THEORY FOR CIRCULAR CYLINDRICAL SHELLS

In structural engineering, equilibrium equations usually come from taking the forces on a differential element which, being infinitely small, represents a point on the structure. Thus the differential lengths, dy and $r\,d\theta$ in Fig. 2-1a, are artificial and must always cancel out in the resulting equations; they are merely a means to the end of equilibrium. There are other ways to get the same equilibrium equations, such as by the principle of virtual work which does not make use of the differential element at all. Here, we use that element because it seems easier for the engineer to visualize derivations by having a physical picture of the forces rather then by using the more abstract ideas of virtual work or the tensor calculus.

For circular cylindrical shells, the membrane theory equilibrium equations come from taking the sum of forces in three orthogonal directions: θ circumferentially, y vertically, and z radially (see Fig. 2-1a). The radius r is constant with y, and the

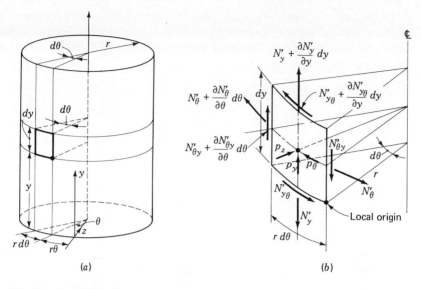

(a) (b)

Fig. 2-1 Definitions for cylindrical shells.

surface loadings are p_θ, p_y, and p_z. Starting with the θ direction, we find from the differential element in Fig. 2-1b that the total circumferential force on the positive face—i.e., that face a positive distance $r\,d\theta$ from the local origin—is equal to

$$\left(N'_\theta + \frac{\partial N'_\theta}{\partial \theta}\,d\theta\right) dy\,\cos\,d\theta$$

where N'_θ defines the magnitude of the stress resultant on the negative face and $(\partial N'_\theta / \partial \theta)\,d\theta$ defines the change in that magnitude from one face to the next. The prime denotes membrane theory values. The $\cos\,d\theta$ expresses the fact that the tangent to the surface at the positive face is not in the same plane as the tangent at the local origin on negative face. Because $d\theta$ is a small angle, we can always take $\cos d\theta \approx 1$ in these derivations. The stress resultant is in units of force per unit length of shell middle surface so that when multiplied by a middle-surface length dy, the result is in units of force.

The circumferential force on the negative face is merely $-N'_\theta\,dy$ so that the net circumferential force acting on the element is

$$\frac{\partial N'_\theta}{\partial \theta}\,d\theta\,dy \qquad\qquad\qquad (a)$$

because the two terms in $N'_\theta\,dy$ cancel out once $\cos d\theta$ is taken as unity.

The in-plane, shear stress resultant on the positive face gives

$$\left(N'_{y\theta} + \frac{\partial N'_{y\theta}}{\partial y}\,dy\right) r\,d\theta$$

and on the negative face $-N'_{y\theta} r\, d\theta$, where the curved length of the element is $r\, d\theta$ and the angular difference between these forces and the tangent plane at the local origin is negligible as before. Thus the total contribution of the shear to the circumferential equilibrium is

$$\frac{\partial N'_{y\theta}}{\partial y} dy\, r\, d\theta \tag{b}$$

The N'_y and the $N'_{\theta y}$ terms are orthogonal to the θ direction and hence do not enter into this equilibrium expression. The only other term is the surface loading p_θ, a pressure, whose total force on the element is

$$p_\theta\, dy\, r\, d\theta \tag{c}$$

where the element area is $dy\, r\, d\theta$. The first equilibrium equation is formed by summing the terms (a), (b), and (c) and canceling the common differential terms $d\theta\, dy$.

The second equilibrium equation follows the same type of derivation now taken in the y direction. The vertical forces give

$$\frac{\partial N'_y}{\partial y} dy\, r\, d\theta \tag{d}$$

the vertical in-plane shears give

$$\frac{\partial N'_{\theta y}}{\partial \theta} d\theta\, dy \tag{e}$$

and the pressure loading

$$p_y\, dy\, r\, d\theta \tag{f}$$

The equation results from adding (d), (e), and (f) and canceling again $d\theta\, dy$.

Finally the third equation comes from considering the components of the four membrane theory stress resultants N'_θ, $N'_{y\theta}$, N'_y, and $N'_{\theta y}$ in the radial or z direction. In this simple case of the cylinder, only N'_θ has such a component. The force on the positive face acts at an angle of $d\theta$ from the tangent plane at the local origin. Thus the component of that force in the z direction is

$$\left(N'_\theta + \frac{\partial N'_\theta}{\partial \theta} d\theta \right) dy\, \sin d\theta$$

By small angles for which $\sin d\theta \approx d\theta$, that term reduces to

$$N'_\theta\, dy\, d\theta + \frac{\partial N'_\theta}{\partial \theta} d\theta\, dy\, d\theta$$

As in the previous two equations, we shall cancel the terms $dy\, d\theta$, but now there remains a $d\theta$ in the second term. Therefore, since $d\theta \ll 1$ by definition, we can neglect any term in a derivation which, after canceling common differential lengths, is still left with an extra one. This is what is meant by neglecting higher-order terms. Thus the total contribution of N'_θ is

$$N'_\theta \, dy \, d\theta \tag{g}$$

Also there is the influence of the radial pressure

$$p_z \, dy \, r \, d\theta \tag{h}$$

It is easy to see that N'_y and $N'_{\theta y}$ play no role because they are exactly vertical, but to show why $N'_{y\theta}$ is neglected, we need to consider again the influence of small angles. The shear component (b) acts at an angle $d\theta/2$ from the tangent plane to the origin. Thus in the z direction it will have a component

$$\frac{\partial N'_{y\theta}}{\partial y} \, dy \, r \, d\theta \, \frac{d\theta}{2}$$

This term contains an extra $d\theta$ which, being of a higher order than (g), can be neglected.

The final results are three equations of equilibrium in four unknowns N'_θ, N'_y, $N'_{y\theta}$, and $N'_{\theta y}$:

$$\frac{\partial N'_\theta}{\partial \theta} + r \frac{\partial N'_{y\theta}}{\partial y} + p_\theta r = 0$$

$$r \frac{\partial N'_y}{\partial y} + \frac{\partial N'_{\theta y}}{\partial \theta} + p_y r = 0 \tag{2-1}$$

$$\frac{N'_\theta}{r} + p_z = 0$$

From the definition of the stress resultants, we have

$$N'_{\theta y} = \int_{-h/2}^{h/2} \tau'_{\theta y} \, dz \tag{i}$$

$$N'_{y\theta} = \int_{-h/2}^{h/2} \tau'_{y\theta} \, (1 - z/r) \, dz \tag{j}$$

Since in the theory of elasticity $\tau'_{y\theta} = \tau'_{\theta y}$, $N'_{y\theta} \neq N'_{\theta y}$. However, in thin shell theory $z/r \ll 1$ so that the small term is dropped in (j), leaving $N'_{y\theta} = N'_{\theta y}$. In this way the three equilibrium equations (2-1) describe a statically determinate structure with three unknowns and, therefore, general expressions for the stress resultants are easily derived from (2-1).

Beginning with the third equation, we find

$$N'_\theta = -p_z r \tag{2-2a}$$

The first of (2-1) can be written as

$$\frac{\partial N'_{y\theta}}{\partial y} = -\frac{1}{r} \frac{\partial N'_\theta}{\partial \theta} - p_\theta \tag{2-2b}$$

and the second of (2-1) as

$$\frac{\partial N'_y}{\partial y} = -\frac{1}{r}\frac{\partial N'_{\theta y}}{\partial \theta} - p_y \tag{2-2c}$$

Upon integration, we obtain

$$N'_\theta = -p_z r$$

$$N'_{y\theta} = -\int \left(\frac{1}{r}\frac{\partial N'_\theta}{\partial \theta} + p_\theta\right) dy + f_1(\theta) \tag{2-3}$$

$$N'_y = -\int \left(\frac{1}{r}\frac{\partial N'_{\theta y}}{\partial \theta} + p_y\right) dy + f_2(\theta)$$

where $f_1(\theta)$ and $f_2(\theta)$ depend upon boundary conditions. Eqs. (2-3) can now be solved for various common loadings.

Gravity Loads

Where there are only dead loads or axisymmetric gravity live loads, the first of Eqs. (2-1) disappears and the second can be rewritten directly in terms of N'_y only by expressing the total vertical load as a single, axial load R equilibrated by $2\pi r N'_y$; then Eqs. (2-1) become

$$N'_y = \frac{R}{2\pi r}$$

$$N'_\theta = -p_z r \tag{2-1a}$$

For the vertical walls under vertical axisymmetrical loading only, $p_z = 0$ and $p_y = -q_y$; then

$$R = -2\pi r \int_y^H q_y \, dy$$

If $q_y = $ a constant q, then $R = -2\pi r q(H - y)$ and

$$N'_y = -q(H - y)$$
$$N'_\theta = 0$$

while if q_y decreases with y say from q_H at the top to q_0 at the bottom for the dead load of a tapered wall, then

$$q_y = q_0(1 - \alpha y)$$

where

$$\alpha = (q_0 - q_H)/q_0 H \quad \text{and} \quad R = -2\pi r q_0 \left[(H - y) - \frac{\alpha}{2}(H^2 - y^2)\right]$$

giving from Eqs. (2-1a)

$$N'_y = -q_0\left[(H - y) - \frac{\alpha}{2}(H^2 - y^2)\right]$$

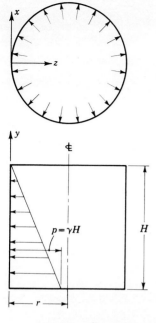

Fig. 2-2 Cylindrical shells under axisymmetrical pressure loadings.

Internal Pressure Loading

For this axisymmetrical case, $p_\theta = p_y = 0$ and $p_z = pf(y)$ (see (Fig. 2-2). Since p_z is independent of θ, Eqs. (2-3) reduce to

$$N'_\theta = -rpf(y)$$

$$N'_{y\theta} = f_1(\theta)$$

$$N'_y = - \int \left(\frac{1}{r} \frac{\partial}{\partial \theta} f_1(\theta) \right) dy + f_2(\theta)$$

where $f_1(\theta)$ and $f_2(\theta)$, which represent stress resultants independent of N'_θ and y, can be taken as zero. The effects of $f_1(\theta)$, a shear force at the edges, and $f_2(\theta)$, a vertical edge force or the dead weight of a roof, can be superimposed on the results of the analysis for pressure loads.

Since the y axis is directed upward from the bottom, then for pressure due to a head of liquid H ft with density γ lb/ft^3,

$$p_z = -\gamma(H - y)$$

and

$$\begin{aligned} N'_\theta &= \gamma(H - y)r \\ N'_{y\theta} &= N'_y = 0 \end{aligned} \tag{2-4}$$

Seismic Loading

As an approximate analysis for earthquake (seismic) loading, we can take a horizontal loading in one direction, say $\theta = 0$, which is equal to the dead weight times

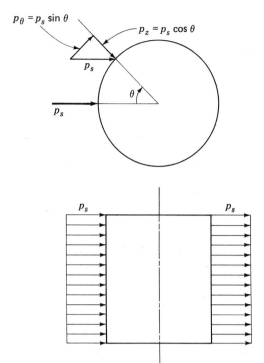

Fig. 2-3 Cylindrical shells under seismic-type loadings.

the ratio a/g, where a is the horizontal seismic acceleration and g is the acceleration of gravity. Thus the loading will be

$$p_\theta = p_s \sin \theta$$
$$p_y = 0$$
$$p_z = p_s \cos \theta$$

where vertical seismic acclerations are not considered here and where $p_s = p_d a/g$ (see Fig. 2-3). If the density of concrete is taken as γ_c, then $p_d = h\gamma_c$.

The Eqs. (2-3) can now be integrated as before where

$$N'_\theta = -p_s r \cos \theta \qquad\qquad (2\text{-}5a)$$

and

$$N'_{y\theta} = -\int (p_s \sin \theta + p_s \sin \theta)\, dy + f_1(\theta)$$
$$= -2 p_s y \sin \theta + f_1(\theta)$$

at $y = H$, $N'_{y\theta} = 0$, from which

$$f_1(\theta) = 2 p_s H \sin \theta$$

and

$$N'_{y\theta} = 2 p_s (H - y) \sin \theta \qquad\qquad (2\text{-}5b)$$

Finally

$$N'_y = - \int 2 \frac{p_s}{r} (H - y) \cos \theta \, dy + f_2(\theta)$$

$$= - \frac{2p_s}{r} \left(Hy - \frac{y^2}{2} \right) \cos \theta + f_2(\theta)$$

at $y = H$, $N'_y = 0$; thus

$$f_2(\theta) = \frac{p_s}{r} H^2 \cos \theta$$

and

$$N'_y = \frac{p_s}{r} (H - y)^2 \cos \theta \qquad\qquad (2\text{-}5c)$$

Eqs. (2-5) show that for the distribution of stress resultants, the shear varies linearly with y and the vertical stress parabolically, both as in a cantilever beam under uniform load. In fact, the same results may be obtained directly from statics without (2-3) by using the familiar beam formulas

$$v = \frac{VQ}{Ib}$$

$$\qquad\qquad (2\text{-}6)$$

$$f = \frac{Mx}{I}$$

in which

$$v = \frac{N'_{y\theta}}{h}$$

$$V = 4 \int_y^H \int_0^{\pi/2} p_s r \, d\theta \, dy = 2p_s \pi r (H - y)$$

$$Q = 2 \int_0^\theta hr^2 \cos \theta \, d\theta = 2hr^2 \sin \theta$$

$$\qquad\qquad (2\text{-}7)$$

$$I = 4 \int_0^{\pi/2} hr^3 \cos^2 \theta \, d\theta = \pi hr^3$$

$$b = 2h$$

$$f = \frac{N'_y}{h}$$

$$M = p_s \pi r (H - y)^2$$

$$x = r \cos \theta$$

Introducing (2-7) into (2-6), we obtain

$$N'_{y\theta} = \frac{2p_s \pi r (H - y) \, 2hr^2 \sin \theta}{2\pi hr^3} = 2p_s (H - y) \sin \theta$$

$$N'_y = \frac{p_s \pi r (H - y)^2 r \cos \theta}{\pi r^3} = \frac{p_s}{r} (H - y)^2 \cos \theta$$

which are identical to (2-5). N'_y plotted on a straight-line projection of the cross section is a straight line, and the component of shear in the direction of the load $N'_{y\theta}$ $\sin \theta$ is a parabola, just as in a beam cross section.

Wind Loads

In general, wind velocity causes external pressure on structures that is well estimated by

$$p_{30} = 0.0026 \, V_{f30}^2$$

where V_{f30} is the velocity measured at 30 ft above ground in open country for the fastest mile of wind; i.e., the mean wind velocity over a time that corresponds to a movement of air of 1 mi. Thus $V_{f30} = 60$ mph implies an average wind velocity for 1 min and $V_{f30} = 120$ mph for ½ min. Values of V_{f30} can be found from Isotach charts.[1] At any point on a cylindrical shell, the pressure can be expressed by

$$p_z = H_\theta K_z G p_{30}$$
$$p_\theta = p_y = 0$$

where H_θ gives the circumferential distribution of pressure, K_z gives the vertical distribution, and G is a gust factor which depends upon the dynamic response of the structure to wind turbulence. We first consider H_θ.

Wind acts on cylindrical shells by exerting an external pressure over an arc length of about $\pm 35°$ from the windward meridian and then suction rising to a peak at about $\pm 70°$ and dropping to a nearly constant value around the leeward half of the shell. Table 2-1 gives values for this circumferential distribution in terms of θ and in terms of a Fourier series which approximates that curve reasonably well as

$$H_\theta = \sum_{n=0}^{7} A_n \cos n\theta$$

The Niemann values in Table 2-1 come from low velocity wind measurements recorded by Niemann,[2] whereas the design values come from analysis of higher velocity measurements recorded by Sollenberger and Scanlan[3] and by Niemann[4] and are probably more representative of design wind loadings.

Furthermore the wind pressure varies in the vertical direction from zero at ground level to a maximum value at some elevation called the gradient height z_g above which the wind velocity is practically constant. This variation depends upon the roughness of the terrain. For example, where the surface is continuously interrupted by buildings, as in a city, the wind pressures near the ground are lower and it takes a greater distance to reach the gradient height, while in open country the gradient height comes much lower. One recommended guide[5] for this vertical distribution gives

$$K_z = 2.64 \left(\frac{z}{z_g} \right)^{2\alpha} \tag{k}$$

where for

Cities (Exposure A) $\alpha = 1/3$, $z_g = 1500$ ft
Rolling country or shrubs (Exposure B) $\alpha = 1/4.5$, $z_g = 1200$ ft
Open country (Exposure C) $\alpha = 1/7$, $z_g = 900$ ft

The coefficient 2.64 in (k) is derived from the condition that $K_z = 1.0$ for $z = 30$ ft in Exposure C, typical of locations (such as airports) where such wind records have been recorded.

Finally the determination of G depends upon a dynamic analysis of the thin shell, and such analyses have shown that values of $G \leq 1.0$ are reasonable under certain conditions for design wind pressures, so that we shall take $G = 1$ here.[6]

Table 2-1

Angle from windward meridian, θ degrees	H_θ coefficients	n	A_n
	Design values*		
0	1.0	0	-0.2636
15	0.8	1	0.3419
30	0.3	2	0.5418
45	-0.3	3	0.3872
60	-0.8	4	0.0525
75	-1.0	5	-0.0771
90	-0.8	6	-0.0039
105	-0.5	7	0.0341
120	-0.4		
135	-0.4		
150	-0.4		
165	-0.4		
180	-0.4		
	Niemann values†		
0	1.0	0	-0.3923
15	0.8	1	0.2602
30	0.2	2	0.6024
45	-0.5	3	0.5046
60	-1.2	4	0.1064
75	-1.3	5	-0.0948
90	-0.9	6	-0.0186
105	-0.4	7	0.0468
120	-0.4		
135	-0.4		
150	-0.4		
165	-0.4		
180	-0.4		

* Measurements from Ref. 4.
† Measurements from Ref. 2.

To derive membrane theory stress resultants, it is simplest to consider only one term at a time in the Fourier series and then obtain a complete solution by superposition. Also for the vertical distribution it is a simple approximation to assume either a sine or a uniform distribution for preliminary design, where the pressure at the top of the cylinder p_H comes from (k) with

$$p_H = 2.64 \left(\frac{H}{Z_g}\right)^{2\alpha} p_{30}$$

We begin by making the conservative assumption that p_z is constant with y so that

$$p_{zn} = p_H A_n \cos n\theta$$

from which (with $p_\theta = p_y = 0$)

$$N'_{\theta n} = -p_H r A_n \cos n\theta \tag{2-8a}$$
$$N'_{y\theta n} = -p_H y A_n n \sin n\theta + f_{1n}(\theta)$$

At $y = H$, $N'_{y\theta n} = 0$ again so that

$$f_{1n}(\theta) = p_H H A_n n \sin n\theta$$

and

$$N'_{y\theta n} = p_H (H - y) A_n n \sin n\theta \tag{2-8b}$$

Then

$$N'_{yn} = -\frac{p_H}{r}\left(Hy - \frac{y^2}{z}\right) A_n n^2 \cos n\theta + f_{2n}(\theta)$$

At $y = H$, $N'_{yn} = 0$ once more, and hence

$$f_{2n}(\theta) = \frac{p_H H^2}{2r} A_n n^2 \cos n\theta$$

and

$$N'_{yn} = \frac{p_H}{2r}(H - y)^2 A_n n^2 \cos n\theta \tag{2-8c}$$

If now we consider a sine variation for p_z,

$$p_z = p_H \sin \frac{\pi y}{2H} \sum_0^7 A_n \cos n\theta \tag{1}$$

Then Eqs. (2-8) become

$$N'_{\theta n} = -p_H r \sin \frac{\pi y}{2H} A_n \cos n\theta$$

$$N'_{y\theta n} = p_H \frac{2}{\pi} H \cos \frac{\pi y}{2H} A_n n \sin n\theta \tag{2-9}$$

$$N'_{yn} = \frac{p_H}{2r} \frac{8H^2}{\pi^2}\left(1 - \sin \frac{\pi y}{2H}\right) A_n n^2 \cos n\theta$$

When the top of such a cylinder is open, the wind blowing over it will cause an internal suction which is normally approximated as constant with θ and as having the same vertical variation as the external p_z. Thus the effect of such an internal suction is to increase the value of A_0 in Table 2-1. A discussion of structural behavior under wind loads comes in Sec. 2-9 where several examples are worked out.

2-3   STRESS RESULTANTS FROM THE MEMBRANE THEORY FOR CONICAL SHELLS

Following the same derivation as for cylindrical shells, we express equilibrium in the θ direction for the differential element of Fig. 2-4. The influence of N'_θ is as before:

$$\frac{\partial N'_\theta}{\partial \theta} \, d\theta \, dy \tag{a}$$

The effect of $N'_{y\theta}$ is different for cones because the positive y face is longer than the negative one; therefore,

$$-N'_{y\theta} r_0 \, d\theta + \left(N'_{y\theta} + \frac{\partial N'_{y\theta}}{\partial y} \, dy \right) \left(r_0 + \frac{\partial r_0}{\partial y} \, dy \right) d\theta$$

from which we find

$$\frac{\partial N'_{y\theta}}{\partial y} r_0 \, d\theta \, dy + N'_{y\theta} \frac{\partial r_0}{\partial y} d\theta \, dy = \frac{\partial (N'_{y\theta} r_0)}{\partial y} d\theta \, dy \tag{b}$$

where the higher-order term, $dy \, dy \, d\theta$, is neglected as before. Again the terms in N'_y do not enter since they are orthogonal to θ; but terms in $N'_{\theta y}$ do enter because the

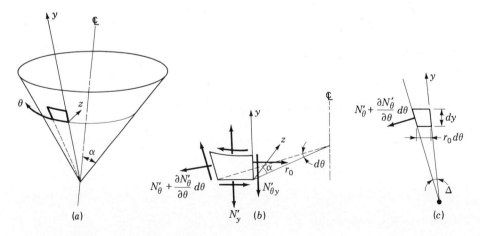

Fig. 2-4   Definitions for conical shells.

positive θ face is now sloped with respect to its negative face by the angle (Fig. 2-4c)

$$\Delta = \frac{\partial r_0}{\partial y} \, d\theta$$

Thus $N'_{\theta y}$ on the positive face has a component in the θ direction of

$$N'_{\theta y} \, dy \sin \Delta \approx N'_{\theta y} \frac{\partial r_0}{\partial y} \, d\theta \, dy \qquad (c)$$

where $\sin \Delta \approx \Delta$ and where the higher-order term in $N'_{\theta y}$ is again neglected. Finally the pressure loading term is

$$p_\theta r_0 \, d\theta \, dy \qquad (d)$$

The equation results from adding (a) through (d).

In the y direction we next find for the N'_y terms

$$-N'_y r_0 \, d\theta + \left(N'_y + \frac{\partial N'_y}{\partial y} \, dy \right) \left(r_0 + \frac{\partial r_0}{\partial y} \, dy \right) d\theta$$

which following (b) reduces to

$$\frac{\partial (N'_y r_0)}{\partial y} \, d\theta \, dy \qquad (e)$$

The $N'_{\theta y}$ give the same result as for the cylinder because the influence of the angle Δ comes as $\cos \Delta$ which can be taken as one.

$$\frac{\partial N'_{\theta y}}{\partial \theta} \, d\theta \, dy \qquad (f)$$

Because the two y faces of the element are parallel, the $N'_{y\theta}$ will have no component in the y direction; but the N'_θ on the positive θ face will have a component in the y direction:

$$-N'_\theta \, dy \sin \Delta = -N'_\theta \frac{\partial r_0}{\partial y} \, d\theta \, dy \qquad (g)$$

This influence is negative, as shown in Fig. 2-4c. Finally the pressure loading term gives

$$p_y r_0 \, d\theta \, dy \qquad (h)$$

The equilibrium equation results from adding (e) through (h).

The last membrane theory equilibrium equation for conical shells comes from considering the z direction in which, as for cylinders, only the N'_θ and p_z have effects; thus $N'_\theta \, d\theta \, dy$ is the influence of the circumferential stress resultant in the radial r_0 direction. Since z is at an angle of α to that r_0 direction, the component of N'_θ will be

$$N'_\theta \, d\theta \, dy \cos \alpha \qquad (i)$$

and the pressure loading term is

$$p_z r_0 \, d\theta \, dy \qquad (j)$$

When these two terms are combined we find

$$N'_\theta \cos \alpha + p_z r_0 = 0 \qquad (k)$$

but since $r_0/\cos \alpha = r_2$, the principal radius of curvature, we can rewrite (k) in terms only of that r_2. Thus, the three equations of equilibrium for conical shells are

$$\frac{\partial N'_\theta}{\partial \theta} + N'_{\theta y} \frac{\partial r_0}{\partial y} + \frac{\partial (N'_{y\theta} r_0)}{\partial y} + p_\theta r_0 = 0$$

$$\frac{\partial (N'_y r_0)}{\partial y} - N'_\theta \frac{\partial r_0}{\partial y} + \frac{\partial N'_{\theta y}}{\partial \theta} + p_y r_0 = 0 \qquad (2\text{-}10)$$

$$\frac{N'_\theta}{r_2} + p_z = 0$$

or, with (Fig. 2-5a) $r_0 = y \sin \alpha$, $\partial r_0/\partial y = \sin \alpha$, $r_2 = y \tan \alpha$, and $N'_{y\theta} = N'_{\theta y}$, we obtain for Eqs. (2-10)

$$N'_\theta = -p_z y \tan \alpha$$

$$\frac{1}{y^2} \frac{\partial (N'_{y\theta} y^2)}{\partial y} = \frac{\partial N'_{y\theta}}{\partial y} + \frac{2N'_{y\theta}}{y} = -\frac{1}{y \sin \alpha} \frac{\partial N'_\theta}{\partial \theta} - p_\theta \qquad (2\text{-}11)$$

$$\frac{1}{y} \frac{\partial (N'_y y)}{\partial y} = \frac{\partial N'_y}{\partial y} + \frac{N'_y}{y} = -\frac{1}{y \sin \alpha} \frac{\partial N'_{y\theta}}{\partial \theta} + \frac{N'_\theta}{y} - p_y$$

These equations can now be solved for a variety of loadings and for two basic configurations: first as a water tank–type shell (Fig. 2-5a) and second as a cooling tower–type one (Fig. 2-5b). For the former, gravity, internal pressure, and wind are considered; while for the latter, gravity and wind.

Gravity Loads

For a constant dead load of q psf, $p_z = -q \sin \alpha$, $p_y = -q \cos \alpha$, $p_\theta = 0$. There will be no variations with respect to θ, and $N'_{y\theta} = N'_{\theta y} = 0$; thus Eqs. (2-11) reduce to (for the tank–type shell)

$$N'_\theta = -p_z y \tan \alpha$$

$$\frac{d(N'_y y)}{dy} = -p_z y \tan \alpha - p_y y$$

or

$$N'_y y = -\int_y^L (p_z \tan \alpha + p_y) y \, dy$$

and

$$N'_y = \frac{q}{y} \left(\frac{\sin^2 \alpha}{\cos \alpha} + \cos \alpha \right) \int_y^L y \, dy = \frac{qy^2}{2y \cos \alpha} + \frac{C}{y} \qquad (l)$$

With the full support assumed at the base of the cone, $N'_y = 0$ at $y = L$ (Fig. 2-5a); hence

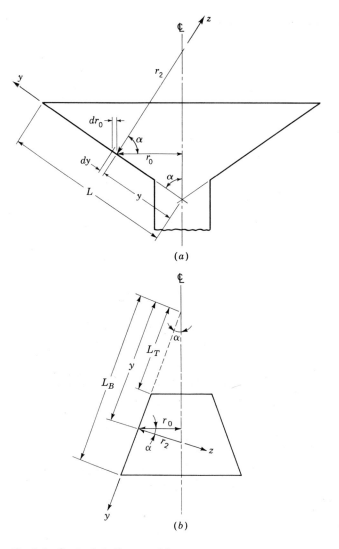

Fig. 2-5 Conical shell geometries.

$$C = - \frac{qL^2}{2 \cos \alpha}$$

and

$$N'_y = - \frac{q}{2} \frac{L^2 - y^2}{y \cos \alpha}$$

$$N'_\theta = qy \tan \alpha \sin \alpha$$

(2-12)

Equations (2-12) could have been derived directly from vertical equilibrium, as in

Eqs. (2-1a) where

$$N_y' 2\pi y \sin \alpha \cos \alpha - R = 0$$

and with $R = -2\pi q \sin \alpha \int_y^L y \, dy = -\pi q(L^2 - y^2) \sin \alpha$,

$$N_y' = -\frac{q}{2} \frac{L^2 - y^2}{y \cos \alpha}$$

Where $y = 0$, (2-12) indicate that $N_y' = \infty$. This case cannot, of course, occur, and equilibrium is established by the vertical reaction.

$$R_v = -R = \pi q L^2 \sin \alpha$$

Stress resultants are plotted in Fig. 2-6b for the conical shell shown in Fig. 2-6a.

For the conical cooling tower–type shell (Fig. 2-5b), $p_z = q \sin \alpha$, $p_y = q \cos \alpha$, $p_\theta = 0$; hence Eq. (l) becomes $-(qy^2/2y \cos \alpha) + C/y$, and with $N_y' = 0$ at $y = L_T$,

$$C = \frac{qL_T^2}{2 \cos \alpha}$$

and

$$N_y' = -\frac{q}{2} \frac{y^2 - L_T^2}{y \cos \alpha} \tag{2-12a}$$

with N_θ' the same as (2-12) except with the reversed sign.

Internal Pressure

For an internal pressure $p_z = -\gamma(L - y) \cos \alpha$, $p_\theta = p_y = 0$, and (2-11) become (for the geometry in Fig. 2-5a)

$$N_\theta' = \gamma(L - y)y \tan \alpha \cos \alpha$$

$$N_y' = \frac{\gamma}{y} \sin \alpha \int_y^L (L - y)y \, dy$$

$$= \frac{\gamma \sin \alpha}{y} \left(\frac{Ly^2}{2} - \frac{y^3}{3} \right) + \frac{C}{y}$$

Again $N_y' = 0$ at $y = L$; hence

$$C = -\gamma \sin \alpha \frac{L^3}{6}$$

and thus (2-11) result in

$$N_y' = -\frac{\gamma \sin \alpha}{2y} \left(\frac{L^3}{3} - Ly^2 + \frac{2y^3}{3} \right) \tag{2-13}$$

$$N_\theta' = \gamma \sin \alpha (L - y)y$$

Stress resultants are plotted in Fig. 2-6c for the same shell as before, and an elevated conical water tank built in Örebro, Sweden, is shown in Fig. 2-7.[7]

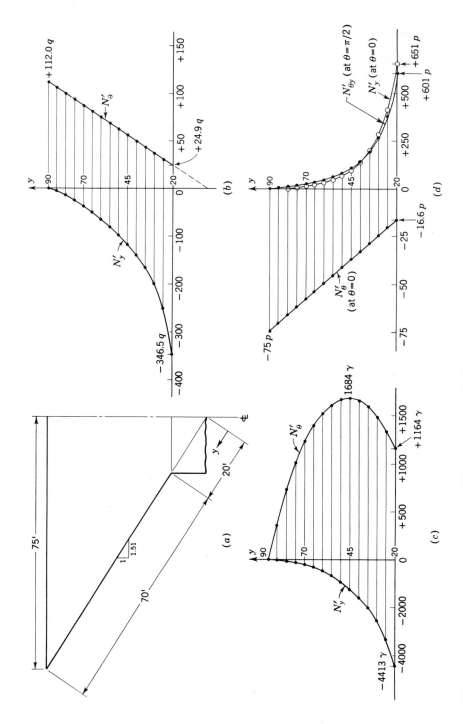

Fig. 2-6 Stress resultants in a conical water tank. (*a*) Geometry. (*b*) Dead load. (*c*) Water load. (*d*) Wind load n = 1.

Fig. 2-7 Elevated water tower, Örebro, Sweden. (Courtesy of Kurt Eriksson.)

For a tank supported tangentially only at the top, N'_θ is the same as in (2-13). $N'_y = 0$ at $y = 0$ and $C = 0$; hence

$$N'_y = \frac{\gamma y}{2}\left(L - \frac{2}{3}y\right)\sin\alpha$$

External Wind Pressure

Let us examine the effect of a pressure distribution $p_{zn} = pA_n\cos\alpha\cos n\theta$, $p_\theta = p_y = 0$, where p is the horizontal pressure at the top of the cone (Fig. 2-5a). The first of Eqs. (2-11) becomes

$$N'_{\theta n} = -pyA_n\cos n\theta\sin\alpha$$

from which the second of (2-11) is solved to give

$$\frac{1}{y^2}\frac{\partial(N'_{y\theta n}y^2)}{\partial y} = -\frac{pyA_n(n\sin n\theta)}{y}$$

$$N'_{y\theta n}y^2 = -pA_n n\sin n\theta\int y^2\,dy$$

$$N'_{y\theta n} = -\frac{py^3 A_n n\sin n\theta}{3y^2} + \frac{C}{y^2}$$

at $y = L$, $N'_{y\theta} = 0$, and thus

$$C = \frac{pL^3 A_n n \sin n\theta}{3}$$

and

$$N'_{y\theta n} = \frac{p}{3}\frac{L^3 - y^3}{y^2} A_n n \sin n\theta$$

Finally from the last of (2-11) we obtain

$$\frac{1}{y}\frac{\partial(N'_{yn}y)}{\partial y} = -\frac{p}{3}\frac{L^3 - y^3}{y^3}\frac{A_n n^2 \cos n\theta}{\sin \alpha} - pA_n \cos n\theta \sin \alpha$$

$$N'_{yn}y = -\frac{pA_n \cos n\theta}{\sin \alpha}\int \left(\frac{L^3 - y^3}{3y^2}n^2 + y \sin^2 \alpha\right) dy$$

$$N'_{yn} = -\frac{pA_n \cos n\theta}{y \sin \alpha}\left[\left(-\frac{L^3}{3y} - \frac{y^2}{6}\right)n^2 + \frac{y^2}{2}\sin^2 \alpha\right] + \frac{C}{y}$$

where $y = L$, $N'_y = 0$; hence

$$C = \frac{pA_n \cos n\theta}{\sin \alpha}\left(-\frac{L^2 n^2}{2} + \frac{L^2}{2}\sin^2 \alpha\right)$$

and

$$N'_{yn} = \frac{pA_n \cos n\theta}{\sin \alpha}\left[\left(\frac{L^3}{3y^2} + \frac{y^2}{6y} - \frac{L^2}{2y}\right)n^2 + \frac{L^2 - y^2}{2y}\sin^2 \alpha\right]$$

Thus (2-11) result in

$$N'_{\theta n} = -pyA_n \cos n\theta \sin \alpha$$

$$N'_{\theta y n} = \frac{p}{3}\frac{L^3 - y^3}{y^2}A_n n \sin n\theta \qquad (2\text{-}14)$$

$$N'_{yn} = \frac{pA_n \cos n\theta}{\sin \alpha}\left(\frac{2L^3 + y^3 - 3L^2y}{6y^2}n^2 + \frac{L^2 - y^2}{2y}\sin^2 \alpha\right)$$

Stress resultants are plotted in Fig. 2-6d again for the same shell as before, except for the case of $n = 1$ only. For a complete design, Eqs. (2-14) should be solved for a wind distribution such as shown in Table 2-1. Eqs. (2-11) can be used to study seismic loading when $p_\theta = p_s \sin \theta$, $p_y = -p_s \sin \alpha \cos \theta$, and $p_z = p_s \cos \alpha \cos \theta$.

2-4 STRESS RESULTANTS FROM THE MEMBRANE THEORY FOR HYPERBOLIC SHELLS

Figure 2-8 shows a surface formed by rotating a hyperbola about the y axis. Based on cartesian coordinates, the equation of the surface is

$$\frac{x^2 + z^2}{a^2} - \frac{y^2}{b^2} = 1 \qquad (2\text{-}15)$$

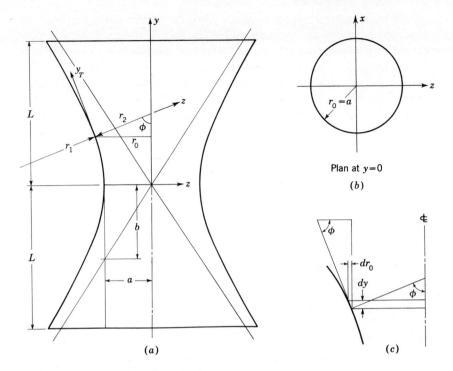

Fig. 2-8 Definitions for hyperbolic shells.

where $x^2 + z^2 = r_0^2$, b/a equals the slope of the asymptotes to the hyperbola, and $a = r_0$ at $y = 0$. Again $r_0 = r_2 \sin \phi$.

If we intersect the surface with a vertical plane parallel to the xy plane but at $z = a$, then the intersection curve will be

$$\frac{x^2 + a^2}{a^2} - \frac{y^2}{b^2} = 1$$

or

$$\frac{x^2}{a^2} = \frac{y^2}{b^2}$$

which is the straight line (where a is the throat radius)

$$y = \frac{b}{a} x$$

which has the same slope as the asymptotes. Any vertical plane tangent to the waist circle (where $r_0 = a$) will intersect the surface with straight lines of the same slope. Hence this cylinder can be obtained by rotating about an axis a nonparallel straight line. This property permits the construction of this doubly curved surface to be made entirely by straight elements. Roof structures having this same property are described in Chap. 7.

The membrane theory for hyperbolic shells yields the same equations as found for cones, Eqs. (2-10), except for two differences: The first difference is that the meridian is curved with a changing radius r_1, and therefore the coordinate y is commonly changed to ϕ, the angle of slope with respect to the horizontal (see Fig. 2-8c). Thus the meridional differential length dy becomes $r_1\,d\phi$ and the derivatives with respect to ∂y change to $r_1\,\partial\phi$ and all subscripts y change to ϕ. The second difference arises in the third of Eqs. (2-10), where $N'_y = N'_\phi$ now enters because of the meridional curvature r_1. The term in N'_ϕ follows from a derivation similar to that in Sec. 2-2 for (g). Thus the equations are

$$\frac{\partial N'_\theta}{\partial\theta} + N'_{\theta\phi}\frac{\partial r_0}{r_1\,\partial\phi} + \frac{\partial(N'_{\phi\theta}r_0)}{r_1\,\partial\phi} + p_\theta r_0 = 0$$

$$\frac{\partial(N'_\phi r_0)}{r_1\,\partial\phi} - N'_\theta\frac{\partial r_0}{r_1\,\partial\phi} + \frac{\partial N'_{\theta\phi}}{\partial\theta} + p_\phi r_0 = 0 \qquad (2\text{-}16)$$

$$\frac{N'_\phi}{r_1} + \frac{N'_\theta}{r_2} + p_z = 0$$

The expressions for r_1 and r_2 for a hyperboloid can be derived from the basic equations for curvature

$$\frac{1}{r_1} = \frac{-d^2r_0/dy^2}{[1 + (dr_0/dy)^2]^{3/2}} \qquad (a)$$

where y is a vertical coordinate. Also,

$$r_2 = r_0/\sin\phi \qquad (b)$$

where, from Fig. 2-8 and Eq. (2-15) we can find

$$r_0 = \frac{a}{b}\,(b^2 + y^2)^{1/2} \qquad (c)$$

$$\frac{dr_0}{dy} = \frac{a}{b}\,\frac{y}{(b^2 + y^2)^{1/2}} = \frac{a^2}{b^2}\,\frac{y}{r_0} \qquad (d)$$

$$\frac{d^2r_0}{dy^2} = \frac{ab}{(b^2 + y^2)^{3/2}} \qquad (e)$$

Introducing (c), (d), and (e) into (a), we find

$$r_1 = -a^2b^2\left(\frac{r_0^2}{a^4} + \frac{y^2}{b^4}\right)^{3/2} \qquad (f)$$

and then introducing into (b) both

$$\sin\phi = \frac{1}{\sqrt{1 + \cot^2\phi}}$$

and

$$\cot\phi = \frac{dr_0}{dy} \qquad (g)$$

as well, which comes from Fig. 2-8c, we obtain

$$r_2 = a \sqrt{1 + \left(\frac{1}{b^2} + \frac{a^2}{b^4}\right) y^2} \qquad (h)$$

from which we see that

$$r_1 = -\frac{b^2}{a^4} r_2^3 \qquad (i)$$

Gravity Loads[8]

Consider the vertical dead load, where $p_y = -q \sin \phi$ and $p_z = -q \cos \phi$. Note that the tangential axis y_T along which p_y is measured is not the same as the vertical y axis. The first of Eqs. (2-1a) becomes

$$N'_\phi = -\frac{R}{2\pi r_2 \sin^2 \phi} = -\frac{1}{2\pi r_2 \sin^2 \phi} \int_0^\phi q 2\pi r_0 r_1 \, d\phi$$

$$= -\frac{1}{r_2 \sin^2 \phi} \int_0^\phi q r_1 r_2 \sin \phi \, d\phi$$

and with $r_1 \, d\phi = ds = dy/\sin \phi$,

$$dy = r_1 \sin \phi \, d\phi$$

so that, eliminating all polar terms, we find

$$N'_\phi = -\frac{r_2}{r_0^2} \int_0^y q r_2 \, dy \qquad (2\text{-}17a)$$

Substituting (i) into the third of Eqs. (2-16), we get

$$N'_\theta = -p_z r_2 + \frac{a^4}{b^2 r_2^2} N'_\phi$$

or

$$N'_\theta = q r_0 \cot \phi - \frac{a^4}{b^2 r_0^2 r_2} \left(\int q r_2 \, dy + C_0 \right) \qquad (j)$$

The constant of integration represents a load applied to the top free edge. Putting (g) and (d) into (j), we find

$$N'_\theta = q y \frac{a^2}{b^2} + \frac{a^4}{b^2 r_0^2 r_2} \left(\int q r_2 \, dy + C_0 \right) \qquad (2\text{-}17b)$$

These equations can be integrated analytically if q (and hence h) is a constant, but it is easier to perform the integrations numerically, especially since the large-scale hyperboloids used for cooling towers do not have constant thicknesses. Dividing a tall hyperboloid into only four approximately conical sections gives a close estimate of dead-load stress resultants; furthermore, analysts commonly use computer-based numerical solutions which do not integrate Eqs. (2-17) either. Figure

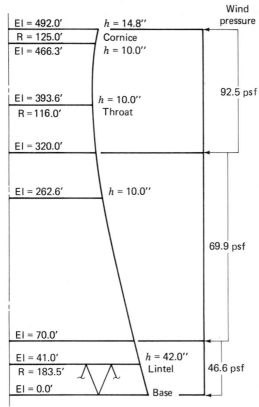

Fig. 2-9 Dimensions of the Trojan Tower.

2-9 gives the dimensions and loading for the 492-ft-high cooling tower (Fig. 2-10) built by Research-Cottrell Inc. for the Trojan Nuclear Plant of the Portland General Electric Company and completed in 1974.[9] These dead-load values come from a finite-element analysis and from an approximate numerical analysis of Eqs. (2-17) gotten by dividing the tower into four bands and using

$$N'_{\phi j} = - \frac{\sum\limits_{i=1}^{j} r_{0i} q_i \, \Delta S_i}{r_{0j} \sin \phi_j} \tag{2-18a}$$

$$N'_{\theta j} = - \frac{r_{0j}}{\sin \phi_j} \left(\frac{N'_{\phi j}}{r_{1j}} - q_j \cos \phi_j \right) \tag{2-18b}$$

(see Table 2-2) where the q_i are the average unit weights of each band of meridional length Δs_i and average horizontal radii r_{0i}.

Table 2-2 summarizes the results in which the load $p_z = -q \cos \phi$, $r_2 = r_0/\sin \phi$, and $\Delta R_s = 2\pi r_0(av) \, q \, \Delta S$. The value of r_1 can be computed by finite differences as

Fig. 2-10 Trojan Tower, Oregon, 1974. (Courtesy Research-Cottrell, Inc.)

Table 2-2 TROJAN TOWER

Band	r_0 (av)	q (av)	ΔS, ft	$r_0 q \, \Delta S$, kips	r_0, ft	ϕ, degrees	y, ft	Fy, ft-kips
1	120.5	0.125	98.4	1,660*	116	90	385	640,000
2	123.5	0.125	131.0	2,000	131	100.6	287	574,000
3	142.5	0.190	110.8	3,000	154	103.3	167	500,000
4	169.0	0.390	110.8	7,300	183.5	104.2	56	410,000
Σ	. . .	. . .	. . .	13,960	. . .	. . .	. .	2,124,000

* $5.0 \times 1.67 \times 0.15 \times 125 = 156^k$ weight of ring is included in Band 1.

follows (where $\partial r_0 / \partial y \ll 1$):

$$\frac{1}{r_1} \approx - \left(\frac{d^2 r_0}{dy^2} \right)_y \approx - \frac{1}{(\Delta y)^2} (r_{02} - 2r_{01} + r_{00})$$

so that, for example, at the throat, $\Delta y = 20.968$ ft, $r_{02} = 116.64$, $r_{01} = 116.015$, and $r_{00} = 116.300$, and thus

$$\frac{1}{r_1} = - \frac{1}{20.97^2} (232.94 - 232.030) = - \frac{1}{481} \text{ ft}^{-1}$$

Thus at the throat the stress resultants will be

$$N'_{\phi 1} = -\frac{1660}{116 \times 1.0} = -14.3 \text{ kips/ft } (-15.0)$$

$$N'_{\theta 1} = -116 \left[\frac{-14.3}{-481} - 0.125(0)\right] -- 3.45 \text{ kips/ft } (-3.65)$$

while at the base

$$N'_{\phi 4} = -\frac{13960}{183.5 \times 0.966} = -79 \text{ kips/ft } (-75.6)$$

$$N'_{\theta 4} = -\frac{183.5}{0.966}\left[\frac{-79}{\infty} + 0.525\,(0.245)\right] = -24.5 \text{ kips/ft } (-2.4)$$

The values in parentheses are from a finite-element solution discussed in Ref. 9. The discrepancies in N'_ϕ are mainly due to approximations in R. The large difference in $N'_{\theta 4}$ is due to the base restraint, which is not accounted for in the membrane theory. Otherwise the correspondence is good and acceptable for preliminary design.

Internal Pressure

Let us consider the case of a liquid pressure in that part of the hyperboloid above the throat, where

$$p_z = -(L - y)\gamma$$
$$p_y = 0$$
$$p_x = 0$$

Therefore

$$R = 2\pi \int r_0 p_z \cos \phi r_1 \, d\phi + C$$

$$= 2\pi \int r_2 p_z \cos \phi \, dy + C$$

$$= 2\pi \left(\frac{a}{b}\right)^2 \int p_z y \, dy + C$$

$$= -2\pi \left(\frac{a}{b}\right)^2 \gamma \int (L - y)y \, dy + C$$

$$= -2\pi \left(\frac{a}{b}\right)^2 \gamma \left(\frac{Ly^2}{2} - \frac{y^3}{3}\right) + C$$

$R = 0$ when $y = L$; hence $C = 2\pi\,(a/b)^2\gamma L^3/6$, from which

$$R = -2\pi \left(\frac{a}{b}\right)^2 \gamma \left(\frac{Ly^2}{2} - \frac{y^3}{3} - \frac{L^3}{6}\right)$$

and from an equation like (2-17a) for the loaded part of the hyperboloid

$$-2\pi r_0 \sin \phi \, N'_\phi - R = 0$$

and since $\sin \phi = r_0/r_2$, then $N'_\phi = -(r_2/r_0^2)(R/2\pi)$ so that

$$N'_\phi = \frac{r_2 a^2 \gamma}{r_0^2 b^2} L^3 \left(\frac{y^2}{2L^2} - \frac{y^3}{3L^3} - \frac{1}{6} \right) \tag{2-19a}$$

and from the third of Eqs. (2-16),

$$N'_\theta = \frac{a^6 \gamma}{r_0^2 r_2 b^4} L^3 \left(\frac{y^2}{2L^2} - \frac{y^3}{3L^3} - \frac{1}{6} \right) + (L-y)\gamma r_2 \tag{2-19b}$$

Eduardo Torroja designed an elevated storage tank of hyperboloid form out of brick in the 1930s. [10]

Seismic Loads

As with the cylinder, the hyperboloid carries seismic loadings primarily as a cantilever with a closed circular cross section. The general membrane theory approach would be to solve Eqs. (2-16) for $p_\theta = p_s \sin \theta$, $p_\phi = p_s \cos \theta \cos \phi$, and $p_z = p_s \cos \theta \sin \phi$ which reduce to the loads used for cylinders when $\phi = 90°$. This approach is more complicated than one using the cantilever beam formulas (2-6) and (2-7), especially where the integrals of (2-7) are solved numerically by summing up conical bands of merdional length Δs as with the dead load.

The formulas for the seismic are the same as those given for cylinders, Eqs. (2-5), except for the changes in r and the influence of N'_ϕ on N'_θ. In general, we compute M_ϕ and I_ϕ and find $N'_\phi/h = M_\phi r/I_\phi$. Table 2-2 gives the moment of 2,124,000 ft-kips which must be multiplied by 2π and the coefficient c for seismic intensity which we take here to be 0.2. Thus

$$M_\phi = 2\pi(0.2)\, 2{,}124{,}000 = 2{,}680{,}000 \text{ ft-kips}$$

at the base $I_\phi = \pi r^3 h = \pi(183.5)^3 h = 1.91 \times 10^6 \, h$, so that the maximum

$$N'_\phi = \frac{M_\phi r h}{I_\phi} = \frac{2{,}680{,}000 \times 183.5}{1.91 \times 10^6} = 25.7 \text{ kips/ft}$$

which compares closely with the finite-element results. [11]

Wind Loads

Just as with cylinders, the wind loads are distributed around a hyperboloid, as given in Table 2-1. As with the conical shell, the membrane theory equations can now be solved for each term of the series representation of the loading. Because of the double curvature, this solution is complicated and cannot be expressed simply as was done by Eqs. (2-14) for cones. Solutions are available using computer-based analyses, but they are mainly used for preliminary design. We can get some idea of the size of the stress resultants by replacing the hyperboloid by an equivalent cone and using Eqs. (2-14).

While the membrane theory gives good results for axisymmetric and seismic loads, it is not always reliable for wind loadings. The reason is that in the series representation of the loading, the terms with higher n give progressively poorer results. Physically this means that loads oscillating around the circumference tend to bend the shell, and these bending effects, while in themselves small, are large enough to influence the stress resultants significantly. Fortunately the terms in higher n have smaller coefficients (see Table 2-1) so that the summed results have provided reasonable results for the cooling towers built in the United States up to 1980. A more detailed discussion of bending in shell walls under wind loading appears in Sec. 2-9.

2-5 DISPLACEMENTS FROM THE MEMBRANE THEORY FOR CIRCULAR CYLINDRICAL SHELLS LOADED SYMMETRICALLY WITH RESPECT TO THEIR AXES

Figure 2-11 shows the displacements for the differential element of a circular cylinder under axisymmetrical loads. The origin of the local or surface coordinates displaces vertically by a distance v and radially by w; there is no displacement in the θ direction because of axial symmetry. At a distance dy from the origin, the vertical displacement is changed by the quantity $(dv/dy)\, dy$ and the radial displacement by $(dw/dy)\, dy$. Thus the vertical strain ϵ_y, defined by the change in length of the dy side, is

$$\epsilon_y = \frac{v + (dv/dy)\, dy - v}{dy} = \frac{dv}{dy} \qquad (a)$$

and the circumferential strain ϵ_θ, defined by the change in length of the $r\, d\theta$ side, is

$$\epsilon_\theta = \frac{(r - w)\, d\theta - r\, d\theta}{r\, d\theta} = -\frac{w}{r} \qquad (b)$$

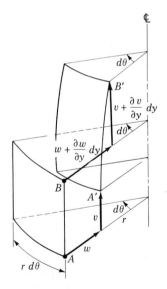

Fig. 2-11 Displacements for a cylindrical shell.

where the negative sign arises because a displacement w in the positive z direction leads to a negative or compressive strain. There is no shearing strain in the element because of axial symmetry. The rotation of the dy side arises because of the change in radial displacement

$$\phi_y = \frac{w + (dw/dy)\, dy - w}{dy} = \frac{dw}{dy} \qquad (c)$$

whereas the rotation of the $r\, d\theta$ side and the twist of the element are both zero again because of axial symmetry.

Having the strain-displacement and rotation-displacement expressions, we next must determine the stress-strain values which we take from elementary elasticity as

$$\epsilon_\theta = \frac{1}{E}\,[\sigma_\theta - \nu(\sigma_y + \sigma_z)] \qquad (d)$$

$$\epsilon_y = \frac{1}{E}\,[\sigma_y - \nu(\sigma_\theta + \sigma_z)] \qquad (e)$$

$$\gamma_{\theta y} = \frac{1}{G}\,\tau_{\theta y} \qquad (f)$$

For thin shells $\sigma_z \to 0$. G is the modulus of rigidity and by symmetry $\gamma_{\theta y} = 0$.

Recalling the definition of stress resultants from Chap. 1 where $\sigma'_\theta = N'_\theta/h$ and $\sigma'_y = N'_y/h$, we can write (d) and (e) as

$$\epsilon_\theta = \frac{1}{Eh}\,(N'_\theta - \nu N'_y)$$

$$\epsilon_y = \frac{1}{Eh}\,(N'_y - \nu N'_\theta) \qquad (2\text{-}20)$$

or in terms of stress resultants

$$N'_\theta = \frac{Eh}{1 - \nu^2}\,(\epsilon_\theta + \nu\epsilon_y)$$

$$N'_y = \frac{Eh}{1 - \nu^2}\,(\epsilon_y + \nu\epsilon_\theta) \qquad (2\text{-}21)$$

Introducing (a) and (b) into (2-21), we obtain

$$N'_\theta = \frac{Eh}{1 - \nu^2}\left(-\frac{w}{r} + \nu\frac{dv}{dy}\right)$$

$$N'_y = \frac{Eh}{1 - \nu^2}\left(\frac{dv}{dy} - \nu\frac{w}{r}\right) \qquad (2\text{-}22)$$

Therefore, for a liquid pressure $p_z = -\gamma(H - y)$, from (2-4)

$$N'_\theta = \gamma r(H - y), \qquad N'_{y\theta} = N'_y = 0$$

and the strain

$$\epsilon'_\theta = \frac{N'_\theta}{Eh} = \frac{\gamma r}{Eh}(H - y) = -\frac{w}{r}$$

Thus

$$w = -\frac{\gamma r^2}{Eh}(H - y) \tag{2-23a}$$

and from (c) with $h = $ constant,

$$\phi_y = \frac{\gamma r^2}{Eh} \tag{2-23b}$$

Where h has a linear variation

$$h = h_0(- \alpha y)$$

$$w = -\frac{\gamma r^2(H - y)}{Eh_0(1 - \alpha y)} \tag{2-24a}$$

$$\phi_y = \frac{1 - H\alpha}{1 - 2\alpha y + \alpha^2 y^2}\frac{\gamma r^2}{Eh_0} \tag{2-24b}$$

2-6 BENDING THEORY FOR CIRCULAR CYLINDERS UNDER AXISYMMETRICAL LOADING

The equilibrium equations (2-1) need to be modified to include shearing stress resultants Q_θ and Q_y as well as stress couples M_θ, M_y, $M_{y\theta}$, and $M_{\theta y}$. The twisting couples $M_{y\theta} = M_{\theta y}$ for the same reasons that the $N_{y\theta} = N_{\theta y}$ described in (i) and (j) of Sec. 2-2. Figure 2-12 shows these additional quantities. The first of (2-1) is influenced by the component of Q_θ on the positive face in the θ direction:

$$-Q_\theta \, dy \, d\theta \tag{a}$$

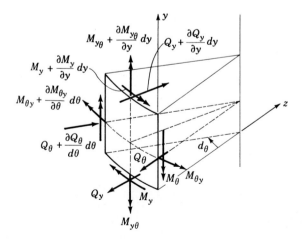

Fig. 2-12 Bending moments and shears in a cylindrical shell.

The second of (2-1) is not influenced by Q_y because there is no meridional curvature, whereas the third of (2-1) is influenced by both Q_θ and Q_y, the total radial effect of which gives for Q_θ

$$\left(Q_\theta + \frac{\partial Q_\theta}{\partial \theta} d\theta\right) dy - Q_\theta\, dy = \frac{\partial Q_\theta}{\partial \theta} d\theta\, dy \tag{b}$$

and for Q_y

$$\left(Q_y + \frac{\partial Q_y}{\partial y} dy\right) r\, d\theta - Q_y r\, d\theta = \frac{\partial Q_y}{\partial y} r\, d\theta\, dy \tag{c}$$

These two terms are added to the third of (2-1). In addition, there will be two equations of moment equilibrium: one about the θ axis and one about the y axis. For example, about the θ axis we find

$$M_y r\, d\theta - \left(M_y + \frac{\partial M_y}{\partial y} dy\right) r\, d\theta = -\frac{\partial M_y}{\partial y} r\, d\theta\, dy \tag{d}$$

$$-M_{\theta y}\, dy + \left(M_{\theta y} + \frac{\partial M_{\theta y}}{\partial \theta} d\theta\right) dy = \frac{\partial M_{\theta y}}{\partial \theta} d\theta\, dy \tag{e}$$

$$\left(Q_y + \frac{\partial Q_y}{\partial y} dy\right) dy\, r\, d\theta = Q_y r\, d\theta\, dy \tag{f}$$

and similarly about the y axis. When these terms are collected together, we obtain the general equilibrium equations for circular cylindrical shells.

$$\frac{\partial N_\theta}{\partial \theta} + \frac{\partial N_{y\theta}}{\partial y} r - Q_\theta + p_\theta r = 0$$

$$\frac{\partial N_y}{\partial y} r + \frac{\partial N_{\theta y}}{\partial \theta} + p_y r = 0$$

$$\frac{\partial Q_\theta}{\partial \theta} + \frac{\partial Q_y}{\partial y} r + N_\theta + p_z r = 0 \tag{2-25}$$

$$-\frac{\partial M_y}{\partial y} r + \frac{\partial M_{\theta y}}{\partial \theta} + Q_y r = 0$$

$$+\frac{\partial M_\theta}{\partial \theta} + \frac{\partial M_{y\theta}}{\partial y} r - Q_\theta r = 0$$

For axisymmetrical loading,

$$N_{y\theta} = N_{\theta y} = M_{y\theta} = M_{\theta y} = 0$$

and thus (2-25) reduce to

$$\frac{dN_y}{dy} r + p_y r = 0$$

$$\frac{dQ_y}{dy} r + N_\theta + p_z r = 0 \tag{2-26}$$

$$-\frac{dM_y}{dy} r + Q_y r = 0$$

We observe that the first of (2-26) is independent of the second and third; i.e., axisymmetrical loads directed along the straight-line generators (p_y) are carried directly by N_y stress resultants and do not affect the other stress resultants and stress couples. If N_y is set to zero, the bending in circular cylinders due to axisymmetrical loading can be described by two equilibrium equations with three unknowns, Q_y, N_θ, and M_y. To solve for these unknowns, we must utilize the expressions (2-22) for stress resultants and give moments in terms of displacements.

$$N_y = K \left(\frac{dv}{dy} - \nu \frac{w}{r} \right) \tag{g}$$

$$N_\theta = K \left(-\frac{w}{r} + \nu \frac{dv}{dy} \right) \tag{h}$$

$$M_y = -D \frac{d\phi_y}{dy} \tag{i}$$

where $K = Eh/(1 - \nu^2)$ and $D = Eh^3/12(1 - \nu^2)$, the extensional and flexural rigidities, respectively. With $N_y = 0$, (g) results in

$$\frac{dv}{dy} = \nu \frac{w}{r}$$

which, when substituted into (h), gives

$$N_\theta = -K(1 - \nu^2)\frac{w}{r} = -\frac{Ehw}{r} \tag{2-27a}$$

and with $\phi_y = dw/dy$,

$$M_y = -D \frac{d^2w}{dy^2} \tag{2-27b}$$

Equations (2-26) are now reduced to a single equation by substituting

$$\frac{dM_y}{dy} = Q_y$$

into the second of (2-26) to give

$$r \frac{d^2M_y}{dy^2} + N_\theta + p_z r = 0 \tag{j}$$

Expressions (2-27) are now introduced into (j) to give

$$\frac{d^2}{dy^2} \left(D \frac{d^2w}{dy^2} \right) + \frac{Ehw}{r^2} = p_z \tag{2-28}$$

which is the general expression for axisymmetrical deformation in circular cylindrical shells. The solution to (2-28) can be easily made for the case where the shell has a constant thickness and hence a constant D. It is a logical practice to increase the wall thickness with depth since liquid pressure and, hence, membrane ring value N'_θ increases with head. It has been shown[12] that approximate calculations based on a constant thickness give results close to those obtained by solving (2-28) directly. For the case of constant thickness (2-28) becomes

$$D \frac{d^4w}{dy^4} + \frac{Eh}{r^2} w = p_z \tag{2-29}$$

Equation (2-29) is of the same form as the equation for a beam on an elastic foundation, usually written as[12]

$$EI \frac{d^4w}{dy^4} + kw = p_z \tag{2-29a}$$

in which the load intensity p_z is the action, and the beam bending stiffness, $EI(d^4w/dy^4)$, and foundation stiffness, kw, are the reactions. We can make a physical analogy directly between (2-29) and (2-29a) by noting that k corresponds to Eh/r^2; hence we may designate k_r as the ring stiffness

$$k_r = \frac{Eh}{r^2}$$

as the equivalent foundation modulus of the cylinder. We shall not use this coefficient directly in the analysis, but it does clearly illustrate the parameters which control the deformations of a cylinder. With

$$\beta^4 = \frac{Eh}{4r^2D} = \frac{3(1 - v^2)}{r^2h^2} \tag{2-30}$$

Eq. (2-29) becomes

$$\frac{d^4w}{dy^4} + 4\beta^4w = \frac{p_z}{D} \tag{2-31}$$

For the solution we find

$$w = e^{\beta y}(C_1 \cos \beta y + C_2 \sin \beta y)$$
$$+ e^{-\beta y}(C_3 \cos \beta y + C_4 \sin \beta y) + f(y) \tag{2-32}$$

$f(y)$ is the particular solution of (2-31); in this case it is the membrane solution

$$w = \frac{p_z r^2}{Eh}$$

and C_1, C_2, C_3, and C_4 are constants of integration dependent upon the edge conditions of the cylinder. In fact, the four constants refer to two displacements at each of the two edges, one displacement due to a shear force Q_0 and the other due to a moment M_0. Since each of these axisymmetrical forces is in equilibrium by itself, it can be concluded from the principle of Saint-Venant that its effect on the cylinder is confined to the region near the edges. If the cylinder is long enough so that the effect of Q_0 and M_0 on one edge is negligible at the other edge, then each edge can be treated independently. The problem is thereby considerably simplified, and Eq. (2-32) can be solved with reference to only two constants.

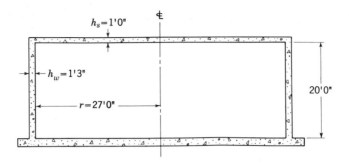

Fig. 2-13 Reinforced concrete tank.

On the basis of this discussion, we can observe that the quantity $e^{\beta y}$ increases with y. It will not lead to a localized effect unless C_1 and C_2 are zero, which must therefore be the case, so that Eq. (2-32) becomes

$$w = e^{-\beta y}(C_3 \cos \beta y + C_4 \sin \beta_y) \tag{2-33}$$

Figure (2-13) shows a typical liquid storage tank, in which the constant-thickness cylindrical wall is built into both the roof and the base. Each edge of the cylinder is treated separately based on the simplifications leading to (2-33). Figure (2-14) shows the errors and corrections both for the wall-base and wall-roof analyses. The former case is described here to illustrate the procedure. We can formulate the analysis in the usual four steps (see Sec. 1-13).

Primary System   The base of the wall is assumed to be free to slide and to rotate under the liquid pressure loads.

Errors   A translation and a rotation occur at the base (Eqs. 2-23).

Corrections $X_1 = -Q_0 = 1$ and $X_2 = M_0 = 1$ are applied at the bottom edge and the resulting translations (D_{11}, D_{12}) and rotations (D_{21}, D_{22}) are computed.*

Compatibility   The usual simultaneous equations are written and solved for X_1 and X_2, the correct values to remove the errors.

To obtain a solution, we need to develop expressions for D_{11}, $D_{12} = D_{21}$, and D_{22}. First we shall determine the translation due to Q_0 and to M_0. At the free edge

$$M_y = -D \left(\frac{d^2 w}{dy^2} \right)_{y=0} = M_0 \tag{k}$$

and from (2-26)

$$Q_y = \left(\frac{dM_y}{dy} \right)_{y=0} = -D \left(\frac{d^3 w}{dy^3} \right)_{y=0} = Q_0 \tag{l}$$

---

*X_1 is taken as $-Q_0$ so that D_{11} (Eq. 2-36) will be positive.

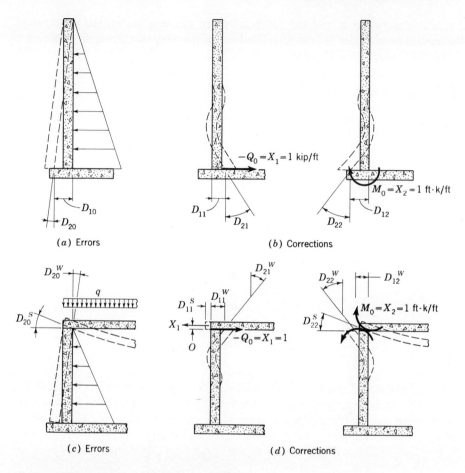

(a) Errors (b) Corrections

(c) Errors (d) Corrections

Fig. 2-14 Edge effects on cylindrical shells.

Differentiating (2-33) successively, we obtain

$$\frac{d^2w}{dy^2} = 2\beta^2 e^{-\beta y}(C_3 \sin \beta y - C_4 \cos \beta y)$$

$$\frac{d^3w}{dy^3} = 2\beta^3 e^{-\beta y}[C_3(\cos \beta y - \sin \beta y) + C_4(\sin \beta y + \cos \beta y)] \qquad (m)$$

from which, where $y = 0$,

$$M_0 = 2\beta^2 D C_4$$

$$C_4 = \frac{M_0}{2\beta^2 D}$$

$$Q_0 = -2\beta^3 D(C_3 + C_4) \qquad (n)$$

$$C_3 = -\frac{1}{2\beta^3 D}(Q_0 + \beta M_0)$$

The expressions for w and the rotation dw/dy can now be written in terms of the applied edge forces M_0 and Q_0. The only values which are generally of interest in the shell analysis are those at $y = 0$.

$$(w)_{y=0} = -\frac{1}{2\beta^3 D}(\beta M_0 + Q_0)$$

$$\left(\frac{dw}{dy}\right)_{y=0} = \frac{1}{2\beta^2 D}(2\beta M_0 + Q_0)$$

(2-34)

The distribution of forces over the shell can be obtained by differentiating (2-33) and substituting in the proper values of C_3 and C_4. Table 2-3 (from pages 470 – 473 of Ref. 12) gives a summary of the equations for displacements and forces in terms of edge loads M_0 and Q_0 and four functions of βy values which are included in the table. The four forces in the shell can be found from these equations as follows:

$$N_\theta = -\frac{Ehw}{r}$$

$$M_\theta = \nu M_y$$

$$M_y = -D\frac{d^2 w}{dy^2}$$

(2-35)

$$Q_y = -D\frac{d^3 w}{dy^3}$$

In the analysis of shells built monolithically with flexible roofs or floors, it will be valuable to have expressions for displacements in the same form as was used for the force method of analysis in frames. If $-Q_0 = X_1 = 1$ and $M_0 = X_2 = 1$, then

$$D_{11} = \frac{1}{2\beta^3 D}$$

$$D_{12} = -\frac{1}{2\beta^2 D}$$

(2-36)

$$D_{22} = \frac{1}{\beta D}$$

Table 2-4 gives a summary of Eqs. (2-35) and (2-36) where at $y = 0$ for $Q = -X_1 = -1$, $w = D_{11}$ and $dw/dy = D_{21}$, while for $M_0 = 1$, $w = D_{12}$ and $dw/dy = D_{22}$. In using Table 2-4, we recall that $4\beta^4 D = Eh/r^2$ so that, for example, $w_{y=0} = -2\beta(r^2/Eh)Q_0 = (2\beta/4\beta^4 D)Q_0 = 1/(2\beta^3 D)$ for $X_1 = 1$.

2-7 ANALYSIS OF A CIRCULAR CYLINDRICAL WALL CONNECTED TO A CIRCULAR ROOF SLAB

Some storage tanks are built without covers, but most have a permanent roof. Figure 2-15 shows a circular slab built monolithically with the wall—the system which will be considered in this section. Table 2-5 gives the results of circular-slab

Table 2-3

$$w = -\frac{1}{2\beta^3 D}[\beta M_0\psi(\beta y) + Q_0\theta(\beta y)]$$

$$\frac{dw}{dy} = \frac{1}{2\beta^2 D}[2\beta M_0\theta(\beta y) + Q_0\phi(\beta y)]$$

$$\frac{d^2w}{dy^2} = -\frac{1}{2\beta D}[2\beta M_0\phi(\beta y) + 2Q_0\zeta(\beta y)]$$

$$\frac{d^3w}{dy^3} = \frac{1}{D}[2\beta M_0\zeta(\beta y) - Q_0\psi(\beta y)]$$

$$\phi(\beta y) = e^{-\beta y}(\cos \beta y + \sin \beta y)$$
$$\psi(\beta y) = e^{-\beta y}(\cos \beta y - \sin \beta y)$$
$$\theta(\beta y) = e^{-\beta y} \cos \beta y$$
$$\zeta(\beta y) = e^{-\beta y} \sin \beta y$$

Table of functions*

βy	ϕ	ψ	θ	ζ
0	1.0000	1.0000	1.0000	0
0.1	0.9907	0.8100	0.9003	0.0903
0.2	0.9651	0.6398	0.8024	0.1627
0.3	0.9267	0.4888	0.7077	0.2189
0.4	0.8784	0.3564	0.6174	0.2610
0.5	0.8231	0.2415	0.5323	0.2908
0.6	0.7628	0.1431	0.4530	0.3099
0.7	0.6997	0.0599	0.3798	0.3199
0.8	0.6354	−0.0093	0.3131	0.3223
0.9	0.5712	−0.0657	0.2527	0.3185
1.0	0.5083	−0.1108	0.1988	0.3096
1.1	0.4476	−0.1457	0.1510	0.2967
1.2	0.3899	−0.1716	0.1091	0.2807
1.3	0.3355	−0.1897	0.0729	0.2626
1.4	0.2849	−0.2011	0.0419	0.2430
1.5	0.2384	−0.2068	0.0158	0.2226
1.6	0.1959	−0.2077	−0.0059	0.2018
1.7	0.1576	−0.2047	−0.0235	0.1812
1.8	0.1234	−0.1985	−0.0376	0.1610
1.9	0.0932	−0.1899	−0.0484	0.1415
2.0	0.0667	−0.1794	−0.0563	0.1230
2.1	0.0439	−0.1675	−0.0618	0.1057
2.2	0.0244	−0.1548	−0.0652	0.0895
2.3	0.0080	−0.1416	−0.0668	0.0748
2.4	−0.0056	−0.1282	−0.0669	0.0613
2.5	−0.0166	−0.1149	−0.0658	0.0492
2.6	−0.0254	−0.1019	−0.0636	0.0383
2.7	−0.0320	−0.0895	−0.0608	0.0287
2.8	−0.0369	−0.0777	−0.0573	0.0204
2.9	−0.0403	−0.0666	−0.0534	0.0132

Table 2-3 (Continued)

βy	ϕ	ψ	θ	ς
3.0	−0.0423	−0.0563	−0.0493	0.0071
3.1	−0.0431	−0.0469	−0.0450	0.0019
3.2	−0.0431	−0.0383	−0.0407	−0.0024
3.3	−0.0422	−0.0306	−0 0364	−0.0058
3.4	−0.0408	−0.0237	−0.0323	−0.0085
3.5	−0.0389	−0.0177	−0.0283	−0.0106
3.6	−0.0366	−0.0124	−0.0245	−0.0121
3.7	−0.0341	−0.0079	−0.0210	−0.0131
3.8	−0.0314	−0.0040	−0.0177	−0.0137
3.9	−0.0286	−0.0008	−0.0147	−0.0140
4.0	−0.0258	0 0019	−0.0120	−0.0139
4.1	−0.0231	0.0040	−0.0095	−0.0136
4.2	−0.0204	0.0057	−0.0074	−0.0131
4.3	−0.0179	0.0070	−0.0054	−0.0125
4.4	−0.0155	0.0079	−0.0038	−0.0117
4.5	−0.0132	0.0085	−0.0023	−0.0108
4.6	−0.0111	0.0089	−0.0011	−0.0100
4.7	−0.0092	0.0090	0.0001	−0.0091
4.8	−0.0075	0.0089	0.0007	−0.0082
4.9	−0.0059	0.0087	0.0014	−0.0073
5.0	−0.0046	0.0084	0.0019	−0.0065
5.1	−0.0033	0.0080	0.0023	−0.0057
5.2	−0.0023	0.0075	0.0026	−0.0049
5.3	−0.0014	0.0069	0.0028	−0.0042
5.4	−0.0006	0.0064	0.0029	−0.0035
5.5	0.0000	0.0058	0.0029	−0.0029
5.6	0.0005	0.0052	0.0029	−0.0023
5.7	0.0010	0.0046	0.0028	−0.0018
5.8	0.0013	0.0041	0.0027	−0.0014
5.9	0.0015	0.0036	0.0026	−0.0010
6.0	0.0017	0.0031	0.0024	−0.0007
6.1	0.0018	0.0026	0.0022	−0.0004
6.2	0.0019	0.0022	0.0020	−0.0002
6.3	0.0019	0.0018	0.0018	0.0001
6.4	0.0018	0.0015	0.0017	0.0003
6.5	0.0018	0.0012	0.0015	0.0004
6.6	0.0017	0.0009	0.0013	0.0005
6.7	0.0016	0.0006	0.0011	0.0006
6.8	0.0015	0.0004	0.0010	0.0006
6.9	0.0014	0.0002	0.0008	0.0006
7.0	0.0013	0.0001	0.0007	0.0006

* Taken from S. P. Timoshenko and S. Woinowsky-Krieger, *Theory of Plates and Shells,* 2d ed., McGraw-Hill Book Company, New York, 1959.

Table 2-4 FORCES AND DISPLACEMENTS IN CIRCULAR CYLINDERS LOADED BY EDGE FORCES UNIFORM AROUND A PARALLEL CIRCLE

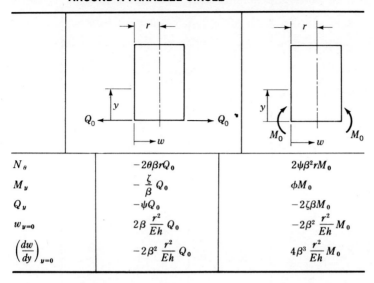

N_θ	$-2\theta\beta r Q_0$	$2\psi\beta^2 r M_0$
M_y	$-\dfrac{\zeta}{\beta} Q_0$	ϕM_0
Q_y	$-\psi Q_0$	$-2\zeta\beta M_0$
$w_{y=0}$	$2\beta \dfrac{r^2}{Eh} Q_0$	$-2\beta^2 \dfrac{r^2}{Eh} M_0$
$\left(\dfrac{dw}{dy}\right)_{y=0}$	$-2\beta^2 \dfrac{r^2}{Eh} Q_0$	$4\beta^3 \dfrac{r^2}{Eh} M_0$

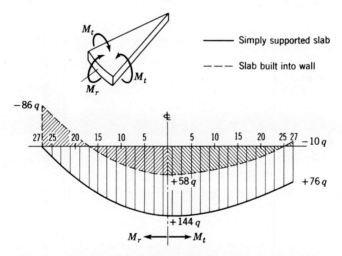

——— Simply supported slab

– – – Slab built into wall

Fig. 2-15 Roof-slab moments in foot-kips per foot under uniform vertical load q in kips per square foot.

analysis. For a large-diameter tank and more than one column, an approximate analysis has commonly been used. [13]

The general analysis is presented and illustrated by the tank shown in Fig. 2-13. A 40-psf load is exerted on the roof by snow, and the tank is filled with water.

Primary System   The roof slab is assumed to be freely supported on the wall. Slab moments due to the roof load are computed from Table 2-5; wall stress

Table 2-5 EXPRESSIONS FOR SYMMETRICAL BENDING OF CIRCULAR PLATES*

	General formulas	$\overset{M_a}{\curvearrowright}$	$\overset{q}{\downarrow\downarrow\downarrow\downarrow\downarrow}$	$\overset{q}{\Downarrow}$
w	$\dfrac{qr^4}{64D} + C_1\dfrac{r^2}{4} + C_2\log\dfrac{r}{a} + C_3$	$\dfrac{M_a}{2D(1+\nu)}(a^2 - r^2)$	$\dfrac{q}{64D}(a^2 - r^2)^2$	$\dfrac{q}{64D}(a^2 - r^2)\left(\dfrac{5+\nu}{1+\nu}a^2 - r^2\right)$
$\dfrac{dw}{dr}$	$\dfrac{qr^3}{16D} + C_1\dfrac{r}{2} + C_2\dfrac{1}{r}$	$-\dfrac{M_a r}{D(1+\nu)}$	$-\dfrac{qr}{16D}(a^2 - r^2)$	$-\dfrac{qr}{16D}\left(\dfrac{3+\nu}{1+\nu}a^2 - r^2\right)$
$\dfrac{d^2w}{dr^2}$	$\dfrac{3qr^2}{16D} + \dfrac{C_1}{2} - C_2\dfrac{1}{r^2}$	$-\dfrac{M_a}{D(1+\nu)}$	$-\dfrac{q}{16D}(a^2 - 3r^2)$	$-\dfrac{q}{16D}\left(\dfrac{3+\nu}{1+\nu}a^2 - 3r^2\right)$
C_1		$-\dfrac{2M_a}{D(1+\nu)}$	$-\dfrac{qa^2}{8D}$	$-\dfrac{qa^2}{8D}\dfrac{3+\nu}{1+\nu}$
C_2		0	0	0
C_3		$\dfrac{M_a a^2}{2D(1+\nu)}$	$\dfrac{qa^4}{64D}$	$\dfrac{qa^4}{64D}\dfrac{5+\nu}{1+\nu}$
M_r	$-D\left(\dfrac{d^2w}{dr^2} + \dfrac{\nu}{r}\dfrac{dw}{dr}\right)$	$+M_a$	$\dfrac{q}{16}[a^2(1+\nu) - r^2(3+\nu)]$	$\dfrac{q}{16}(3+\nu)(a^2 - r^2)$
M_t	$-D\left(\dfrac{1}{r}\dfrac{dw}{dr} + \nu\dfrac{d^2w}{dr^2}\right)$	$+M_a$	$\dfrac{q}{16}[a^2(1+\nu) - r^2(1+3\nu)]$	$\dfrac{q}{16}[a^2(3+\nu) - r^2(1+3\nu)]$
M_a		$+M_a$	$-\dfrac{qa^2}{8}$	0
$M_{\mathbb{C}}$		$+M_a$	$\dfrac{qa^2}{16}(1+\nu)$	$\dfrac{qa^2}{16}(3+\nu)$

* For a derivation of these values see Ref. 12, chapter 3. a is the outside radius of the plate.

couples and stress resultants are computed for lateral loadings by considering the proper base restraint (see Sec. 2-8). The slab moments are shown in Fig. 2-15.

Errors A slab rotation occurs at the wall; for a uniform load it would be, where $r = a$ and $v = 1/6$,

$$D_S = \frac{Eh_s^3}{12(1 - v^2)} = \frac{E(1)^3}{12(35/36)} = 0.0857E$$

$$-\frac{dw}{dr} = D_{20}^S = \frac{qa^3}{16D_S} \frac{12}{7} = \frac{q(27)^3(3)}{28(0.0857E)} = 24,600\ \frac{q}{E}$$

For a tank filled with liquid there would be no translation at the top, but a rotation occurs, shown as negative in Fig. 2-14c.

$$\frac{dw}{dy} = D_{20}^W = -\frac{\gamma r^2}{Eh} = -\frac{\gamma(27)^2}{1.25E} = -584\ \frac{\gamma}{E}$$

Corrections A unit moment and a unit shear are applied at the junction of roof and wall; with

$$\beta^4 = \frac{3(1 - v^2)}{r^2 h^2} = \frac{3(1 - 1/36)}{(27)^2(1.25)^2} = 0.00257\ \text{ft}^{-4}$$

$$\beta^3 = 0.0114\ \text{ft}^{-3}$$
$$\beta^2 = 0.0507\ \text{ft}^{-2}$$
$$\beta = 0.225\ \text{ft}^{-1}$$

the resulting displacements in the wall are, from (2-36),

$$D_W = \frac{Eh_W^3}{12(1 - v^2)} = \frac{E(1.25)^3}{12(35/36)} = 0.167E$$

$$D_{11}^W = \frac{1}{2\beta^3 D_W} = \frac{1}{2(0.0114)0.167E} = \frac{262}{E}$$

$$D_{12}^W = -\frac{1}{2\beta^2 D_W} = -\frac{1}{2(0.0507)0.167E} = -\frac{59.2}{E}$$

$$D_{22}^W = \frac{1}{\beta D_W} = \frac{1}{(0.225)0.167E} = \frac{26.7}{E}$$

In the roof slab, the unit horizontal force which produces no rotation leads to a translation of

$$D_{11}^S = \frac{r}{Eh}(1 - v) = \frac{27}{E}\frac{5}{6} = \frac{22.5}{E}$$

and the moment, which gives no translation, leads to a rotation, from Table 2-5, of

$$D_{22}^S = \frac{r}{D_s(1 + v)} = \frac{27}{0.0857(1.167)E} = \frac{272}{E}$$

Compatibility We shall consider first the effect of the roof load, for which the compatibility equations are

$$X_1(262 + 22.5) + X_2(-59.2) = 0$$
$$X_1(-59.2) + X_2(26.7 + 272) + 24,600q = 0$$

from the first equation

$$X_1 = 0.204X_2$$

and then from the second

$$X_2 = -24,600 \frac{q}{286.6} = -86.0q = M_0$$

and   (a)

$$X_1 = 0.204X_2 = -17.5q = -Q_0$$

Substituting these values into the expressions in Table 2-3 and thence into Eqs. 2-35, we obtain, for example, N_θ and M_y which are plotted in Fig. 2-16. Note again that $Q_0 = -X_1$.

The solution is somewhat simpler if the effect of the horizontal displacement of the roof slab is neglected; thus with $D_{11}^S = 0$,

$$X_1 = 0.226X_2$$

$$X_2 = -24,600 \frac{q}{285.3} = -86.5q \qquad (b)$$

$$X_1 = 0.226X_2 = -19.5q$$

If we neglect X_1 altogether, we obtain

$$X_1 = 0$$
                                                                                      (c)
$$X_2 = -24,600 \frac{q}{298.7} = -83.0q$$

The effect of X_1 is obviously not great in this case. Since (a) and (b) differ only slightly, we may consider (b) as close enough for design purposes, in which case a direct solution may be obtained by moment distribution.

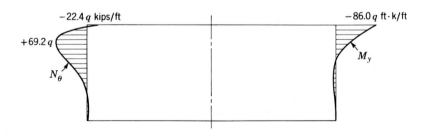

Fig. 2-16 Wall built into roof slab under roof load q in kips per square foot.

The stiffness factor of a cylindrical wall is computed from the value of rotation determined in Sec. 2-6. The stiffness factor is the moment which gives a unit rotation ($dw/dy = 1$ at $y = 0$) with no translation ($w = 0$ at $y = 0$):

$$(w)_{y=0} = - \frac{1}{2\beta^3 D} (\beta M_0 + Q_0) = 0 \tag{2-37}$$

$$\left(\frac{dw}{dy}\right)_{y=0} = \frac{1}{2\beta^2 D} (2\beta M_0 + Q_0) = 1 \tag{2-38}$$

From (2-37)

$$Q_0 = -\beta M_0$$

and from (2-38)

$$2\beta M_0 - \beta M_0 = 2\beta^2 D$$

from which

$$M_0 = K_W = 2\beta D \tag{2-39}$$

or, using the values for β and D, we find

$$K_W = \frac{E h_W^3}{6(1 - v^2)} \sqrt[4]{\frac{3(1 - v^2)}{r^2 h_W^2}}$$

Separating parameters and including H, the wall height, we obtain

$$K_W = C_W \frac{E h_W^3}{H} \tag{2-40}$$

where

$$C_W = \frac{\sqrt[4]{12(1 - v^2)}}{6(1 - v^2)} \sqrt{\frac{H^2}{2rh_W}} = 0.32 \sqrt{\frac{H^2}{2rh_W}}$$

for $v = 1/6$. These values of C_W are given in Table 2-6 for a wide range of $H^2/2rh_W$ ratios.

Table 2-6 STIFFNESS OF CYLINDRICAL WALL; NEAR EDGE HINGED, FAR EDGE FREE* ($K_W = C_W \times E h_W^3/H$)

$H^2/2rh_W$	Coefficient	$H^2/2rh_W$	Coefficient
0.4	0.139	5	0.713
0.8	0.270	6	0.783
1.2	0.345	8	0.903
1.6	0.399	10	1.010
2.0	0.445	12	1.108
3.0	0.548	14	1.198
4.0	0.635	16	1.281

* From Ref. 13.

For the roof slab, we obtain the stiffness factor from Table 2-5,

$$K_S = \frac{D(1+v)}{r} = C_S \frac{Eh_s^3}{r} \tag{2-41}$$

where

$$C_S = \frac{1}{12(1-v)} = \frac{1}{10} \text{ for } v = 1/6$$

For this example

$$\frac{H^2}{2rh_W} = 5.92 \approx 6.0$$

which, from Table 2-5, gives

$$K_W = \frac{0.78(1.25)^3}{20} E = 0.0761E$$

and from (2-41)

$$K_S = 0.1E/27 = 0.00370E$$

The fixed end moments are, for the slab load,

$$M^F = -\frac{qr^2}{8} = -91.1q$$

and for the wall load, with $D_{11}^S = 0$,

$$M^F = \frac{584\gamma}{26.7 - 13.4} = 44.0\gamma$$

obtained by solving the compatibility equations with $D_{22}^S = 0$.

The wall moment due to slab load will now be

$$M_0 = -91.1q \frac{0.0761}{0.0798} = -86.6q$$

and due to wall load

$$M_0 = 44.0\gamma \frac{0.0037}{0.0798} = 2.04\gamma$$

which can be obtained directly from the compatibility equations where

$$X_1 = 0.226X_2$$
$$X_2 = \frac{584\gamma}{285.3} = 2.05\gamma$$

2-8   ANALYSIS OF A CIRCULAR CYLINDRICAL WALL CONNECTED TO A FOUNDATION

We may distinguish five different types of restraint to the base of a cylindrical tank (Fig. 2-17):

1. *Free-sliding,* for which the membrane theory gives a complete solution.
2. *Hinged,* wherein the effect of a radial shear at the base must be super-imposed upon the membrane theory solution.
3. *Fixed,* in which both a moment and shear occur at the base.
4. *Partially fixed,* where a rotation of the base slab is considered, hence reducing the fixing moment determined in type 3.
5. *Partially sliding,* in which some shear and possibly some moment occur, but of less magnitude than in types 2 or 3. This case has been incorporated into designs of prestressed tanks to avoid wall bending under initial prestressing and yet to obtain a monolithic structure in service. Analysis of this system is not presented here.

By successively studying the example of Fig. 2-13 for each type of restraint, we shall illustrate the analyses.

Free-Sliding (No Restraint)

The results of the membrane theory, with expression (2-4), are plotted in Fig. 2-17. There are no bending moments in this case.

Hinged Restraint

The analysis proceeds in the normal way:

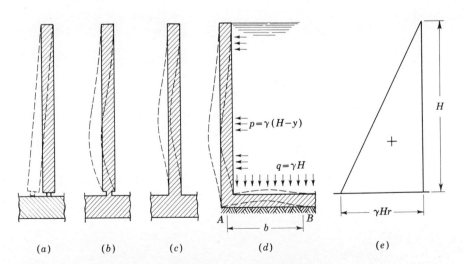

$$p = \gamma(H - y)$$

$$q = \gamma H$$

(a) (b) (c) (d) (e)

Fig. 2-17   Base restraints for cylindrical wall. (a) Free-sliding. (b) Hinged. (c) Fixed. (d) Partially fixed. (e) Ring tension for free-sliding wall.

Primary System   The free-sliding wall.

Error   From Eq. (2-23a),

$$D_{10} = w = -\frac{\gamma r^2 H}{Eh} = -\frac{\gamma(27)^2 20}{E(1.25)} = -\frac{11,700\gamma}{E}$$

Correction   From (2-36), see Fig. 2-14b,

$$D_{11} = \frac{1}{2\beta^3 D} = \frac{262}{E}$$

Compatibility

$$X_1 D_{11} + D_{10} = 0$$

$$X_1 = -\frac{D_{10}}{D_{11}} = \frac{11,700\gamma}{262} = 44.7\gamma$$

and where $\gamma = 0.0624$ lb/ft^3,

$$X_1 = 2.78 \text{ kips/ft}$$

The distribution throughout the shell of N_θ and M_y is obtained by substituting X_1 for Q_0 into the equations of Table 2-3 and combining the results with the membrane theory results. The resulting N_θ and M_y are plotted in Fig. 2-18.

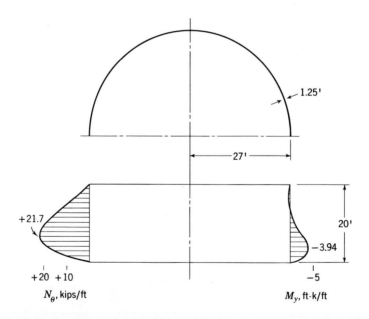

Fig. 2-18   Wall hinged at base.

Fixed Restraint

The same type of analysis used for hinged restraint results in the following:

Primary System   The free-sliding wall.

Errors   See Fig. 2-14a.

$$D_{10} = w = -11,700 \frac{\gamma}{E}$$

$$D_{20} = \frac{dw}{dy} = \frac{\gamma r^2}{Eh} = 584 \frac{\gamma}{E}$$

Corrections   Same as in Sec. 2-7; see Fig. 2-14b.

$$D_{11} = \frac{262}{E}$$

$$D_{12} = -\frac{59.2}{E}$$

$$D_{22} = \frac{26.7}{E}$$

Compatibility

$$262X_1 - 59.2X_2 - 11,700\gamma = 0$$
$$-59.2X_1 + 26.7X_2 + 584\gamma = 0$$

from which

$$X_2 = 155\gamma$$
$$X_1 = 79.8\gamma$$

or with $\gamma = 0.0624$

$$X_2 = 9.65 \text{ ft-k/ft}$$
$$X_1 = 4.96 \text{ kips/ft}$$

The results are plotted in Fig. 2-19.

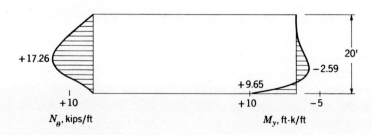

Fig. 2-19   Wall fixed at base.

Partially Fixed Restraint

The wall is rigidly connected to the floor slab and the supporting foundation is assumed to be immovable. Two sets of forces act on the floor slab (Fig. 2-17): an edge bending moment M_0 and a vertical pressure q. The cylinder wall is restrained from translation and rotation by the slab. The lateral load tends to rotate the wall outward at the base and hence to rotate the plate upward. As the plate rotates upward off of the foundation, the vertical pressure tends to push it back down. Such a case has been treated rigorously on page 308 of Ref. 12, and a good approximation has been given on page 455 of Ref. 8. The approximate analysis will now be developed. If that part of the slab which lifts off the support is confined to the region near the edge, then it can be considered as a series of beams of constant width spanning some distance b and acted upon by an end moment M and a vertical load q. If the actual plate is considered, the width of a strip would not be constant but would vary from a unit value at the edge to zero at the center.

With this approximation, two moment-area equations can be written to express the facts that there must be no relative deflection between points A and B and that there must be no rotation at B.

$$\phi_A = \frac{1}{3D_S} Mb - \frac{qb^3}{24D_S}$$

$$\phi_B = \frac{1}{6D_S} Mb - \frac{qb^3}{24D_S} = 0$$

from which

$$b = 2\sqrt{\frac{M}{q}} \tag{2-42}$$

and

$$\phi_A = \frac{1}{3D_S}\sqrt{\frac{M^3}{q}} = X_2 D_{22}^S \tag{2-43}$$

The value of b must be small relative to r, so that D_S will be reasonably constant for the length b. M represents an unknown moment at the wall base which will be less than the fixed-base moment. The analysis now proceeds as usual.

Primary System   This is again the free-sliding wall.

Errors   Same as in type 3.

Corrections   These include the flexibility of the floor slab. We shall begin by including the axial displacement of the floor slab; hence as in Sec. 2-7 with $h_S = 1.0$ ft,

$$D_{11}^S = \frac{r}{Eh}(1 - \nu) = \frac{22.5}{E}$$

$$D_{12}^S = 0$$

and from (2-43)

$$X_2 D_{22}^S = \frac{1}{3D_S}\sqrt{\frac{(X_2)^3}{q}} = \frac{1}{(0.0857)3E}\sqrt{\frac{(X_2)^3}{q}}$$

$$= \frac{3.92}{E}\sqrt{\frac{(X_2)^3}{q}}$$

and the total corrections become

$$D_{11} = \frac{262}{E} + \frac{22.5}{E} = \frac{284.5}{E}$$

$$D_{12} = -\frac{59.2}{E} + 0 = -\frac{59.2}{E}$$

$$D_{22} = \frac{26.7}{E} + \frac{3.92}{EX_2}\sqrt{\frac{(X_2)^3}{q}}$$

Compatibility

$$284.5X_1 - 59.2X_2 - 11,700\gamma = 0$$

$$-59.2X_1 + 26.7X_2 + 3.92\sqrt{\frac{(X_2)^3}{q}} + 584\gamma = 0$$

where $q = H\gamma = 20\gamma$

$$284.5X_1 - 59.2X_2 = +11,700\gamma$$

$$-59.2X_1 + 26.7X_2 + 0.876\sqrt{\frac{(X_2)^3}{\gamma}} = -584\gamma$$

With $\gamma = 0.0624$ kip/ft^3

$$
\begin{aligned}
284.5X_1 - 59.2X_2 &= +730 \\
-59.2X_1 + 26.7X_2 + 3.5\sqrt{(X_2)^3} &= -36.4 \\
+59.2X_1 - 12.3X_2 &= +151.8 \\
14.3X_2 + 3.5\sqrt{(X_2)^3} &= +115.4
\end{aligned}
$$

$$X_2 = 5.2 \text{ ft-k/ft}$$

$$X_1 = \frac{730 + 308}{284.5} = 3.65 \text{ kips/ft}$$

$$b = 2\sqrt{\frac{X_2}{q}} = 2\sqrt{5.2/1.25} = 4.08 \text{ ft}$$

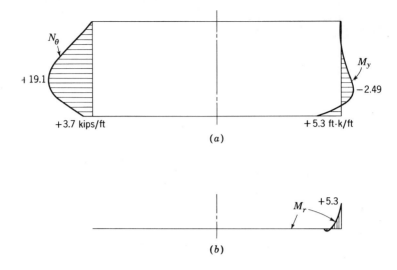

(a)

(b)

Fig. 2-20 (a) Wall partially fixed at base. (b) Moments in the base slab.

The average width of the trapezoidal strip of slab is less than the 12 in. used at the edge.

$$t_{av} = 24.96/27 = 0.925 \text{ ft}$$

If we use a new $D_S = 0.0857E (0.925) = 0.0794E$,

$$X_2 = 5.3 \text{ ft-k/ft}$$
$$X_1 = 3.67 \text{ kips/ft}$$
$$b = 4.12 \text{ ft} \approx 4.08 \text{ ft}$$

2-9 BENDING BEHAVIOR OF SHELL WALLS UNDER WIND

The study of bending in shell walls under wind loads can often best be done by using numerical methods adapted for the computer. Here we shall make use of results from a finite-element program for axisymmetrical shells. Many such programs are available.[14-16] This section begins with a study of cylinders and concludes with a comparative study of cylinders, cones, and hyperboloids.

Cylinders

The cylinder studied (Fig. 2-21) is taken as free at the top, simply supported at the base, of constant thickness, and loaded by a wind pressure uniformly distributed from top to bottom while varying in the circumferential direction following the design values in Table 2-1. Simply supported means that $u = v = w = M_y = 0$ at

$y = 0$. Also, in the finite-element program the rotation of the shell at its base about a tangent line is also zero, i.e., $\partial w/\partial y = 0$ at $y = 0$.

Tables 2-7a and b give results from membrane and bending theories for harmonic loadings given in Table 2-1. Table 2-7a gives results considered to be more representative of high wind velocity loadings, whereas Table 2-7b gives values obtained from a wind distribution found for lower wind velocities. The principal difference in results is that the distribution (Table 2-7b) with the higher external suction ($H_\theta = -1.3$ at $\theta = 75°$) leads to vertical stress resultants which are about 1.25 times those found for the lower suction ($H_\theta = -1.0$ at $\theta = 75°$). A second difference of some importance is the

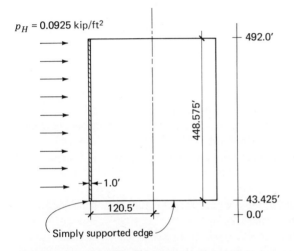

Fig. 2-21   Wind on a cylindrical shell.

Table 2-7a   CYLINDER UNDER WIND LOAD (Stress resultants at $\theta = 0°$)

n	A_n	n^2	$A_n n^2$	N'_{yn}* $y = 0$	$N'_{\theta m}$* $y = 100$†	N_{yn}* $y = 0$	$N_{\theta m}$* $y = 100$†	$M_{\theta n}$‡ $y = 100$†
0	0.2364§	0	0	0	−2.63	0	−2.63	0
1	0.3419	1	0.3419	26.4	−3.81	26.1	−3.81	0
2	0.5418	4	2.1672	167.4	−6.05	163.0	−6.03	−0.72
3	0.3872	9	3.4848	269.1	−4.31	182.0	−4.11	−3.73
4	0.0525	16	0.8400	64.9	−0.59	14.4	−0.48	−0.87
5	−0.0771	25	−1.9275	−148.9	0.86	−12.4	0.52	1.59
6	−0.0039	36	−0.1404	−10.8	0.04	−0.4	0.02	0.09
7	0.0341	49	1.6709	129.0	−0.38	2.3	−0.06	−0.74
Σ	1.5138	....	6.4369	497.1	−16.9	375.1	−16.6	−4.38

* In kips per foot.

† In feet.

‡ In kips.

§ Taken from Table 2-1, design values, but including an internal suction of 0.5.

Table 2-7b   CYLINDER UNDER WIND LOAD

n	A_n	n^2	$A_n n^2$	N'_{yn}*	$N'_{\theta n}$*	N_{yn}*	$N_{\theta n}$*	$M_{\theta n}$†
0	0.1077‡	0	0	0	−1.20	0	−1.21	0
1	0.2602	1	0.2602	20.10	−2.90	19.82	−2.91	0
2	0.6024	4	2.4096	186.10	−6.72	180.7	−6.72	−0.97
3	0.5046	9	4.5414	350.74	−5.62	236.9	−5.30	−5.88
4	0.1064	16	1.7024	131.48	−1.19	29.07	−0.93	−2.10
5	−0.0948	25	−2.3700	−183.04	1.06	−15.17	−0.58	2.26
6	−0.0186	36	−0.6696	−51.71	0.21	−1.90	−0.06	0.47
7	0.0468	49	2.2932	177.11	−0.52	3.174	−0.06	−1.10
Σ	1.5147	....	7.9132	630.77	−16.89	452.6	−16.50	−6.4

* In kips per foot.

† In kips.

‡ Taken from Table 2-1, Niemann values, but including an internal suction of 0.5.

increase by about 46 percent in M_θ for the case of higher external suction. The coefficients for $n > 1$ are all higher in that case, meaning that the ring loading is more oscillatory and hence the bending is increased. The case of higher external suction corresponds to a more streamlined wind flow giving less overall drag (measured by the lower A_1 of Table 2-7b), but the structural consequence is a more severe redistribution from the simple $n = 1$ case.

Focusing now only on Table 2-7a, we observe that the axisymmetrical load $n = 0$ gives no vertical stress at the base and a compression ring-stress-resultant equal to $N'_{\theta 0} = 0.0925\,(120.5)\,(0.2364) = +2.63$ kips/ft. from Eq. (2-2a). The finite-element solution gives the same value as the membrane theory, but only at some distance from the base. Near the base the shell must move inward under the axisymmetrical compression, but the restraint there (taken simply supported in this case) prevents that motion by introducing a horizontal edge shear Q_y which in turn introduces bending near the edge. Thus to get a true comparison of the two methods, it is necessary to compare N_θ at some distance from the edge disturbance. We take $y = 100$ ft in this case, and the table shows that there $N'_{\theta 0} = N_{\theta 0}$.

The solution for $n = 1$ is similar to seismic loading and can be obtained directly from the beam equation (2.6), where for a hollow circular cross section, $I = \pi h r^3$, $x = r$, and M is the cantilever moment of the load $p_H A_1 \cos \theta$ applied around the circumference and over the height $y = H$; i.e.,

$$M_p = 4 \int_0^H \int_0^{\pi/2} (p_H A_1 \cos \theta) \cos \theta \, r \, d\theta (H - y) \, dy \qquad (a)$$

which for $y = 0$ gives $M = p_H \pi r (H^2/2) A_1$ and for $p_H = 0.0925$ kip/ft², $r = 120.5$ ft and $H = 448.6$ ft, and $A_1 = 0.3419$ results in $M = 1,200,000$ ft-kips. So that with $I = 5.5 \times 10^6$ ft⁴. $N'_{y1} = (1,200,000 \times 120.5 \times 1)/(5.5 \times 10^6) = 26.3$ kips/ft, essentially the same value shown in Table 2-7a both from the membrane theory and from the solution including bending.

However, the wind is not distributed as $n = 1$ but rather as pressures and suctions mostly on the windward half of the tower; this means that higher terms need to be used to account for the true distribution. These terms do not add any additional load to the cylinder; rather they redistribute the load obtained from $n = 1$ in a manner consistent with pressures measured on full-scale towers. The use of a Fourier series to redistribute the loading is both mathematically convenient and physically enlightening. Physically it shows that the cylinder will not behave as a cantilever beam where its beam stress resultant would be only 26 kips/ft, but rather as a kind of deep beam where its actual maximum stress resultant can be over 10 times greater but its lever arm between maximum tension and maximum compression is very much smaller. Figure 2-22 shows the N_y values at the base plotted over the cross section on a projected horizontal to give a comparison with the $n = 1$ solution. The total wind-load moment is the same for both results ($n = 1$ and $n = 0$-7); only the circumferential distribution of load is different.

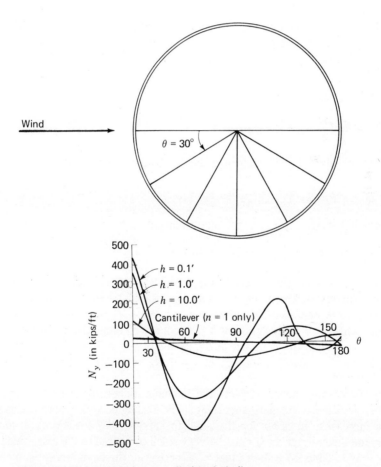

Fig. 2-22 N_y from wind on a cylindrical shell.

Table 2-7a also shows how the membrane theory solution begins to break down for higher values of n. This is so because of the flexural stiffness of the shell in the circumferential direction. To see this requires study of a vertical slice of the shell, as shown in Fig. 2-23. The membrane theory requires that for each harmonic load, the slice defined by an arc of $2\theta_n$ is held in vertical equilibrium by the N'_{yn} at the base and the $N'_{\theta yn}$ along the vertical edges (Fig. 2-23), giving the equation

$$-2\int_0^{\theta_n} N'_{yn} r\, d\theta + 2\int_0^H N'_{\theta yn}\, dy = 0$$

$$-2\int_0^{\theta_n} \frac{p_H}{r} A_n n^2 \frac{H^2}{2} \cos n\theta r\, d\theta + 2\int_0^H p_H A_n n(H-y)\sin n\theta_n\, dy = 0$$

$$-p_H A_n nH^2 + p_H A_n nH^2 = 0$$

The horizontal loads $p_H A_n \cos n\theta \cos \theta$ are equilibrated by the shear $N'_{y\theta n}$ at the base. The N'_{yn} forces alone equilibrate the lateral-load cantilever bending moment, as we have already seen in Eq. (a) for $n = 1$. A similar equation holds for $n > 1$, as shown in Fig. 2-23 where for each harmonic, the moment M_n on the vertical slice of shell must be fully balanced by the moment of the N'_{yn} values about the chord AB. For any n this moment will be

$$M_{AB} = 2\int_0^{\theta_n} N'_{yn} r(\cos\theta - \cos\theta_n) r\, d\theta \tag{b}$$

where, for p_H constant with y, at $y = 0$,

$$N'_{yn} = \frac{p_H}{r}\frac{H^2}{2} A_n n^2 \cos n\theta \tag{c}$$

Substituting (c) into (b) and performing the integration yields

$$M_{AB} = -p_H r A_n H^2 \frac{n}{1-n^2}\cos\theta_n \tag{d}$$

For the lateral loading, Eq. (a) is modified to include $n > 1$:

$$M_p = 4p_H r A_n \int_0^H \int_0^{\pi/2} \cos n\theta\cos\theta\, d\theta\,(H-y)\, dy \tag{e}$$

which, upon integration, yields exactly Eq. (d). Thus the membrane theory means that the shell carries each harmonic $n > 1$ as a vertical cantilever of depth $d_n = r(1 - \cos\theta_n)$. Thus as θ_n decreases, d_n decreases and hence N'_{yn} increases. As shown in Table 2-7a, when $n = 3$, the value of N'_{yn} becomes 10 times the value for $n = 1$, even though A_3 is about the same size as A_1. By the time $n = 7$, a value of $A_7 = 1/10$ A_1 still yields an N'_{y7} about 5 times N'_{y1}. At $n = 7$, $d_n = 0.025r = 3$ ft compared to a radius of 120.5 ft.

But the membrane theory neglects the bending stiffness of the shell, and this stiffness becomes more and more important as the loading becomes more oscillatory, i.e., as n increases. In order for the shell to carry $n = 7$ loading as a vertical cantilever, the flexural rigidity D must be very small compared to the axial rigidity $K = Eh/(1 - \nu^2)$. Table 2-7a shows that as n increases, the circumferential moment $M_{\theta n}$ becomes increasingly greater as compared to $N_{\theta n}$. In other words, some of the load is transferred circumferentially by out-of-plane shear and bending before it is transferred vertically by in-plane shear and bending (N_y).

Table 2-8 shows how the N_{yn} change as the shell flexural rigidity increases with respect to its axial rigidity. The values for $n = 0$ and $n = 1$ do not change since they do not depend upon the out-of-plane bending stiffness in the circumferential direction. But the values for $n \geq 2$ decrease as the shell gets stiffer flexurally, until in the limit the wind load is carried entirely by $n = 1$; i.e., the ring flexural stiffness is so great that the circumferential pressure variation plays no role in N_{yn}, but only the overall loading, expressed by M_p, influences N_y. In effect a circumferentially stiff ring means that the tower behaves like a cantilever beam because the fundamental idea of beam theory is that the cross section itself remains undeformed in its plane after bending. Figure 2-22 emphasizes this behavior by showing how the circumferential variations in N_y approach beam behavior as the ring stiffness increases.

The cylinder with $h = 0.1$ ft was given an $E = 5.19 \times 10^6$ kips/ft^2 or 10 times the value of E used for the reference $h = 1.0$-ft cylinder where $E = 5.19 \times 10^5$ kips/ft^2 $= 3.6 \times 10^6$ psi $= 57,000 \sqrt{f_c'}$ for $f_c' = 4000$ psi. The cylinder with $h = 10.0$ ft had $E = 5.19 \times 10^4$ kips/ft^2. In this way the axial rigidity of the three cylinders remained the same, $k = Eh/(1 - \nu^2)$, while the flexural rigidity varied as h^2.

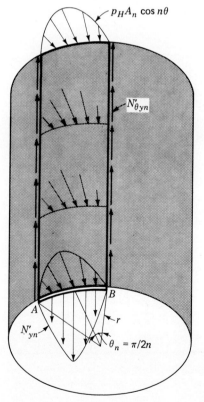

Fig. 2-23 Equilibrium of a cylindrical slice.

Table 2-8 CYLINDER UNDER WIND LOAD (Influence of bending stiffness on harmonics)

n	$N'_{yn}{}^*$	$N_{yn}{}^*$ $h = 0.1$ ft	$N_{yn}{}^*$ $h = 1.0$ ft	$N_{yn}{}^*$ $h = 10$ ft
0	0	0	0	0
1	26.4	26.2	26.1	25.2
2	167.4	166.1	163.0	73.3
3	269.1	265.9	182.0	15.1
4	64.9	61.2	14.4	0.8
5	−148.9	−113.4	−12.4	−0.5
6	−10.8	−4.9	−0.4	0.0
7	129.0	30.5	2.3	0.1
Σ	497.1	431.7	375.1	113.9

* In kips per foot.

Table 2-9 shows how the maximum values of N_y, N_θ, and M_θ change as the shell flexural rigidity increases. The very thin shell behaves much like the membrane theory, except that small moments M_θ do give significant shell stresses

$$\sigma_\theta = \frac{6M_\theta}{h^2} = \frac{6 \times 0.3}{0.1^2} = 180 \text{ kips/ft}^2 \text{ (1250 psi)}$$

For the 1-ft-thick cylinder, the maximum bending gives an even higher stress $\sigma_\theta = 234$ kips/ft² (1620 psi), whereas for the very thick cylinder, the stress drops to $\sigma_\theta = 16.1$ kips/ft² (112 psi). The moments increase with height because there is no restraint at the top where the shell acts more and more like a ring.

Indeed for the thick cylinder under harmonic loading of $n \geq 2$, the values at the least restrained part, the top, are very close to those for a ring under harmonic loading. The ring equations come directly from Eqs. (2-25) when we take all $\partial/\partial y = 0$,

Table 2-9 CYLINDER UNDER WIND (Influence of wind on maximum values)

y^*	Membrane theory		$h = 0.1$ ft			$h = 1.0$ ft			$h = 10$ ft		
	$N_y\dagger$	$N_\theta\dagger$	$N_y\dagger$	$N_\theta\dagger$	$M_\theta\ddagger$	$N_y\dagger$	$N_\theta\dagger$	$M_\theta\ddagger$	$N_y\dagger$	$N_\theta\dagger$	$M_\theta\ddagger$
0	497.1	−16.9	431.7	−1.2	−0.1	375.1	23.7	−0.5	113.9	9.1	−3.1
100	300.2	−16.9	258.6	−16.8	−0.1	222.2	−16.6	−4.4	52.7	−14.3	−67.6
350	24.7	−16.9	19.9	−16.7	−0.3	15.8	−15.0	−28.5	1.7	−7.4	−223.3
450	0	−16.9	0	−17.6	−0.3	0	−23.2	−39.0	0	−7.1	−268.4

* In feet.
† In kips per foot.
‡ In kips.

$p_y = p_\theta = 0$, and $p_z = p_H A_n \cos n\theta$ so that

$$\frac{dN_{\theta n}}{d\theta} - Q_{\theta n} = 0 \tag{f}$$

$$\frac{dN_{\theta yn}}{d\theta} = 0 \tag{g}$$

$$\frac{dQ_{\theta n}}{d\theta} + N_{\theta n} + r p_H A_n \cos n\theta = 0 \tag{h}$$

$$\frac{dM_{\theta yn}}{d\theta} = 0 \tag{i}$$

$$-\frac{dM_{\theta n}}{d\theta} + Q_{\theta n} r = 0 \tag{j}$$

from which we see that $N_{\theta yn}$ and $M_{\theta yn}$ can be dropped and Eqs. (f), (h), and (j) represent a statistically determinate system. Differentiating (f) and putting the result into (h) yields

$$\frac{d^2 N_{\theta n}}{d\theta^2} + N_{\theta n} + r p_H A_n \cos n\theta = 0$$

for which the solution $N_{\theta n} = N_{\theta n} \cos n\theta$ yields

$$-N_{\theta n} n^2 + N_{\theta n} + r p_H A_n = 0$$

$$N_{\theta n} = \frac{1}{n^2 - 1} p_H A_n r \tag{k}$$

and with $Q_{\theta n} = Q_{\theta n} \sin n\theta$, Eq. (f) gives

$$Q_{\theta n} = -\frac{n}{n^2 - 1} p_H A_n r \tag{l}$$

Finally with $M_{\theta n} = M_{\theta n} \cos n\theta$, Eq. (j) gives

$$M_{\theta n} = \frac{1}{n^2 - 1} p_H A_n r^2 \tag{m}$$

Table 2-10 compares the results of $M_{\theta n}$ with those from the finite-element solution where $p_H r^2 = 0.0925 \times 125^2 = 1448$ kips. Except for $n = 2$ where the ring $M_{\theta n}$ is 36 percent high, the correspondence is close, showing that the top of the cylinder behaves very nearly as a free ring. The radius of both ring and cylinder is taken as $r = 120 + h/2 = 125$ ft.

Cylinders, Cones, and Hyperboloids

Table 2-11 compares the behavior of the cylinder in Fig. 2-21 ($h = 1$ ft) with that of a cone (Fig. 2-24) and a hyperboloid (Fig. 2-25) of similar dimensions. The dimensions are chosen to represent simplified versions of the Trojan Tower (Fig. 2-10).

Table 2-10 CYLINDER AND RING BENDING

	$M_{\theta n} = p_H A_n r^2/(n^2 - 1)$			
n	A_n	$1/(n^2 - 1)$	$M_{\theta m}$, ring*	$M_{\theta n}$*
0	0.2364	....	....	−0.18
1	0.3419	....	....	−0.005
2	0.5418	1/3	261.5	192.2
3	0.3872	1/8	70.1	75.09
4	0.0525	1/15	5.1	5.15
5	−0.0771	1/24	−4.7	−4.72
6	−0.0040	1/35	−0.2	−0.16
7	0.0341	1/48	1.0	1.04
Σ	....	....	332.8	268.4

* In kips.

Table 2-11 HARMONIC COMPARISON OF FORM

	Cylinder			Cone			Hyperboloid		
n	N_{yn}* $y = 0$	$N_{\theta n}$* $y = 100$‡	$M_{\theta n}$† $y = 100$‡	N_{yn}* $y = 0$	$N_{\theta n}$* $y = 100$‡	$M_{\theta n}$† $y = 100$‡	$N_{\phi n}$* $y = 0$	$N_{\theta n}$* $y = 100$‡	$M_{\theta n}$† $y = 100$‡
0	0.0	−2.63	0	0.0	−3.6	0	0.0	−3.4	0
1	26.1	−3.81	0	12.2	−5.2	0	12.7	−4.7	0
2	163.0	−6.03	−0.7	76.9	−8.1	−0.2	62.3	−6.5	−0.2
3	182.0	−4.11	−3.7	100.3	−5.8	−1.1	63.9	−4.2	−0.7
4	14.4	−0.48	−0.9	11.2	−0.8	−0.4	7.8	−0.6	−0.2
5	−12.4	0.52	1.6	−12.8	1.0	0.9	−9.7	0.9	0.7
6	−0.4	0.02	0.1	−0.5	0.0	0.1	−0.5	0.0	0.1
7	2.3	−0.06	−0.7	3.4	−0.3	−0.7	3.4	−0.3	−0.7
Σ	375.1	−16.3	−4.4	190.7	−22.7	−1.4	139.9	−18.7	−1.0

* In kips per foot.
† In kips.
‡ In feet.

For the cone the same type of slice defined by $2\theta_n$ (see Fig. 2-23) again carries each harmonic of wind load, except that the arc length increases as the cone horizontal radius increases. Again the N'_{yn} and $N'_{\theta yn}$ provide vertical equilibrium, and the $N'_{y\theta n}$ and p_H loads along with the horizontal components of the $N'_{\theta yn}$ provide horizontal equilibrium. The moment about the base chord AB is provided solely by the N'_{yn}, except that the base radius is now larger than before, thus tending to reduce the N'_{yn} required.

The hyperboloid slice has the same behavior as the cone, except that the moment about AB now has a component of the $N'_{\theta\phi n}$ acting in the same rotational sense as the $N'_{\phi n}$. Thus we should expect that $N'_{\phi n}$ will be less for $n \geq 2$ in the hyperboloid than

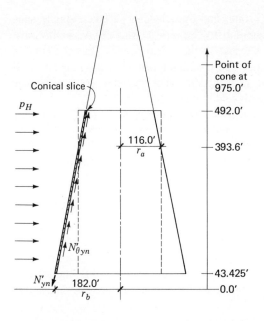

Fig. 2-24 Wind on a conical shell.

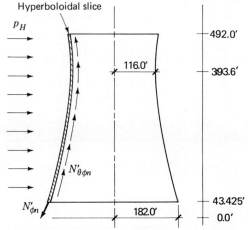

Fig. 2-25 Wind on a hyperbolic shell.

the N'_{yn} in the cone or cylinder, and Table 2-11 shows this to be the case.* The meridional curvature of the hyperboloid allows the $N'_{\phi n}$ to be reduced thanks to the overall bending resistance afforded by the in-plane shears at $\theta = \theta_n$. The high peaks of N'_{yn} found in the cylinder and the cone are smoothed out in the hyperboloid. This basic difference indicates one reason why the doubly curved shell is better suited to carrying wind loads than the singly curved cones and cylinders.

* The discussion here refers to membrane theory values to emphasize the predominantly in-plane behavior of these shells. Table 2-11 gives results from the bending theory analysis which shows the same trends as would results from a membrane theory analysis.

Table 2-11 compares the cylinder, cone, and hyperboloid, all taken simply supported and having constant thicknesses of 1 ft. The $n = 1$ values of N_y and N_θ are smaller for the cone and hyperboloid because of the greater radii. Thus the cone is really not as efficient as it seems because its maximum tension is $190.7/12.2 = 15.6$ times its cantilever beam tension of 12.2 kips/ft². The cylinder tension ratio is $375.1/26.2 = 14.3$, whereas the hyperboloid is only 11.

Table 2-12 shows that the cylinder and cone have much higher M_θ at the top than the hyperboloid, and Table 2-13 compares results for the simplified hyperboloid with those for the Trojan Tower, where the increased thickness does result in higher M_θ. These moments, however, lead to lower stresses than do the smaller ones in the simplified hyperboloid.

One further comparison illustrates the special ability of hyperboloids to carry wind loadings. Figure 2-26 shows the circumferential distribution of meridional stress at $y = 0$ for banded loading on each form. Band A corresponds to the 92.5 psf load only on the top 172 ft of the tower, Band B is a load of 69.9 psf only on the lower 250 ft, and Band C a load of 46.6 psf on the bottom 29 ft (see Fig. 2-9). Whereas for both the cylinder and the cone, Band A gives higher N_y at the base than Band B, the reverse is true for the N_ϕ in the hyperboloid. This remarkable behavior arises because the

Table 2-12 FORM COMPARISON OF STRESS VALUES

$y*$	Cylinder			Cone			Hyperboloid		
	N_y†	N_θ†	M_θ‡	N_y†	N_θ†	M_θ‡	N_ϕ†	N_θ†	M_θ‡
0	375.1	23.7	0.5	190.7	11.5	0.6	139.9	6.3	−0.4
~100	222.2	−16.3	−4.4	135.0	−22.7	−1.4	108.0	−18.7	−1.0
~350	15.8	−15.0	−28.5	15.6	−17.3	−15.1	21.1	−10.4	−6.7
~450	0	−23.2	−39.0	0	−11.7	−30.9	0	−22.5	−1.1

* In feet.
† In kips per foot.
‡ In kips.

Table 2-13 INFLUENCE OF VARYING THICKNESS ON HYPERBOLOIDS

$y*$	$h = 1.0$ ft				h = variable and flexible columns			
	N_ϕ†	N_θ†	M_θ‡	$h*$	N_ϕ†	N_θ†	M_θ‡	$h*$
0	139.9	6.3	−0.4	1.0	116.0	46.0	−30.0	3.50
~100	108.0	−18.7	−1.0	1.0	95.9	−13.2	−9.0	1.83
~350	21.1	−10.4	−6.7	1.0	21.2	−10.8	−3.2	0.83
~450	0	−22.5	−1.1	1.0	0	−35.7	0.3	1.67§

* In feet.
† In kips per foot.
‡ In kips.
§ Ring 5 ft × 1.67 ft.

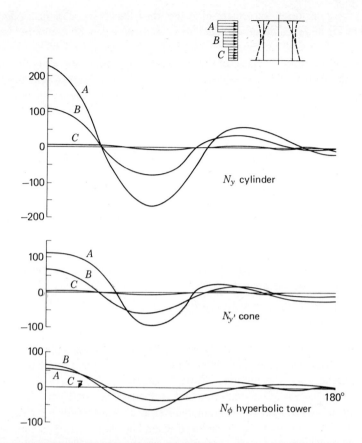

Fig. 2-26 N_y or N_ϕ for banded loads.

straight-line generators of the hyperboloid tend to distribute the higher-up load away from the windward meridian. Flügge had discussed this behavior,[17] and Elms demonstrated it experimentally.[18] It means that hyperboloids are able to carry higher wind loads with less meridional tension than cones and cylinders.* At the same time such behavior means that these shells will be subject to a reverse problem under differential settlements. Here that influence will propagate up to the throat and beyond and requires special care in design.[19]

2-10 THERMAL GRADIENTS IN SHELL WALLS

Temperature can have two kinds of effects: one, an overall expansion or contraction which only causes stresses where the shell is restrained at its base, and

---

* The Trojan Tower is an offset hyperboloid and its generators are thus not quite straight lines, but the behaviour is still evident (see Ref. 9).

the other, a change in temperature through the shell thickness. The first effect is often negligible for design of large cooling towers but can easily be computed based on the axisymmetrical shell bending theory in Sec. 2-6. The second effect can influence cooling tower behavior because of the relatively high inside operating temperatures that can go together with low winter ambient conditions. Except near the shell boundaries, the moments can be calculated on the basis of plate theory[12] from which

$$M_y = M_\theta = \frac{\alpha \Delta T D (1 + \nu)}{h} \qquad (a)$$

where α is the coefficient of thermal expansion usually taken as 6.5×10^{-6} in. per inch per degree Fahrenheit, ΔT the temperature gradient across the shell thickness in degrees Fahrenheit, $D = Eh^3/12(1 - \nu^2)$ = the flexural rigidity of the shell (E is the modulus of elasticity in kips per square foot), ν is Poissons ratio taken as about 0.15. For example, a 7-in.-thick shell with $\Delta T = 50°$ will have a moment

$$M_y = M_\theta = 5.43 \text{ ft-k/ft}$$

whereas the ultimate moment capacity for such a shell with 0.175 percent steel in each face in each direction is approximately

$$M_u = 0.00175 \times 12 \times 7 \times 60/12 \,(5.5 - 0.1)$$
$$= 3.96 \text{ ft-k/ft}$$

Hence the shell must crack under a gradient of 50°F, and when it does, its flexural rigidity immediately drops because D can no longer be based upon the full uncracked h. When the shell cracks, the moment drops substantially. For the 7-in. shell described in Ref. 20, the cracked-section moment is about 0.38 ft-k/ft (see Sec. 9-3).

This cracking, while reducing the bending moments, also decreases the circumferential stiffness and thus needs to be considered when buckling capacity is estimated. For example, an approximate buckling calculation for the Trojan Tower estimated a safety factor of about 19 compared to about 6 when cracking was considered (Ref. 9).

REFERENCES

1. Thom, H. C. S., "New Distribution of Extreme Winds in the United States," *J. Struct. Div.*, ASCE, vol. 94, no. ST7, July 1968, pp. 1787–1801; and E. Simiu, and R. H. Scanlan, *Wind Effects on Structures*, John Wiley & Sons, New York, 1978.
2. Niemann, H. J., H. L. Peters, and W. Zerna, "Naturzugkühlturme in Wind," *Beton and Stahlbetonbau*, vol. 67, no. 6, 1972.
3. Sollenberger, N. J. and Scanlan, R. H., "Pressure Differences across the Shell of a Hyperbolic Natural Draft Cooling Tower," *Proceedings of the International Conference on Full Scale Testing of Wind Effects*, London and Ontario, Canada, June 1974.
4. Sollenberger, N. J., R. H. Scanlan, and D. P. Billington, "Wind Loading and Response of Cooling Towers," *J. Struct. Div.*, ASCE, March 1980.
5. "Reinforced Concrete Cooling Tower Shells: Practice and Commentary," Rep. ACI-ASCE Committee on Concrete Shell Design and Construction, ACI-ASCE Committee 334, *J. ACI*, January 1977, Proc. vol. 24, no. 1, pp. 22–31.
6. Billington, D. P., and J. F. Abel, "Design of Cooling Towers for Wind," *Proceedings of the National Structural Engineer Conference on Methods of Structural Analysis*, ASCE, Madison, Wis., 1976, pp. 242–267.

7. Erikson, Kurt, "Prestressed-Concrete Water Tower in Örebro, Sweden," *Civ. Eng.*, October 1958, pp. 68–73.
8. Girkmann, K., *Flächentragwerke*, 5th ed., Springer-Verlag OHG, Vienna, 1959.
9. Cole, P. P., J. F. Abel, and D. P. Billington, "Buckling of Cooling-Tower Shells: Bifurcation Results," *J. Struct. Div.*, ASCE, vol. 10, no. ST6, June 1975, pp. 1205–1222.
10. Torroja, E., *The Structures of Eduardo Torroja*, New York, 1958, pp. 19–22.
11. Abel, J. F., and D. P. Billington, "Effect of Shell Cracking on Dynamic Response of Concrete Cooling Towers," *IASS World Congress on Spatial Enclosures*, Montreal, July 1976, pp. 895–904.
12. Timoshenko, S. P., and S. Woinowsky-Krieger, *Theory of Plates and Shells*, 2d ed., McGraw-Hill Book Company, New York, 1959.
13. "Circular Concrete Tanks without Prestressing," Portland Cem. Assoc., Chicago (c 1945).
14. Brombolich, L. J., and P. L. Gould, "Finite Element Analysis of Shells of Revolution," *Proceedings of the Conference on Application of the Finite Element Method in Civil Engineering*, Vanderbilt University Press, Nashville, November 1969, pp. 279–307.
15. Cole, P. P., J. F. Abel, and D. P. Billington, "Buckling of Hyperbolic Cooling-Tower Shells," *Dep. Civ. and Geol. Eng. Res. Rep.*, no. 73-SM-2, July 1973.
16. Bushnell, D. and S. Smith, "Stresses and Buckling of Nonuniformly Heated Cylindrical and Conical Shells," *AIAA J.*, vol. 9, no. 12, December 1971, pp. 2314–2321.
17. Flügge, W., *Stresses in Shells*, 2d ed., Springer-Verlag, New York, 1973, pp. 71–79.
18. Elms, D. G., D. P. Billington, and R. Mark, "Stresses in Hyperboloids under Edge Loadings," *J. Eng. Mech. Div.*, ASCE, vol. 91, no. EM4, August 1965, pp. 89–114.
19. Abel, J. F., S-C Wu, and D. P. Billington, "The Effects of Uplift on Cooling Towers," *Dep. Civ. Eng. Res. Rep.*, no. 78-SM-5, Princeton University, N.J.; and J. A. Dumitrescu and D. P. Billington, "Shell-Foundation Interaction for Hyperbolic Cooling Towers," *Dep. Civ. Eng. Res. Rep.*, no. 80-SM-28, Princeton University, N.J., October 1980.
20. Larrabee, R. D., D. P. Billington, and J. F. Abel, "Thermal Loading of Thin-Shell Concrete Cooling Towers," *J. Struct. Div.*, ASCE, vol. 100, no. ST12, December 1974, pp. 2367–2383.

3
ANALYSIS OF DOMES

3-1 DEFINITIONS

The types of shells considered in this chapter under the title of *domes* are defined as thin shells in the form of surfaces of revolution, which serve primarily for roof structures. A surface of revolution is obtained by the rotation of a plane curve about an axis lying in the plane of the curve. This curve is called the *meridian* and its plane the *meridian plane.* The differential element used in the shell analysis is cut from the surface by two adjacent meridian planes and by two adjacent parallel circles, planes normal to the axis of rotation that pass through the surface. Figure 3-1 shows the element and defines the coordinate system to be used.

By the above definitions, domes can be classified as shells of positive gaussian curvature and as rotational shells. It is, therefore, to be expected that the membrane theory will give a reasonably accurate basis for design throughout the system except near the edge boundaries. Segments of domed surfaces can be formed by creating boundaries along meridians, as in a band shell (Fig. 3-2*a*); along parallel circles, as in the roof of a cylindrical tank (Fig. 3-2*b*); or along any other type of curved line in the surface (for example, Fig. 3-2*c*). Only the second type of shell system, that which is symmetrical about its axis of revolution, will be considered in this chapter. For the analysis of other types of domes see Ref. 1.

The analysis will be divided into four parts as described in Sec. 1-13:

1. *Primary System* based on the membrane theory.
2. *Errors* at the boundaries due to membrane stress resultants.
3. *Corrections* due to unit edge effects at the boundaries.
4. *Compatibility* which is achieved by computing the size of the edge effects necessary to eliminate the errors.

The next sections will deal successively, therefore, with (1) stress resultants derived from the membrane theory (3-2 to 3-3); (2) boundary deformations derived from the membrane theory (3-4); (3) boundary deformations and stress distributions throughout the dome caused by unit edge effects (3-5); and (4) formulation and solution of compatibility equations (3-6). A discussion of dome design is presented in Chap. 4.

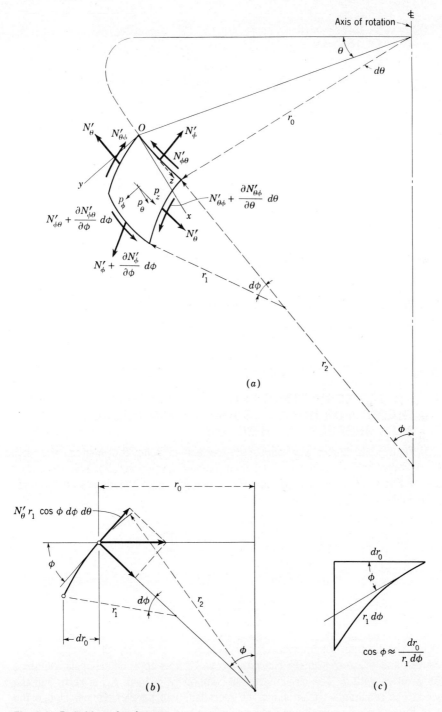

Fig. 3-1 Definitions for domes.

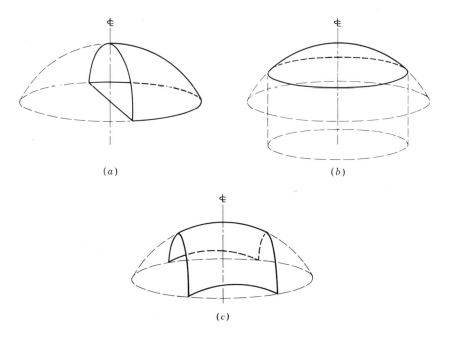

Fig. 3-2 Dome forms.

3-2 STRESS RESULTANTS FROM THE MEMBRANE THEORY FOR DOMES LOADED SYMMETRICALLY WITH RESPECT TO THEIR AXES

Figure 3-1 defines the differential element in polar coordinates. For shell systems which are symmetrical about their axes of revolution, all purely geometric terms involving $\partial\theta$ vanish, and we find for domes the same equations as derived for hyperboloids, i.e., Eqs. (2-16):

$$\frac{\partial N'_\theta}{\partial\theta}r_1 + N'_{\theta\phi}\frac{\partial r_0}{\partial\phi} + \frac{\partial(N'_{\phi\theta}r_0)}{\partial\phi} + p_\theta r_0 r_1 = 0$$

$$\frac{\partial(N'_\phi r_0)}{\partial\phi} - N'_\theta\frac{\partial r_0}{\partial\phi} + \frac{\partial N'_{\theta\phi}}{\partial\theta}r_1 + p_\phi r_0 r_1 = 0 \tag{3-1}$$

$$\frac{N'_\theta}{r_2} + \frac{N'_\phi}{r_1} + p_z = 0$$

When the loading is symmetrical with respect to the axis, all terms involving $\partial\theta$ vanish and the terms in $\partial\phi$ can now be written as total differentials $d\phi$ since nothing varies with θ. The circumferential component of load, p_θ, is zero, and the shear stress resultants vanish along the meridians and parallel circles. Hence the first of Eqs. (3-1) disappears, and the remaining two reduce to

$$\frac{d(N'_\phi r_0)}{d\phi} - N'_\theta \frac{dr_0}{d\phi} + p_\phi r_0 r_1 = 0$$

$$\frac{N'_\theta}{r_2} + \frac{N'_\phi}{r_1} + p_z = 0$$ (3-2)

Thus the membrane theory for axisymmetrical domes under axisymmetrical loading requires the solution of (3-2) for the two membrane stress resultants N'_θ and N'_ϕ. The first of Eqs. (3-2) can be written in another form which permits a simpler solution. From Fig. 3-1c, we observe that

$$\cos \phi \approx \frac{dr_0}{r_1 \, d\phi}$$

or

$$\frac{dr_0}{d\phi} = r_1 \cos \phi$$

which can be substituted into the first of (3-2) to give

$$\frac{d(N'_\phi r_0)}{d\phi} - N'_\theta r_1 \cos \phi + p_\phi r_0 r_1 = 0$$

The effect of N'_θ in the meridian direction could have been directly obtained from Fig. 3-1b by noting that the meridian sides of the element form an angle $d\theta \cos \phi$ with each other. Thus in the meridional direction there will be a component of the hoop stress resultant in the negative direction

$$N'_\theta r_1 \, d\phi \, d\theta \cos \phi$$

which, when the multipliers $d\phi \, d\theta$ are dropped, reduces to

$$N'_\theta r_1 \cos \phi$$

The second equation of (3-2) can be solved for

$$N'_\theta = - \frac{r_0}{\sin \phi} \left(\frac{N'_\phi}{r_1} + p_z \right)$$ (a)

and when (a) is introduced into the first of (3-2) and each term is multiplied by $\sin \phi$, we obtain

$$\sin \phi \frac{d(N'_\phi r_0)}{d\phi} + \sin \phi \frac{r_0}{\sin \phi} \left(\frac{N'_\phi}{r_1} + p_z \right) r_1 \cos \phi + \sin \phi p_\phi r_0 r_1 = 0$$

which, when multiplied by 2π and integrated with respect to ϕ, yields

$$\int_0^\phi \sin \phi \frac{d(N'_\phi r_0)}{d\phi} \, d\phi + \int_0^\phi N'_\phi r_0 \cos \phi \, d\phi$$

$$= - \frac{1}{2\pi} \int_0^\phi (p_\phi \sin \phi + p_z \cos \phi) 2\pi r_0 r_1 \, d\phi$$ (b)

The first integral can be integrated by parts in the form

$$\int u \, dv = uv - \int v \, du$$

where $u = \sin \phi$, $du = \cos \phi \, d\phi$, $dv = [d(N'_\phi r_0)/d\phi] \, d\phi$, and $v = N'_\phi r_0$. The left side of (b) becomes

$$\sin \phi N'_\phi r_0 - \int_0^\phi N'_\phi r_0 \cos \phi \, d\phi + \int_0^\phi N'_\phi r_0 \cos \phi \, d\phi$$

and hence (b) simplifies to

$$N'_\phi = - \frac{1}{2\pi r_0 \sin \phi} \int_0^\phi (p_\phi \sin \phi + p_z \cos \phi)(2\pi r_0) r_1 d\phi \tag{c}$$

The expression $(p_\phi \sin \phi + p_z \cos \phi)$ gives the vertical component of the load. $2\pi r_0$ sums this vertical load over a complete parallel circle, and $\int_0^\phi r_1 \, d\phi$ integrates the vertical load along a meridian. Thus the integral in (c) represents the total vertical load (R in Fig. 3-3) above the parallel circle defined by ϕ. The quantity $N'_\phi 2\pi r_0 \sin \phi$ is easily seen as the total vertical component of N'_ϕ at the parallel circle ϕ. Hence N'_ϕ can be directly written as

$$N'_\phi = - \frac{R}{2\pi r_0 \sin \phi} \tag{3-3a}$$

and from (a)

$$N'_\theta = \frac{R}{2\pi r_1 \sin^2 \phi} - p_z \frac{r_0}{\sin \phi} \tag{3-3b}$$

Several specific loading conditions on domes will now be described to illustrate the use of (3-3).

Spherical Domes

Uniform Load over the Dome Surface   In the important case of the spherical dome of uniform thickness under dead load, $r_1 = r_2 = a$, $p_\phi = q \sin \phi$,

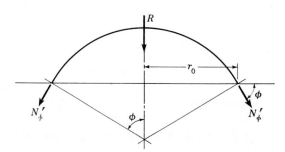

Fig. 3-3   Dome equilibrium.

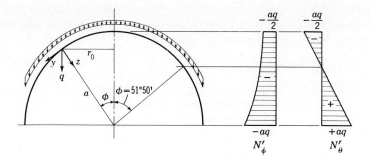

Fig. 3-4   Spherical dome under surface loads.

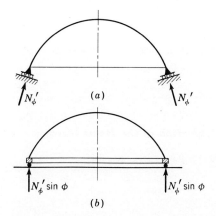

Fig. 3-5   Dome edge conditions.

and $p_z = q \cos \phi$, where q is the dead weight of the shell, and

$$R = 2\pi a^2 q \int_0^{\phi} \sin \phi \, d\phi$$

$$= 2\pi a^2 q (1 - \cos \phi)$$

and (3-3) become

$$N'_{\phi} = -aq \frac{1}{1 + \cos \phi} \tag{3-4}$$

$$N'_{\theta} = aq \left(\frac{1}{1 + \cos \phi} - \cos \phi \right) \tag{3-5}$$

Figure 3-4 shows the distribution of the two membrane stress resultants over a hemisphere. The meridional values, which are compressive, increase from the crown to the edge. The hoop values decrease from a maximum compression at the crown to zero where $\cos \phi = 1/(1 + \cos \phi)$, or about 51°50'; they then become tension and increase to a maximum at the edge.

In many cases the dome cannot be provided with a support tangent to the meridian as in Fig. 3-5a. Where the support is vertical only, as shown in Fig. 3-5b, there will be an incompatibility in the edge force since the support does not supply a horizontal restraint to resist the horizontal thrust $H_\phi = N'_\phi \cos \phi$. A tension ring may be provided at the dome edge to resist the ring force,

$$T_\phi = N'_\phi \, a \sin \phi \cos \phi \qquad\qquad (3\text{-}3c)$$

Since this tension can be large, it is usually necessary to use an edge-stiffening ring with reinforcement. The strain in such a ring is tensile, whereas dome hoop strain is usually compressive and will rarely equal that in the ring. This incompatibility of strains cannot exist, so bending must occur along the meridians. The solution of the dome-ring bending problem is discussed in Chap. 4. Where the dome is supported vertically, the equilibrium of horizontal membrane stress resultants must be obtained within the dome such that

$$\int_0^\phi N'_\theta \, a \, d\phi + T_\phi = 0$$

Where the free edge occurs at $51°50'$, T_ϕ will be a maximum. Where $\phi = 90°$, $T_\phi = 0$, and

$$\int_0^{\pi/2} N'_\theta \, a \, d\phi = 0$$

Uniform Load over a Horizontal Projection of the Dome Surface

$$p_z = p \cos^2 \phi \qquad p_\phi = p \sin \phi \cos \phi \qquad p_\theta = 0$$

In this case $R = p \pi r_0^2 = p a^2 \pi \sin^2 \phi$ and

$$N'_\phi = -\frac{ap}{2} \qquad\qquad (3\text{-}6)$$

$$N'_\theta = -a \left(p \cos^2 \phi - \frac{ap}{2a} \right) = -\frac{ap}{2} \left(2 \cos^2 \phi - 1 \right)$$

$$N'_\theta = -\frac{ap}{2} \cos 2\phi \qquad\qquad (3\text{-}7)$$

The distribution of membrane values shown in Fig. 3-6 indicates that, whereas N'_ϕ is a constant compression, N'_θ varies from compression at the crown to tension at the edge. Zero hoop value occurs where $\cos 2\phi = 0$ or $\phi = 45°$.

Uniform External Pressure   For a uniform pressure p all over the dome

$$p_z = p \qquad p_\phi = 0 \qquad \text{and} \qquad p_\theta = 0$$

$$R = 2\pi a^2 p \int_0^\phi \sin \phi \cos \phi \, d\phi$$

$$= \frac{\pi a^2 p}{2} (2 \sin^2 \phi)_0^\phi$$

$$N'_\phi = -\frac{\pi a^2 p \sin^2 \phi}{2a \sin^2 \phi} = -\frac{ap}{2} \qquad\qquad (3\text{-}8)$$

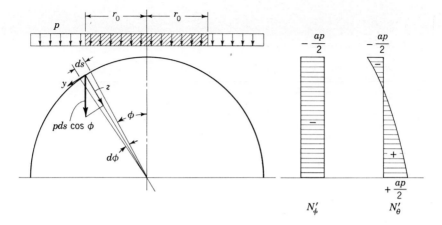

Fig. 3-6 Spherical dome under horizontally uniform loads.

$$N'_\theta = -a \left(p - \frac{ap}{2a} \right) = -\frac{ap}{2} \tag{3-9}$$

Cosine Variation of Pressure In some preliminary blast designs a maximum pressure is assumed at the crown of the dome decreasing to zero at the edge. In such a case a cosine variation could be used:

$$p_z = p \cos \phi \qquad p_\phi = p_\theta = 0$$

$$R = 2\pi a^2 p \int_0^\phi \sin \phi \cos^2 \phi \, d\phi$$

$$= -2\pi a^2 p \left(\frac{\cos^3 \phi}{3} \right)_0^\phi$$

$$= 2\pi a^2 p \left(\frac{1}{3} - \frac{\cos^3 \phi}{3} \right)$$

$$N'_\phi = -\frac{R}{2\pi a \sin^2 \phi} = -\frac{ap}{3 \sin^2 \phi} (1 - \cos^3 \phi)$$

$$= -\frac{ap}{3} \left(1 + \cos \phi - \frac{\cos \phi}{1 + \cos \phi} \right) \tag{3-10}$$

$$N'_\theta = -ap \cos \phi - N'_\phi$$

$$= -\frac{ap}{3} \left(2 \cos \phi - 1 + \frac{\cos \phi}{1 + \cos \phi} \right) \tag{3-11}$$

Variation in Dome Thickness Consider the dome made up of two elements—one of constant thickness h and the other of a regular variation h' which changes as a function of ϕ and is zero at ϕ_0 (see Fig. 3-7). The membrane stress resultants for constant h are given by (3-4) and (3-5); thus it remains to add the

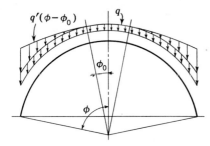

Fig. 3-7 Spherical dome under variable loads.

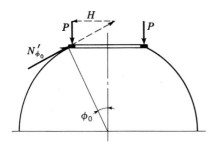

Fig. 3-8 Spherical dome under skylight loads.

effect of h'. Let q' be the increase in weight due to h', and ϕ_0 the parallel circle below which the thickness varies.

$$p_z = q'(\phi - \phi_0) \cos \phi \qquad p_\phi = q'(\phi - \phi_0) \sin \phi \qquad p_\theta = 0$$

$$R = 2\pi a^2 q' \int_{\phi_0}^{\phi} \sin \phi (\phi - \phi_0)\, d\phi$$

$$= 2\pi a^2 q' [\sin \phi - \sin \phi_0 - (\phi - \phi_0) \cos \phi]$$

$$N'_\phi = -\frac{aq'}{\sin^2 \phi} [\sin \phi - \sin \phi_0 - (\phi - \phi_0) \cos \phi] \qquad (3\text{-}12)$$

$$N'_\theta = -aq'(\phi - \phi_0) \cos \phi$$

$$+ \frac{aq'}{\sin^2 \phi} [\sin \phi - \sin \phi_0 - (\phi - \phi_0) \cos \phi]$$

$$= -\frac{aq'}{\sin^2 \phi} [2(\phi - \phi_0) \cos \phi - (\phi - \phi_0) \cos^3 \phi$$

$$- \sin \phi + \sin \phi_0] \qquad (3\text{-}13)$$

Concentrated Load around a Skylight Opening Domes are often made discontinuous near the crown by cutting a circular hole along a parallel circle. The opening is usually covered by a skylight which is not monolithic with the dome but does place a uniform concentrated load P around the free upper edge. P acts vertically and cannot be resisted by the meridional thrust alone, as shown in Fig. 3-8. A horizontal thrust must be supplied,

$$H = \frac{P}{\tan \phi_0}$$

which induces a ring compression into the edge of the shell.

$$C_{\phi_0} = Hr_0 = Ha \sin \phi_0 = aP \cos \phi_0 \qquad (3\text{-}14)$$

C_{ϕ_0} can be large so a stiffening edge ring is usually required around such openings. The membrane values are readily calculated from the conditions that $p_z = p_\phi = p_\theta = 0$ and $R = P2\pi a \sin \phi_0$. Therefore

$$N'_\phi = -\frac{P2\pi a \sin \phi_0}{2\pi a \sin^2 \phi} = -\frac{P \sin \phi_0}{\sin^2 \phi} \qquad (3\text{-}15)$$

$$N'_\theta = \frac{P \sin \phi_0}{\sin^2 \phi} \qquad (3\text{-}16)$$

where $\phi = \phi_0$, $N'_{\phi_0} = -P/\sin \phi_0$ (see Fig. 3-8). Note that the hoop stress resultant is tensile below ϕ_0. Therefore in an edge ring a large compressive ring force is obtained from the membrane theory, whereas just below in the dome a hoop tension is indicated. Since the strain must always be the same at any point, this incompatibility represents an error at the shell boundary and thus bending will occur.

Concentrated Load at the Axis As a limiting case of the skylight load P, the entire load may occur at the crown. In such a case the membrane values do not change.

$$N'_\phi = -\frac{P_c}{2\pi a \sin^2 \phi} \qquad (3\text{-}17)$$

$$N'_\theta = \frac{P_c}{2\pi a \sin^2 \phi} \qquad (3\text{-}18)$$

It will be noted, however, that near the crown $\sin^2 \phi$ becomes small and, hence, N'_ϕ and N'_θ become large; as $\phi \to 0$ they approach infinity. This condition cannot occur, and, in fact, considerable bending must take place in this area to spread out the forces. The problem of such localized loadings can be treated to determine the bending and shearing forces involved (see page 558 of Ref. 1). Finite-element analysis is used in such cases as well.

Conoidal Domes

The same general procedure used for spherical domes is applicable in this case. Only the derivation of dead-load membrane forces will be given.

As shown in Fig. 3-9, this dome consists of the rotation of arc 0-1, which is circular about an axis parallel to a vertical radius of the arc but separated from it by a distance r'.

$$r_1 = a \qquad \text{and} \qquad r_2 = \frac{r_0}{\sin \phi}$$

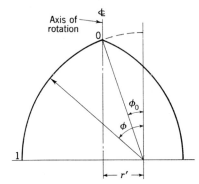

Fig. 3-9 Conoidal dome.

In this case, however, $r_0 = a \sin \phi - r'$, so that $r_2 \neq a$ but varies with ϕ and $r' = a \sin \phi_0$.

$$r_2 = \frac{a \sin \phi - r'}{\sin \phi} = a - \frac{r'}{\sin \phi}$$

$$p_z = q \cos \phi \qquad p_\phi = q \sin \phi \qquad p_\theta = 0$$

and

$$R = 2\pi a q \int_{\phi_0}^{\phi} r_0 \, d\phi$$

$$= 2\pi a^2 q \int_{\phi_0}^{\phi} \sin \phi \, d\phi - 2\pi a q r' \int_{\phi_0}^{\phi} d\phi$$

$$= 2\pi a^2 q (\cos \phi_0 - \cos \phi) - 2\pi a r' q (\phi - \phi_0)$$

$$N'_\phi = -\frac{R}{2\pi r_0 \sin \phi} = -\frac{R}{2\pi a \sin \phi (\sin \phi - \sin \phi_0)}$$

$$= -\frac{2\pi a^2 q [\cos \phi_0 - \cos \phi - \sin \phi_0 (\phi - \phi_0)]}{2\pi a \sin \phi (\sin \phi - \sin \phi_0)}$$

$$= -qa \frac{\cos \phi_0 - \cos \phi - \sin \phi_0 (\phi - \phi_0)}{\sin \phi (\sin \phi - \sin \phi_0)} \qquad (3\text{-}19)$$

$$N'_\theta = -\frac{a \sin \phi - a \sin \phi_0}{\sin \phi} \left\{ q \cos \phi \right.$$

$$\left. - \frac{aq[\cos \phi_0 - \cos \phi - \sin \phi_0 (\phi - \phi_0)]}{a(\sin \phi - \sin \phi_0) \sin \phi} \right\}$$

$$= -\frac{aq}{\sin^2 \phi} \left[(\sin \phi - \sin \phi_0) \sin \phi \cos \phi \right.$$

$$\left. - (\cos \phi_0 - \cos \phi) + \sin \phi_0 (\phi - \phi_0) \right] \qquad (3\text{-}20)$$

Elliptical Domes

For the dome shown in Fig. 3-10, the elliptical arc is described by

$$\frac{x^2}{a^2} + \frac{y^2}{b^2} = 1 \qquad (3\text{-}21)$$

where a is called the major axis, b is the minor axis, and x corresponds to r_0, the radius of the parallel circle; so

$$x = r_0 = a\sqrt{1 - \frac{y^2}{b^2}} = \frac{a}{b}\sqrt{b^2 - y^2} \tag{d}$$

$$\frac{dy}{dx} = \frac{1}{dx/dy} = -\tan\phi = -\frac{b}{ay}\sqrt{b^2 - y^2}$$

and

$$\sin\phi = \frac{dy}{ds} = \frac{dy}{\sqrt{dy^2 + dx^2}} = \frac{dy/dx}{\sqrt{(dy/dx)^2 + 1}}$$

$$= \frac{-\tan\phi}{\sqrt{1 + \tan^2\phi}} = \frac{-b\sqrt{b^2 - y^2}}{\sqrt{b^4 + y^2(a^2 - b^2)}} \tag{e}$$

Similarly

$$\cos\phi = \frac{1}{\sqrt{1 + \tan^2\phi}} = \frac{ay}{\sqrt{b^4 + y^2(a^2 - b^2)}}$$

The two principal radii of curvature are given by the expressions[2]

$$r_1 = \frac{[b^4 + y^2(a^2 - b^2)]^{3/2}}{ab^4} \qquad r_2 = \frac{a}{b^2}\sqrt{b^4 + y^2(a^2 - b^2)} \tag{f}$$

At the crown of the dome, $y = b$ and $r_1 = r_2 = a^2/b$, whereas at the base, $y = 0$, $r_1 = b^2/a$, and $r_2 = a$. The calculation of the dead-load membrane values illustrates the analysis for this type of dome.

$$p_z = q\cos\phi \qquad p_\phi = q\sin\phi \qquad p_\theta = 0$$

$$R = 2\pi q\int_y^b x\,ds$$

where x is given by (d) and

$$ds = dy\sqrt{1 + \left(\frac{dx}{dy}\right)^2}$$

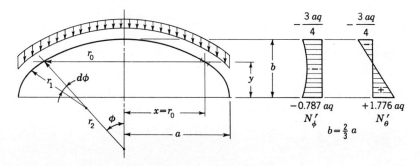

Fig. 3-10 Elliptical dome under surface loads.

with

$$\frac{dx}{dy} = -\frac{ay}{b\sqrt{b^2 - y^2}}$$

$$ds = dy\sqrt{1 + \frac{a^2 y^2}{b^2(b^2 - y^2)}}$$

$$R = 2\pi q \int_y^b \frac{a}{b}\sqrt{b^2 - y^2}\sqrt{\frac{b^2(b^2 - y^2) + a^2 y^2}{b^2(b^2 - y^2)}}\,dy$$

$$R = 2\pi q \frac{a}{b^2}\int_y^b \sqrt{b^4 + y^2(a^2 - b^2)}\,dy$$

$$R = 2\pi a^2 q \left[\frac{1}{2} - \frac{y}{2ab^2}\sqrt{b^4 + y^2(a^2 - b^2)}\right.$$

$$\left. + \frac{b^2}{2a\sqrt{a^2 - b^2}}\log\frac{b(a + \sqrt{a^2 - b^2})}{y\sqrt{a^2 - b^2} + \sqrt{b^4 + y^2(a^2 - b^2)}}\right]$$

This quantity in brackets, evaluated for various ratios of y/b and b/a, has been listed in Table 3-1[3] as a quantity C so that

$$R = 2\pi a^2 q C$$

The membrane values can then be obtained from Eqs. (3-3) and (f):

$$N'_\phi = -\frac{R}{2\pi r_0 \sin\phi} = -\frac{2\pi a^2 q C}{2\pi r_2 \sin^2\phi}$$

$$= -\frac{a^2 q C b^2}{\sin^2\phi\, a\sqrt{b^4 + y^2(a^2 - b^2)}}$$

From (e)

$$N'_\phi = -\frac{a^2 q b^2 C[b^4 + y^2(a^2 - b^2)]}{b^2(b^2 - y^2)a\sqrt{b^4 + y^2(a^2 - b^2)}}$$

$$= -\frac{q a^2}{b}\frac{C}{(b^2 - y^2)/b^2}\frac{\sqrt{b^4 + y^2(a^2 - b^2)}}{ab}$$

Let

$$\sqrt{1 - \frac{a^2 - b^2}{a^2}\left(1 - \frac{y^2}{b^2}\right)} = \frac{\sqrt{b^4 + y^2(a^2 - b^2)}}{ab} = Q$$

Q is also listed in Table 3-1.

Then,

$$N'_\phi = -\frac{q a^2}{b}\frac{C Q}{1 - (y^2/b^2)} \tag{3-22}$$

$$N'_\theta = -r_2\left(p_z + \frac{N'_\phi}{r_1}\right)$$

$$= -r_2 q \cos\phi - \frac{N'_\phi}{r_1}r_2$$

Table 3-1 COEFFICIENTS FOR ELLIPTICAL DOMES

	Values of C^*									Values of $Q\dagger$									
	b/a									b/a									
$g = y/b$	0.2	0.3	0.4	0.5	0.6	0.7	0.8	0.9	1.0	0.2	0.3	0.4	0.5	0.6	0.7	0.8	0.9	1.0	$g = y/b$
0.0	0.547	0.588	0.636	0.690	0.747	0.807	0.870	0.934	1.000	0.200	0.300	0.400	0.500	0.600	0.700	0.800	0.900	1.000	0.0
0.1	0.526	0.558	0.596	0.640	0.687	0.737	0.789	0.844	0.900	0.227	0.315	0.410	0.507	0.605	0.704	0.802	0.901	1.000	0.1
0.2	0.501	0.525	0.554	0.588	0.626	0.666	0.709	0.754	0.800	0.280	0.356	0.440	0.529	0.621	0.714	0.809	0.904	1.000	0.2
0.3	0.469	0.486	0.508	0.534	0.563	0.594	0.627	0.663	0.700	0.356	0.415	0.485	0.563	0.646	0.732	0.820	0.909	1.000	0.3
0.4	0.430	0.441	0.456	0.475	0.496	0.520	0.545	0.572	0.600	0.440	0.485	0.543	0.608	0.680	0.756	0.835	0.917	1.000	0.4
0.5	0.381	0.389	0.399	0.412	0.427	0.443	0.461	0.480	0.500	0.529	0.563	0.608	0.661	0.721	0.786	0.854	0.926	1.000	0.5
0.6	0.323	0.328	0.335	0.343	0.352	0.362	0.374	0.387	0.400	0.621	0.646	0.680	0.721	0.768	0.821	0.877	0.937	1.000	0.6
0.7	0.257	0.260	0.263	0.267	0.272	0.278	0.285	0.292	0.300	0.714	0.732	0.756	0.786	0.821	0.860	0.904	0.950	1.000	0.7
0.8	0.181	0.182	0.183	0.185	0.188	0.190	0.193	0.196	0.200	0.809	0.820	0.835	0.854	0.877	0.903	0.933	0.965	1.000	0.8
0.9	0.095	0.095	0.096	0.096	0.097	0.098	0.098	0.099	0.100	0.904	0.910	0.917	0.926	0.937	0.950	0.965	0.982	1.000	0.9
1.0	0.000	0.000	0.000	0.000	0.000	0.000	0.000	0.000	0.000	1.000	1.000	1.000	1.000	1.000	1.000	1.000	1.000	1.000	1.0

$^*C = \dfrac{1}{2} + \dfrac{1-k^2}{2k}\log(1+k) - \dfrac{g}{2}\sqrt{1-k^2(1-g^2)} - \dfrac{1-k^2}{2k}\log\left[gk + \sqrt{1-k^2(1-g^2)}\right]$, in which $k^2 = 1 - \dfrac{b^2}{a^2}$.

$^\dagger Q = \sqrt{1-k^2(1-g^2)}$.

119

From (f)

$$\frac{r_2}{r_1} = \frac{a^2 b^2}{b^4 + y^2(a^2 - b^2)} = \frac{1}{Q^2}$$

$$N'_\theta = -\frac{aq \cos \phi \sqrt{b^4 + y^2(a^2 - b^2)}}{b^2} + \frac{a^2 q}{b} \frac{C}{[1 - (y^2/b^2)]Q}$$

$$= -\frac{a^2 qy \sqrt{b^4 + y^2(a^2 - b^2)}}{b^2 \sqrt{b^4 + y^2(a^2 - b^2)}} + \frac{a^2 q}{b} \frac{C}{[1 - (y^2/b^2)]Q}$$

$$N'_\theta = -\frac{a^2 q}{b} \left[\frac{y}{b} - \frac{Cb^2}{(b^2 - y^2)Q} \right] \tag{3-23}$$

In both Eqs. (3-22) and (3-23) the values become indeterminate when $y/b = 1.0$, but it is easily shown by plotting results for other values of $y/b < 1.0$ that the terms $CQ/(1 - y^2/b^2)$ and $Cb^2/(b^2 - y^2)Q$ converge to 0.5 as y/b goes to 1.0.

Conical Domes

For this type of dome, as shown in Fig. 3-11, $\phi = \pi/2 - \alpha = $ constant. $r_1 = \infty$, $r_2 = r_0/\cos \alpha$, and $r_0 = y \sin \alpha$.

In the case of dead load for a uniform shell thickness

$$p_z = q \sin \alpha \qquad p_y = q \cos \alpha \qquad p_x = 0$$

$$R = \int_0^y 2\pi r_0 q \, dy = 2\pi q \sin \alpha \int_0^y y \, dy$$

$$= \pi q y^2 \sin \alpha$$

$$N'_\phi = -\frac{R}{2\pi r_0 \sin \phi} = -\frac{\pi q y^2}{2\pi y \cos \alpha} = -\frac{qy}{2 \cos \alpha} \tag{3-24}$$

$$N'_\theta = -r_2 \left(q \sin \alpha + \frac{N'_\phi}{r_1} \right) = -r_2 q \sin \alpha$$

$$= -\frac{r_0}{\cos \alpha} q \sin \alpha = -yq \tan \alpha \sin \alpha \tag{3-25}$$

These results coincide with those from Sec. 2-3 when in Eq. (2-2a) we set $L_T = 0$.

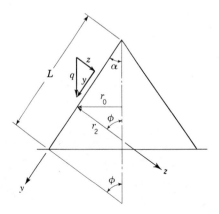

Fig. 3-11 Conical dome.

3-3   INFLUENCE OF DOME GEOMETRY ON MEMBRANE THEORY STRESS RESULTANTS

Using the results from the previous section, we shall compute the dead-load values of N'_ϕ and N'_θ for four types of domes—elliptical, spherical, conoidal, and conical—each with a base radius $r_0 = 160$ ft and a crown rise $b = 80$ ft. Novozhilov gives a similar set of examples having the same proportions.[4]

For the *Elliptical Dome*, $b/a = 0.5$, and following Eqs. (3-22) and (3-23) and Table 3-1, we find the values for N'_ϕ and N'_θ in terms of q as, for example, at $y = 0$

$$N'_\phi = -\frac{160}{0.5} q \; \frac{(0.69)(0.5)}{1} \doteq -110q \text{ kips/ft}$$

$$N'_\theta = -\frac{160}{0.5} q \left[0 - \frac{(0.69)(6400)}{(6400)(0.5)} \right] = 442q \text{ kips/ft}$$

where q is in kips per square foot. The distribution of these values is shown in Fig. 3-12a.

For the *Spherical Dome*, the radius of curvature can be computed from the relationship

$$a^2 = 160^2 + (a - 80)^2$$

from which $a = 200$ ft, and hence the slope at the edge support is 4/3 giving an edge angle $\alpha = 53°8'$. The values of N'_ϕ and N'_θ at the edge come from Eqs. (3-4) and (3-5) which give

$$N'_\phi = -\frac{200}{1 + 0.6} q = -125q \text{ kips/ft}$$

$$N'_\theta = 200q \left(\frac{1}{1 + 0.6} - 0.6 \right) = 5q \text{ kips/ft}$$

The distribution is shown in Fig. 3-12b.

For the *Conoidal Dome*, we take $r' = 40$ ft and $\alpha = 45°$, and thus find a from the relationships

$$a^2 = 40^2 + (a - c)^2$$
$$a^2 = 200^2 + (a - 80 - c)^2$$

where c is the difference between b and the point at which the circular arc has zero slope (top of dashed line in Fig. 3-9). The solution gives $a = 282.84$ ft, $c = 2.84$ ft, $\alpha = 45°$, and $\phi_0 = 8°8'$ from which at $\phi = \alpha$ we get

$$N'_\phi = -282.8q \; \frac{0.99 - 0.707 - 0.141(0.785 - 0.142)}{0.707(0.707 - 0.141)}$$

$$= -136q \text{ kips/ft}$$

$$N'_\theta = -\frac{282.8q}{0.707^2} [(0.707 - 0.141)(0.5) - (0.99 - 0.707)$$

$$+ (0.141)(0.643)] = -51.4q \text{ kips/ft}$$

Fig. 3-12c gives the distribution of these stress resultants from the membrane theory.

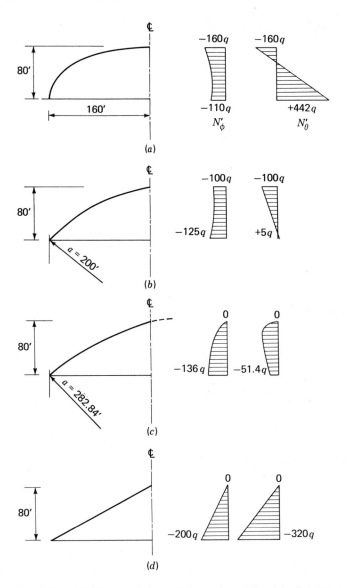

Fig. 3-12 Membrane stress resultants on various domes. (a) Elliptical dome. (b) Spherical dome. (c) Conoidal dome. (d) Conical dome.

Finally for the *Conical Dome,* we have with $r_0 = 160$ ft at the base and $b = 80$ ft, α (the apex half angle) $=$ arctan $160/80 = 63°26'$, the edge slope angle, $\phi = 26°34'$, and L (the meridional length) $= 179$ ft. Therefore at the dome base

$$N'_\phi = -\frac{179q}{2\,(0.447)} = -200q \text{ kips/ft}$$

$$N'_\theta = -179q\,(2)(0.894) = -320q \text{ kips/ft}$$

Figure 3-12d shows the distribution of these membrane theory stress resultants. Using Eq. (3-3c) for the ring force T_ϕ given for spherical domes on vertical supports, we can compare these cases by noting that T_ϕ depends only upon N'_ϕ, r_0, and ϕ at the base; thus Eq. (3-3c) applies to all four domes.

Elliptical dome: $\quad T_\phi = 110q\ (160)\ (0) = 0$
Spherical dome: $\quad T_\phi = 125q\ (160)\ (0.6) = 12,000q$ kips
Conoidal dome: $\quad T_\phi = 136q\ (160)\ (0.707) = 15,400q$ kips
Conical dome: $\quad T_\phi = 200q\ (160)\ (0.894) = 28,800q$ kips

Thus the elliptical dome does not need a tension ring because the shell carries the tension by hoop stresses N'_θ. The flatter the shell in the sense of decreased edge slope, the greater the edge meridional compression and hence the ring tension. As we shall see later, when a concrete edge ring is built to carry T_ϕ, then its continuity with the shell will change the edge values of N'_θ, often drastically, and will introduce bending moments and transverse shearing forces; but the meridional stress resultants N'_ϕ will not usually change very much.

3-4 DISPLACEMENTS FROM THE MEMBRANE THEORY FOR DOMES LOADED SYMMETRICALLY WITH RESPECT TO THEIR AXES

The extensional strains in the meridional and circumferential directions are computed following Fig. 3-13. The total change in radius r_0 owing to the meridional displacement v and the radial displacement w is

$$\Delta r_0 = v \cos \phi - w \sin \phi \qquad (a)$$

and thus the circumferential strain

$$\epsilon_\theta = \frac{2\pi \Delta r_0}{2\pi r_0} = \frac{v}{r_0} \cos \phi - \frac{w}{r_2} \qquad (b)$$

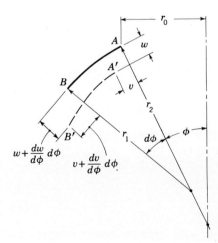

Fig. 3-13 Displacements in domes.

with $\sin \phi / r_0 = 1/r_2$. The meridional length $r_1 \, d\phi$ changes first by $\dfrac{dv}{d\phi} d\phi$ and second under the influence of w. Neglecting the higher-order term in w, we note that an increase in w leads to a decrease in elemental length of $w \, d\phi$ as shown in (b) of Sec. 2-5. Thus the meridional length change

$$\Delta (r_1 d\phi) = \frac{dv}{d\phi} d\phi - w \, d\phi \tag{c}$$

and the corresponding strain

$$\epsilon_\phi = \frac{\Delta (r_1 d\phi)}{r_1 \, d\phi} = \frac{dv}{r_1 \, d\phi} - \frac{w}{r_1} \tag{d}$$

Combining (b) and (d) and solving for v, we obtain

$$\frac{dv}{d\phi} - v \cot \phi = r_1 \epsilon_\phi - r_2 \epsilon_\theta \tag{e}$$

and from Eqs. (2-20) where $y \rightarrow \phi$,

$$\epsilon_\theta = \frac{1}{Eh} (N_\theta' - \nu N_\phi')$$

$$\epsilon_\phi = \frac{1}{Eh} (N_\phi' - \nu N_\theta') \tag{f}$$

When (f) are substituted into (e), the resulting differential equation is

$$\frac{dv}{d\phi} - v \cot \phi = \frac{1}{Eh} [N_\phi'(r_1 + \nu r_2) - N_\theta'(r_2 + \nu r_1)] \tag{3-26}$$

Such an equation can be solved by integration where

$$f(\phi) = \frac{1}{Eh} [N_\phi'(r_1 + \nu r_2) - N_\theta'(r_2 + \nu r_1)]$$

and

$$\frac{dv}{d\phi} - v \cot \phi = f(\phi)$$

The general solution will be

$$v = \sin \phi \left(\int \frac{f(\phi)}{\sin \phi} d\phi + C \right) \tag{g}$$

where C is a constant determined by the support conditions.

Once v is determined, w can be derived from (b):

$$w = v \cot \phi - r_2 \epsilon_\theta$$

$$= v \cot \phi - \frac{r_2}{Eh} (N_\theta' - \nu N_\phi') \tag{h}$$

The meridian rotation is obtained from Fig. 3-13 as

$$\Delta_\phi = \frac{v}{r_1} + \frac{dw}{r_1 \, d\phi}$$

For purposes of analyzing edge disturbances in shells, the values of v, w, and Δ_ϕ are needed only at the edges. Where the edge's supports are tangentially unyielding, $v = 0$ and only w and Δ_ϕ are required.

In the analyses to follow, the values v and w are not used; rather the horizontal movement Δ_H and the rotation Δ_ϕ are used. Δ_H can be derived directly from (f)

$$r_0\epsilon_\theta = \frac{1}{Eh} (N'_\theta - \nu N'_\phi) r_0 = \Delta_H$$

$$\Delta_H = \frac{r_2 \sin \phi}{Eh} (N'_\theta - \nu N'_\phi) \tag{3-27}$$

The meridian rotation at the edge will be, from (h), with $v = 0$,

$$\Delta_\phi = \frac{dw}{r_1 \, d\phi} = \frac{\cot \phi}{r_1} \frac{dv}{d\phi} - \frac{d}{r_1 \, d\phi} \left[\frac{r_2}{Eh} (N'_\theta - \nu N'_\phi) \right] \tag{i}$$

From (3-26), with $v = 0$,

$$\frac{dv}{d\phi} = \frac{1}{Eh} [N'_\phi(r_1 + \nu r_2) - N'_\theta(r_2 + \nu r_1)] \tag{j}$$

From (3-27), we observe that the bracketed term in (i) equals $\Delta_H/\sin \phi$. Thus, substituting (j) into (i), we find

$$\Delta_\phi = \frac{\cot \phi}{r_1 Eh} [N'_\phi(r_1 + \nu r_2) - N'_\theta(r_2 + \nu r_1)] - \frac{d}{r_1 \, d\phi} \left(\frac{\Delta_H}{\sin \phi} \right) \tag{3-28}$$

Where only the horizontal movement is required, it is not necessary to determine v, w, or Δ_ϕ nor to have N'_ϕ and N'_θ as known mathematical functions of ϕ. It is only necessary to compute N'_ϕ and N'_θ at the edge.

As an example, we shall give the dead-load displacements at the edge of a constant-thickness spherical dome, where $r_1 = r_2 = a$, $v = 0$ at the edge, and N'_ϕ and N'_θ are given by (3-4) and (3-5). From (3-27) and (3-28),

$$\Delta_H = \frac{a^2 q}{Eh} \left(\frac{1 + \nu}{1 + \cos \phi} - \cos \phi \right) \sin \phi \tag{k}$$

$$\Delta_\phi = -\frac{aq}{Eh} (2 + \nu) \sin \phi \tag{l}$$

In the same way, for a spherical dome under a uniform load over a horizontal projection of the dome surface where N'_ϕ and N'_θ are given by Eqs. (3-6) and (3-7),

$$\Delta_H = \frac{a^2 p}{2Eh} (-\cos 2\phi + \nu) \sin \phi \tag{m}$$

$$\Delta_\phi = \frac{ap}{2Eh} (3 + \nu) \sin 2\phi \tag{n}$$

and for a spherical dome under uniform external pressure where N'_ϕ and N'_θ come from Eqs. (3-8) and (3-9),

$$\Delta_H = \frac{a^2 p}{2Eh} (-1 + \nu) \sin \phi \qquad (o)$$

$$\Delta_\phi = 0 \qquad (p)$$

For other loadings on spherical domes and for other types of domes, Δ_H can always be computed easily from N'_ϕ and N'_θ, but Δ_ϕ is more complicated, and good approximate values can be obtained by using finite differences for the second term of Eq. (3-28).

3-5 BENDING IN SHELLS OF REVOLUTION UNDER LOADS SYMMETRICAL ABOUT THEIR AXES OF ROTATION

The general equations for equilibrium are the same as Eqs. (3-1), except that shearing terms must now be included and two additional bending equations added to give five equations similar to Eqs. (2-25) for cylinders. Figure 3-14 shows the shear and bending terms. Because of the double curvature, Q_θ will have a component $Q_\theta r_1 \, d\phi \, d\theta$ in the direction θ, and Q_ϕ a component $Q_\phi r_0 \, d\theta \, d\phi$ in the ϕ direction. Furthermore in the z direction the Q_θ and Q_ϕ terms on the negative faces will cancel those on the positive faces, leaving only their differences to be included in the z equilibrium equation.

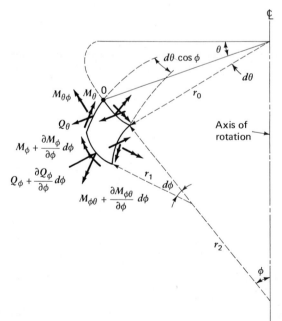

Fig. 3-14 Bending moments and shears in domes.

The bending equations need to be derived more fully; we can do this by reference to Fig. 3-14. For bending around the parallel circle, we find from the vector diagrams that

$$+ M_\phi r_0 \, d\theta - \left(M_\phi + \frac{\partial M_\phi}{\partial \phi} \, d\phi \right) \left(r_0 + \frac{\partial r_0}{\partial \phi} \right) d\theta$$

$$- M_{\theta\phi} r_1 \, d\phi + \left(M_{\theta\phi} + \frac{\partial M_{\theta\phi}}{\partial \theta} \, d\theta \right) r_1 \, d\phi$$

$$+ \left(M_\theta + \frac{\partial M_\theta}{\partial \theta} \, d\theta \right) r_1 \, d\phi \, d\gamma$$

$$+ \left(Q_\phi + \frac{\partial Q_\phi}{\partial \phi} \, d\phi \right) \left(r_0 + \frac{\partial r_0}{\partial \phi} \, d\phi \right) d\theta \, r_1 \, d\phi = 0$$

which, after canceling terms, neglecting higher-order ones, and recognizing that $dy = \cos \phi \, d\theta = \dfrac{\partial r_0}{r_1 \, \partial \phi} \, d\theta$ from Fig. 3-1c, we can reduce to

$$- \frac{\partial (M_\phi r_0)}{\partial \phi} + M_\theta \frac{\partial r_0}{\partial \phi} + r_1 \frac{\partial M_{\theta\phi}}{\partial \theta} + Q_\phi r_0 r_1 = 0$$

A similar equation results for bending around the meridian, so that the final five equations of equilibrium are

$$r_1 \frac{\partial N_\theta}{\partial \theta} + N_{\theta\phi} \frac{\partial r_0}{\partial \phi} + \frac{\partial (N_{\phi\theta} r_0)}{\partial \phi} - Q_\theta r_1 \sin \phi + p_\theta r_0 r_1 = 0$$

$$\frac{\partial (N_\phi r_0)}{\partial \phi} - N_\theta \frac{\partial r_0}{\partial \phi} + r_1 \frac{\partial N_{\theta\phi}}{\partial \theta} - Q_\phi r_0 + p_\phi r_0 r_1 = 0$$

$$r_1 \frac{\partial Q_\theta}{\partial \theta} + \frac{\partial (Q_\phi r_0)}{\partial \phi} + N_\theta r_1 \sin \phi + N_\phi r_0 + p_z r_0 r_1 = 0 \qquad (3\text{-}29)$$

$$- \frac{\partial (M_\phi r_0)}{\partial \phi} + M_\theta \frac{\partial r_0}{\partial \phi} + r_1 \frac{\partial M_{\theta\phi}}{\partial \theta} + Q_\phi r_0 r_1 = 0$$

$$- r_1 \frac{\partial M_\theta}{\partial \theta} + M_{\theta\phi} \frac{\partial r_0}{\partial \phi} - \frac{\partial (M_{\phi\theta} r_0)}{\partial \phi} + Q_\theta r_0 r_1 = 0$$

which, for axisymmetrical loading and with $\partial r_0 / \partial \phi = r_1 \cos \phi$, reduce to

$$\frac{d(N_\phi r_0)}{d\phi} - N_\theta r_1 \cos \phi - Q_\phi r_0 + p_\phi r_0 r_1 = 0$$

$$\frac{d(Q_\phi r_0)}{d\phi} + N_\theta r_1 \sin \phi - N_\phi r_0 + p_z r_0 r_1 = 0 \qquad (3\text{-}30)$$

$$- \frac{d(M_\phi r_0)}{d\phi} + M_\theta r_1 \cos \phi + Q_\phi r_0 r_1 = 0$$

These three equations correspond to the three equations (2-26) and reduce to them when $\phi \to y$, $r_0 \to r$ (a constant), $\phi = 90°$, $r_1 \to \infty$, and $r_1 \, d\phi \to dy$.

Putting the expressions for strain, (b) and (d) from Sec. 3-4, into the formulas for stress resultants, Eqs. (2-21), we find

$$N_\theta = K\left(\frac{v}{r_0}\cos\phi - \frac{w}{r_2} + \frac{v}{r_1}\frac{dv}{d\phi} - \frac{vw}{r_1}\right)$$

$$N_\phi = K\left(\frac{1}{r_1}\frac{dv}{d\phi} - \frac{w}{r_1} + \frac{vv}{r_0}\cos\phi - \frac{vw}{r_2}\right)$$

(3-31)

To determine comparable expressions for the stress couples, we must first develop the changes in curvature for the ϕ and the θ directions. From Fig. 3-13 we see that the meridional rotation at the local origin or upper edge is

$$\Delta_\phi = \frac{v}{r_1} + \frac{dw}{r_1\,d\phi}$$

and at the lower edge that rotation becomes

$$\overline{\Delta}_\phi = \Delta_\phi + \frac{d(\Delta_\phi)}{d\phi}\,d\phi$$

The change in curvature is defined as the difference between rotation at the lower and upper edges divided by the meridional arc length, i.e.,

$$\chi_\phi = \frac{\overline{\Delta}_\phi - \Delta_\phi}{r_1\,d\phi} = \frac{d}{r\,d\phi}\left(\frac{v}{r_1} + \frac{dw}{r_1\,d\phi}\right)$$

(a)

In the circumferential direction, the rotation of the meridian on the negative face of the differential element (the left-hand side in Fig. 3-14) is in the plane perpendicular to the θ direction and hence has no rotational component in the plane perpendicular to the meridian. The rotation of the meridian on the positive face has the same value as that on the negative face because of axial symmetry; but that rotation is now in a plane making an angle of $\cos\phi\,d\theta$ with the ϕ direction. The positive face rotation has a component about the local ϕ axis of $\Delta_\phi\cos\phi\,d\theta$, thus giving a change in curvature in the θ direction of

$$\chi_\theta = \frac{\Delta_\phi\cos\phi\,d\theta - 0}{r_0\,d\theta} = \left(\frac{v}{r_1} + \frac{dw}{r_1\,d\phi}\right)\frac{\cos\phi}{r_0}$$

(b)

Putting (a) and (b) into the general expressions for bending,

$$M_\theta = -D(\chi_\theta + v\chi_\phi)$$
$$M_\phi = -D(\chi_\phi + v\chi_\theta)$$

we get the following equations for the stress couples:

$$M_\theta = -D\left[\left(\frac{v}{r_1} + \frac{dw}{r_1\,d\phi}\right)\frac{\cos\phi}{r_0} + \frac{v}{r_1}\frac{d}{d\phi}\left(\frac{v}{r_1} + \frac{dw}{r_1\,d\phi}\right)\right]$$

$$M_\phi = -D\left[\frac{1}{r_1}\frac{d}{d\phi}\left(\frac{v}{r_1} + \frac{dw}{r_1\,d\phi}\right) + v\left(\frac{v}{r_1} + \frac{dw}{r_1\,d\phi}\right)\frac{\cos\phi}{r_0}\right]$$

(3-32)

The seven equations of (3-30) to (3-32) contain seven unknowns: three stress resultants (N_ϕ, N_θ, and Q_ϕ), two stress couples (M_ϕ and M_θ), and two displacements (v and w).

**Table 3-2 FORCES AND DISPLACEMENTS IN DOMES OF
REVOLUTION LOADED BY EDGE FORCES
UNIFORM AROUND A PARALLEL CIRCLE**

N_ϕ	$\psi H \sin \alpha \cot (\alpha - \psi)$	$-2\zeta\beta M_\alpha \cot (\alpha - \psi)$
N_θ	$2\theta\beta aH \sin \alpha$	$2\psi\beta^2 aM_\alpha$
M_ϕ	$\dfrac{\zeta}{\beta} H \sin \alpha$	ϕM_α
Δ_H	$2\beta \dfrac{a^2}{Eh} H \sin^2 \alpha$	$\dfrac{2\beta^2 a^2 \sin \alpha}{Eh} M_\alpha$
Δ_α	$2\beta^2 \dfrac{a^2}{Eh} H \sin \alpha$	$\dfrac{4\beta^3 a^3}{Eh} M_\alpha$

Note: $\beta^4 = 3(1 - \nu^2)/a^2 h^2$.

These seven equations are normally reduced to two equations (see Chap. 16 of Ref. 1). The resulting two equations can be solved analytically, but such solutions have limited value. In most cases the Geckeler approximation (Sec. 1-4) permits use of the cylinder solution from Chap. 2. Where there is some doubt about the accuracy of that approximation, numerical solutions are useful (Sec. 3-7).

For stress resultants, stress couples, and edge displacements, the expressions for cylinders given in Chap. 2 can be used for domes as long as the dome-edge slope α is taken into account. For an edge force H on a dome, the edge shear $Q_0 = H \sin \alpha$ so that in Table 3-2, Q_0 given in Table 2-4 is replaced by $H \sin \alpha$. Also the radial displacement w, given in Table 2-4, must be modified for horizontal dome-edge displacement $\Delta_H = w \sin \alpha$. Therefore, for example, the expression for Δ_H due to an edge force H will be that given in Table 2-4 multiplied by $\sin^2 \alpha$. These are all given in Table 3-2.

Note that for $H = M_\alpha$, $\Delta_\alpha = \Delta_H$, which must be so because of the reciprocity theorem. Here for an outward thrust there is positive rotation and translation. Table 3-2 gives the expressions of N_ϕ, N_θ, M_ϕ, Δ_α, and Δ_H for each type of edge loading. The complete dome analysis which follows describes how these various expressions are used.

3-6 EXAMPLE OF ANALYSIS

The rigidly supported spherical dome of Fig. 3-15 will be analyzed for uniform gravity load over the dome surface, temperature change, and wind load.

$r_1 = r_2 = a = 94.5$ ft
$h = 4$ in.   constant shell thickness
$\alpha = 28°$

$r_0 = a \sin \alpha = 44.25$ ft   parallel circle radius at springing

$\nu = 1/6 = 0.167$ Poisson's ratio

$t = -10°F$ temperature drop

$q = 50$ psf (D.L.)* $+ 40$ psf (L.L.)† $= 90$ psf

$p = 30$ psf   wind pressure

$$\lambda = \sqrt[4]{3(1 - \nu^2)} \left(\frac{a}{h}\right)^2 = 22 \qquad \beta = \lambda/a = 0.233$$

Uniform Gravity Load Over the Dome Surface

Primary System The stress resultants are computed on the basis of the membrane theory [Eqs. (3-3)] and given in Table 3-4a.

Errors The dome-edge displacements are, from (k) and (l) in Sec. 3-4,

$$\Delta_H = D_{10} = -3300\,\frac{q}{E}$$

$$\Delta_\phi = D_{20} = +295\,\frac{q}{E}$$

Note that D_{20} is positive according to the sign convention given in Fig. 3-15.

Corrections Edge forces X_1 and X_2 are applied (see Fig. 3-15d and e). The displacements of the dome edge due to these forces are given in Table 3-2. Where $H = X_1 = 1$ and $M_\alpha = X_2 = 1$,

$$\Delta_H(H = 1) = D_{11} = \frac{2740}{E}$$

$$\Delta_H(M_\alpha = 1) = D_{12} = \frac{1360}{E}$$

$$\Delta_\alpha(M_\alpha = 1) = D_{22} = \frac{1350}{E}$$

Compatibility The size of the correction forces required are computed by setting up the two equations of compatibility at the dome support:

$$\Sigma\Delta_H = 0 = X_1 D_{11} + X_2 D_{12} + D_{10}$$
$$\Sigma\Delta_\alpha = 0 = X_1 D_{21} + X_2 D_{22} + D_{20}$$

where

$$+2740X_1 + 1360X_2 = +3300q$$
$$+1360X_1 + 1350X_2 = -295q$$

---

* Dead load.
† Live load.

and

$$X_1 = +2.63q = +237 \text{ lb/ft}$$
$$X_2 = -2.87q = -260 \text{ ft-lb/ft}$$

The negative sign indicates that the direction of M_α is opposite to that shown in Fig. 3-15. These values for H and M_α are then substituted into the expressions of Table 3-2 and are shown numerically in Table 3-4a and plotted in Fig. 3-16. Note how rapidly the edge restraint effects are damped out. Table 3-3 gives the influence of edge loadings. Values obtained from a more rigorous analysis as described in Ref. 1, Art. 129, are shown in parentheses.

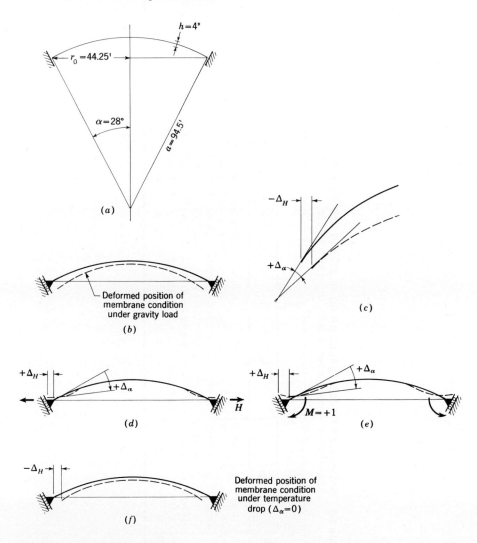

Fig. 3-15   Fixed dome analysis.

Table 3-3 INFLUENCE OF EDGE FORCES ON DOME STRESS RESULTANTS
(Values from a more rigorous analysis are in parentheses)

ψ, deg	N_ϕ X_1		N_ϕ X_2		N_θ X_1		N_θ X_2		M_ϕ X_1		M_ϕ X_2	
0°	+0.881	(+0.883)	0	(0)	+20.6	(+20.49)	+10.3	(+10.5)	0	(0)	+1.00	(+1.00)
1°	+0.346	(+0.347)	−0.234	(−0.246)	+13	(+13.19)	+3.85	(+4.14)	+0.52	(+0.52)	+0.89	(+0.907)
2°	+0.0107	(+0.004)	−0.307	(−0.324)	+6.85	(+7.13)	+0.114	(+0.284)	+0.65	(+0.668)	+0.66	(+0.686)
5°	−0.208	(−0.231)	−0.152	(−0.170)	−1.04	(−1.07)	−1.93	(−2.08)	+0.23	(+0.302)	+0.088	(+0.101)
10°	−0.0037	(−0.0045)	+0.0194	(+0.025)	−0.332	(−0.415)	−0.027	(−0.048)	+0.027	(−0.033)	−0.03	(−0.037)
20°	−0.00119		−0.00144		+0.0016		−0.0037		0		0	
28°	0		0		0		0		0		0	

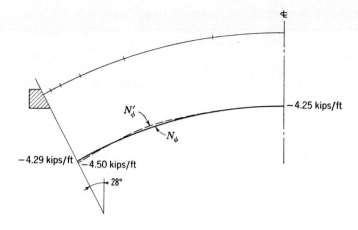

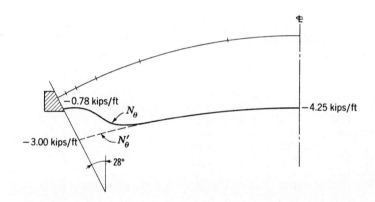

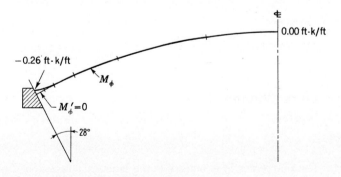

Fig. 3-16   Plots of fixed dome results.

Table 3-4 FORCES IN A SPHERICAL DOME ON FIXED SUPPORTS

(a) Uniform gravity load over dome surface

ψ	0° (edge)	1°	2°	5°	10°	20°	28° (apex)
1. N'_ϕ (membrane)	−4.50	−4.49	−4.48	−4.42	−4.34	−4.25	−4.25
2. N_{ϕ_1} ($H = X_1$)	0.21	0.08	0	−0.05	0	0	0
3. N_{ϕ_2} ($M_\alpha = X_2$)	0.0	0.06	0.08	0.04	−0.01	0	0
4. N_ϕ ($q = 90$ psf)	−4.29	−4.35	−4.40	−4.43	−4.35	−4.25	−4.25
5. N'_θ (membrane)	−3.00	−3.08	−3.16	−3.41	−3.73	−4.15	−4.25
6. N_{θ_1} ($H = X_1$)	4.88	3.10	1.62	−0.25	−0.08	0	0
7. N_{θ_2} ($M_\alpha = X_2$)	−2.66	−1.00	−0.03	0.51	0.08	0	0
8. N_θ ($q = 90$ psf)	−0.78	−0.98	−1.57	−3.15	−3.73	−4.15	−4.25
9. M'_ϕ (membrane)	0	0	0	0	0	0	0
10. M_{ϕ_1} ($H = X_1$)	0	0.123	0.155	0.066	−0.007	0	0
11. M_{ϕ_2} ($M_\alpha = X_2$)	−0.26	−0.229	−0.169	−0.023	0.008	0	0
12. M_ϕ ($q = 90$ psf)	−0.26	−0.106	−0.014	0.043	0.001	0	0

(b) Uniform temperature change over dome surface

ψ	0° (edge)	1°	2°	5°	10°	20°	28° (apex)
1. N'_ϕ (membrane)	0	0	0	0	0	0	0
2. N_{ϕ_1} ($H = X_1$)	0.75	0.29	0.08	−0.18	0	0	0
3. N_{ϕ_2} ($M_\alpha = X_2$)	0	0.20	0.26	0.13	−0.01	0	0
4. N_ϕ ($T = -10°$)	0.75	0.49	0.34	−0.05	−0.01	0	0
5. N_ϕ ($T = -70°$)	5.25	3.44	2.38	−0.35	−0.07	0	0
6. N'_θ (membrane)	0	0	0	0	0	0	0
7. N_{θ_1} ($H = X_1$)	17.5	11.1	5.85	−0.88	−0.28	0	0
8. N_{θ_2} ($M_\alpha = X_2$)	−8.8	−3.3	−0.10	1.64	0.02	0	0
9. N_θ ($T = -10°$)	8.7	7.8	5.75	0.76	−0.26	0	0
10. N_θ ($T = -70°$)	61.0	54.5	40.3	5.31	−1.82	0	0
11. M'_ϕ (membrane)	0	0	0	0	0	0	0
12. M_{ϕ_1} ($H = X_1$)	0	0.41	0.55	0.20	0.02	0	0
13. M_{ϕ_2} ($M_\alpha = X_2$)	−0.85	−0.76	−0.55	−0.07	0.03	0	0
14. M_ϕ ($T = -10°$)	−0.85	−0.32	0	0.13	0.05	0	0
15. M_ϕ ($T = -70°$)	−5.95	−2.24	0	0.91	0.35	0	0

Note: N_ϕ and N_θ are in kips per foot.
M_ϕ is in foot-kips per foot.
Negative sign indicates compression.

Temperature Change

Primary System Under the membrane condition, no forces result from volume change.

Errors The only error in geometry is the lateral displacement Δ_H, given simply as

$$\Delta_H = D_{10} = r_0 T\epsilon = -44.25(10°)6 \times 10^{-6} = -2655 \times 10^{-6}$$

To avoid introducing E into all the other displacements, the temperature movement is expressed as $E\Delta_H$, with $E = 4.32 \times 10^5$ kips/ft².

$$ED_{10} = 4.32 \times 10^5(-2655 \times 10^{-6}) = -1145$$
$$\Delta_\alpha = D_{20} = 0$$

Corrections The same corrections as before are used.

Compatibility

$$+2740X_1 + 1360X_2 = +1145$$
$$+1360X_1 + 1350X_2 = 0$$

from which

$$X_1 = +0.85 \text{ kip/ft}$$
$$X_2 = -0.85 \text{ ft-k/ft}$$

The distributions of N_ϕ, N_θ, and M_ϕ due to these edge values are given in Table 3-4. It is important to note the relative magnitude of the gravity-load values and the volume-change values in a fixed dome:

	Gravity loads	Volume change $(-10°)$
N_ϕ	-4450 lb/ft	$+750$ lb/ft
N_θ	-4250 lb/ft	$+8700$ lb/ft
M_ϕ	-260 ft-lb/ft	-850 ft-lb/ft

Even a 10°F temperature drop will produce a large hoop tension and a moment more than three times the gravity-load moment. Actual temperature drops can often be as high as 70°, so that the tensions and moments due to temperature could control the design. Where temperatures can rise as well as fall, the moments will reverse and the shell must be so reinforced.

Wind Load

Primary System Using the membrane theory, we can solve for wind loading by assuming the pressure to be made up of harmonic components, as in the case of cylinders (Sec. 2-2). Analytic solutions, worked out for relatively simple loading assumptions, are illustrated by Table 3-5 for a loading $p_z = p \cos \theta \sin \phi$ and $p_\phi = p_\theta = 0$ on a spherical shell.

The stress resultants are given in Ref. 5 by

$$N'_\phi = -apK_1 \cos \theta$$
$$N'_{\phi\theta} = -apK_2 \sin \theta \qquad\qquad (3-36)$$
$$N'_\theta = -apK_3 \cos \theta$$

A full derivation of these equations was given in Ref. 6, taken from Ref. 5. These results are useful for very rough calculations, but their main value is as a check on numerical solutions such as those described in Sec. 3-7.

Table 3-5 SPHERICAL DOMES

ϕ	K_0	K_1	K_2	K_3
0	0.0000	0.0000	0.0000	0.0000
10	0.0002	0.0413	0.0419	0.1334
20	0.0036	0.0840	0.0894	0.2580
30	0.0171	0.1187	0.1371	0.3913
40	0.0505	0.1455	0.1900	0.4973
50	0.1124	0.1615	0.2510	0.6045
60	0.2084	0.1605	0.3210	0.7055
70	0.3380	0.1409	0.4125	0.7958
80	0.4947	0.0909	0.5200	0.8943
90	0.6667	0.0000	0.6667	1.0000

Errors A rough approximation for the errors in geometry can be made by assuming[5]

$$\Delta_H = D_{10} = \frac{a \sin \alpha}{Eh} (N'_\theta - \nu N'_\phi)$$

$$= -\frac{a^2 p \sin \alpha}{Eh} (K_3 - \nu K_1) = -\frac{94.5^2 \times 0.469}{E \times 4/12} (0.35)p$$

$$= -4380 \frac{p}{E}$$

and

$$\Delta_\alpha = D_{20} \approx 0$$

The value for Δ_H is derived by considering the maximum membrane stress resultants as constant around the edge and using the same equation, (3-27), which was developed for axisymmetrical displacement. *Corrections* and *Compatibility* are the same as before, and we obtain

$$+2740X_1 + 1360X_2 = +4380p$$
$$+1360X_1 + 1350X_2 = 0$$

from which, with $p = 0.03$ kip/ft^2,

$$X_1 = +3.18 \times 0.03 = +95.5 \text{ lb/ft}$$
$$X_2 = -3.20 \times 0.03 = -96.0 \text{ ft-lb/ft}$$

These values are small and would have little effect on the design in this case.

The complete design of such a dome will be illustrated in Chap. 4 where the practical case of a fluid storage tank covered by a domed roof will be presented.

3-7 FINITE-ELEMENT ANALYSIS

The same problem shown in Fig. 3-15 can be solved directly by the finite-element method where the elements are horizontal slices from the dome. Such idealizations are discussed in Chap. 4.

Table 3-6

	Membrane theory*	4" constant thickness*	Varying thickness†	6" constant thickness*
N_ϕ	−4.50	−4.29	−3.59	−4.25
N_θ	−3.00	−0.78	−0.59	−1.02
M_ϕ	0	−0.26	−0.63	−0.40

* Load taken as 90 psf in all cases.

† By finite-element analysis for actual D.L. + 40 psf L.L.

Table 3-6 gives the edge values for various assumptions. The finite-element analysis gives less N_ϕ because its dead load takes into account the actual shell thickness which over most of the shell is 2½ in. Thus, q should be roughly $1.5 \times 12.5 \approx 19$ psf less and N_ϕ roughly $-(71/90)\,4.29 = -3.4$ kips/ft. The actual N_ϕ is slightly more because of the edge thickening weight.

Although N_ϕ is less, M_ϕ is substantially more in somewhat the same way in which a haunched beam has a greater fixed end moment than a prismatic one. Here, the moments are so small in either case that the increase found through a more accurate idealization in the finite-element analysis would have no influence on design. As we shall see in the next chapter, the same conclusion results when finite-element results are compared to those from classic analysis, even when the dome is flexibly supported by a ring and wall. We thus conclude that these classical solutions developed in 1926 by Geckeler form a sound basis for design when used by experienced engineers.

Other numerical methods exist for solving dome problems of which numerical integration[7] and finite differences are the most used. An example of the latter method was given by Parme in 1950[8] and more recently reworked with simpler formulas.[9] It consists of taking the differential equations and putting them in finite difference form for a series of discrete points on the meridian. If nine points are used, then nine linear simultaneous equations result, the solution to which yields values of the variables at each point. For greater accuracy to be achieved, more points need to be chosen. We have already used a finite difference representation of the curvature for a hyperboloid (see Sec. 2-4).

REFERENCES

1. Timoshenko, S. P., and S. Woinowsky-Krieger, *Theory of Plates and Shells,* 2d ed., McGraw-Hill Book Company, New York, 1959.
2. Girkmann, K., *Flächentragwerke,* 5th ed., Springer-Verlag OHG, Vienna, 1959.
3. "Design of Circular Domes," Portland Cem. Assoc., Chicago.
4. Novozhilov, V. V., *Thin Shell Theory,* Groningen, Netherlands, 1964, p. 127.
5. Dischinger, F., "Schalen und Kuppeln" (Shells and Domes), *Handb. für Eisenbetonbau,* 4th ed., vol. 6, Berlin, 1928.

6. Billington, D. P., *Thin Shell Concrete Structures,* 1st ed., McGraw-Hill Book Company, New York, 1965, pp. 51–55.
7. Goldberg, J. E., J. L. Bogdanoff, and D. L. Alspaugh, "Stresses in Spherical Domes under Arbitrary Loading Including Thermal and Shrinkage Effects," *Proceedings,* Symp. Non-classical Shell Probl., North-Holland Publishing Company, Amsterdam, 1964, pp. 116–138.
8. Parme, A., "Solution of Difficult Structural Problems by Finite Differences," *J. ACI,* vol. 47, no. 3, November 1950, pp. 246–251.
9. Billington, D. P., "Concrete Thin Shells of Revolution," *Concr. Thin Shells,* ACI, Publ. SP-28, 1971, pp. 237–274.

4
ANALYSIS AND DESIGN OF
DOMES AND SHELL WALLS

4-1 INTRODUCTION

Mainly domes supported tangentially to the meridian slope at the edge were considered in Chap. 3. We observed that bending moments are generally small and restricted to a narrow zone at the edge of the dome. In fact, from the analytic example we saw that the dome design is little affected even by fixity, except for some extra bending reinforcement against temperature changes.

It is usually not possible or desirable to supply support tangent to the meridian at the edge. The normal case of dome support is by vertical supports. Figure 4-1 illustrates the general structural problem brought about by this situation. The membrane reaction N'_α cannot be supplied by a vertical support. The lack of lateral restraint gives rise to a horizontal displacement at the dome edge. The resulting dome stresses may be determined by combining the membrane values with those obtained from an analysis of the dome pulled outward by an edge force $H = -N'_\alpha \cos \alpha$. Normally the designer imposes some type of lateral restraint, three types of which are discussed in this chapter (Fig. 4-2): an edge ring, a supporting cylindrical wall, or a combination of ring and wall. The ring support of Fig. 4-2a is often realized in practice by an elastomeric ring pad placed on top of the cylindrical wall. In this case some shear force will be transferred through the pad; the size of that force will depend upon the pad's stiffness, which manufacturers specify. To analyze the interaction of domes, rings, and walls, it will be necessary to study the behavior of circular rings.

All of these analyses will be based on the assumption that there is continuous vertical support with no differential settlement of the ring or wall. Such a condition is generally valid for most dome designs. An important exception is the dome supported only by columns spaced at intervals around the base circumference. A ring will generally be provided around the base between columns. But if this ring is a beam of some depth, it can have little vertical deflection between columns and the assumption of continuous support may still be valid. Often with column-supported domes, the lack of support between columns means that the dome shell must "bridge" the span itself as a deep beam.

Computations for the membrane conditions of column-supported domes have been presented elsewhere[1] and are not treated in this text. In the case of shell walls, such as cooling towers, the lower part of the shell must bridge between the supporting

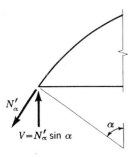

N'_α

$V = N'_\alpha \sin \alpha$

α

Fig. 4-1   Dome edge reactions.

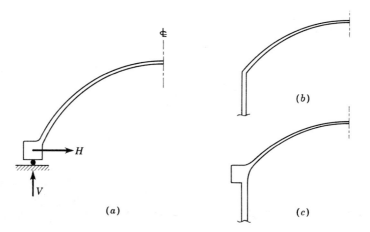

(b)

H

V

(a)

(c)

Fig. 4-2   Dome edge constructions.

columns, and this effect is normally considered by an equivalent deep-beam analysis which will be discussed later in this chapter.

4-2   CIRCULAR RINGS

Figures 4-3 and 4-4 show a circular ring of constant rectangular cross section subjected to two sets of forces.

First, we take a set of horizontal forces, H, uniformly distributed around the circumference and acting radially through the centroid of the cross section (Fig. 4-3). These forces cause a constant circumferential force in the ring equal to

$$T = Hr$$

and a circumferential strain of

$$\epsilon_c = \frac{T}{EA_R}$$

which gives a total change in length of the ring of

$$\Delta_c = \frac{2\pi r}{EA_R} T$$

or a change in radius of

$$\Delta_H = \frac{rT}{EA_R} \tag{a}$$

Expressing (a) in terms of H, we get

$$\Delta_H = \frac{r^2}{EA_R} H \tag{4-1}$$

the horizontal displacement of the ring in the radial direction. Second, we consider the radial moment M_α, uniformly distributed around the circumference, and observe from Fig. 4-4 that on a ring element

$$M_\alpha r\, d\theta - M_x\, d\theta = 0$$

so that

$$M_x = rM_\alpha$$

This moment, acting on the ring cross section, gives ring stresses (see Fig. 4-5)

$$f = \frac{M_x y}{I_R}$$

from which the circumferential strains will be

$$\epsilon_c = \frac{M_x y}{EI_R} \tag{b}$$

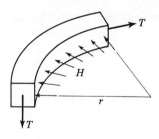

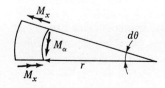

Fig. 4-3 Ring axial tension.

Fig. 4-4 Ring bending moments.

These strains are seen to be linear over the depth d of the ring. The change in length of the ring at any point will be

$$\Delta_c = 2\pi r \frac{M_x y}{EI_R} \tag{c}$$

and the change in radius

$$\Delta_H = \frac{rM_x y}{EI_R}$$

or in terms of M_α

$$\Delta_H = \frac{r^2 y}{EI_R} M_\alpha \tag{4-2}$$

The angular rotation Δ_α can easily be obtained from (4-2) by recognizing that the small angle

$$\Delta_\alpha = \frac{\Delta_H}{y}$$

so that

$$\Delta_\alpha = \frac{r^2}{EI_{R\cdot}} M_\alpha \tag{4-3}$$

Equations (4-1) to (4-3) give the displacements of rings of any shape under uniform horizontal thrust and radial moment.

Equations (4-1) and (4-3) are based on the assumption that the ring width b is small compared to the radius r. In Eq. (b) strains are constant at any level y in the ring, and in Eq. (c) the circumferential elongation at any level y is a constant. This cannot be true since the elongation at any point depends upon the radius to that point, and this varies from $r - b/2$ at the inside of the ring to $r + b/2$ at the outside. Equation (c) assumes an average value r over the full width b. The accurate expression for rotation of a ring of rectangular cross section is given by Timoshenko[2] as

$$\Delta_\alpha = \frac{12rM_\alpha}{Ed^3 \ln(1 + b/r)} \tag{4-3a}$$

For small values of b/r, $\ln(1 + b/r) \approx b/r$ and (4-3a) reduces to (4-3). Consider a ring of width $b = 5$ ft 0 in. and radius $r = 50$ ft 0 in.; then

$$\ln(1 + 0.1) = 0.09531 = (0.9531)\frac{b}{r}$$

or an error of under 5 percent in rotation will occur where approximations of Eq. (4-3) are used. But a ring of 5 ft 0 in. on a radius of 50 ft 0 in. is already beyond the normal practical range for concrete structures, so that we shall consider Eqs. (4-1) to (4-3) as sufficiently accurate for the following discussions. As a further approximation we take r to be the inside radius of the ring (Fig. 4-3) so that it will closely correspond to r_0 of the dome it supports.

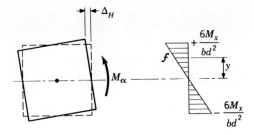

Fig. 4-5 Ring bending stresses.

Because the most frequent ring cross section is the rectangle, Eqs. (4-1) to (4-3) can be more conveniently expressed as

$$\Delta_H = \frac{r^2}{Ebd} H \tag{4-4}$$

$$\Delta_H = \frac{12r^2 y}{Ebd^3} M_\alpha \tag{4-5}$$

$$\Delta_\alpha = \frac{12r^2}{Ebd^3} M_\alpha \tag{4-6}$$

4-3 DOME-RING ANALYSIS

In the dome and ring example,[3] illustrated in Fig. 4-2a, the ring is assumed to be monolithic with the dome and to be free to slide and rotate on an immovable support. With these assumptions we may analyze the structure in the usual way.

Primary System   The stress resultants in the dome are computed on the basis of the membrane theory. The horizontal component of N'_α is held in equilibrium by the ring tension $T = N'_\alpha\, a \sin \alpha \cos \alpha$ (Fig. 4-6a), where the dome radius $a = r/\sin \alpha$.

Errors   There will be four errors: horizontal translation and rotation of the dome edge (Δ^D_H and Δ^D_α, respectively) and of the ring (Δ^R_H and Δ^R_α, respectively). Expressions for the first two, which depend upon the dome loading, have been given in Chap. 3. The expressions for the second two can be easily derived from Eqs. (4-1) to (4-3).

From the membrane theory the statics are satisfied by resolving the meridional thrust N'_α into its components.

$$H = N'_\alpha \cos \alpha$$
$$V = N'_\alpha \sin \alpha$$

where V is taken by the foundation and H by the ring, which will then move outward according to Eq. (4-1).

$$\Delta_H = \frac{r^2}{EA_R} H = \frac{r^2}{EA_R} N'_\alpha \cos \alpha \tag{a}$$

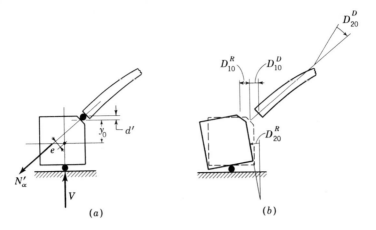

Fig. 4-6 Dome-ring analysis.

Where the line of action of N'_α passes through the center of rotation of the ring, Eq. (a) gives the total Δ_H; but if N'_α does not pass through the center of rotation (see Fig. 4-6), then there will be a moment $M_\alpha = N'_\alpha e$ and an additional horizontal movement from (4-2).

$$\Delta_H = \frac{r^2 y}{EI_R} M_\alpha = \frac{r^2 y N'_\alpha e}{EI_R} \tag{b}$$

Inward translation is taken as positive so that when N'_α is compression (hence negative), (a) becomes negative, and for the moment shown in Fig. 4-6a, (b) will also be negative when N'_α is negative.

Equations (a) and (b) can now be combined to give the total Δ_H^R. Of course it is desired to have the Δ_H^R at the junction of dome and ring as shown in Fig. 4-6a. This will be at some small distance below the "corner" of the ring, which, with h_D as dome-edge thickness, is

$$d' = \frac{h_D}{2} \cos \alpha$$

and

$$y_0 = \frac{d}{2} - d'$$

d' is generally small compared to d, and the succeeding equations can be simplified if d' is taken as zero.

The total Δ_H^R can now be written

$$\Delta_H^R = D_{10}^R = \left(\frac{r^2}{EA_R} \cos \alpha + \frac{r^2 y_0 e}{EI_R} \right) N'_\alpha$$

and the total rotation, from Eq. (4-3), is

$$\Delta_\alpha^R = D_{20}^R = -\frac{r^2 e}{EI_R} N_\alpha'$$

Where the line of action of N_α' passes through the center of rotation of the ring, $e = 0$ and $\Delta_\alpha^R = 0$. These errors can be summarized for a rectangular ring as follows:

$$D_{10}^R = \left(\cos \alpha + \frac{12 y_0 e}{d^2}\right) \frac{r^2 N_\alpha'}{Ebd} \tag{4-7}$$

$$D_{20}^R = -\frac{12 r^2 e N_\alpha'}{Ebd^3} \tag{4-8}$$

where N_α' is taken as negative and e as positive in the direction shown on Fig. 4-6a.

Corrections Figure 4-10 shows the two correction forces needed, designated X_1 and X_2. The dome displacements due to X_1 and X_2 have been given in Chap. 3, and the ring displacements come directly from Eqs. (4-1) to (4-3)

Consider first the force X_1 which is applied at y_0. The three ring equations are written (with X_1 positive inward):

$$\Delta_H = \frac{r^2}{A_R E} X_1 \tag{c}$$

$$\Delta_H = \frac{r^2 y_0^2}{EI_R} X_1 \tag{d}$$

$$\Delta_\alpha = -\frac{r^2 y_0}{EI_R} X_1 \tag{e}$$

Combining Eqs. (c) and (d) and setting $X_1 = 1$, we get

$$D_{11}^R = \left(\frac{1}{A_R} + \frac{y_0^2}{I_R}\right) \frac{r^2}{E} \tag{4-9}$$

and from (e), with $X_1 = 1$,

$$D_{21}^R = -\frac{r^2 y_0}{EI_R} \tag{4-10}$$

Now let us consider the force X_2 alone and write the basic ring equations again. Equation (4-1) does not apply and (4-2) and (4-3) become

$$\Delta_H = -\frac{r^2 y_0}{EI_R} X_2$$

$$\Delta_\alpha = \frac{r^2}{EI_R} X_2$$

which, for $X_2 = 1$, become

$$D_{12}^R = -\frac{r^2 y_0}{EI_R} = D_{21}^R \tag{4-11}$$

$$D_{22}^R = \frac{r^2}{EI_R} \tag{4-12}$$

These displacements are now summarized for the rectangular cross section:

$$D_{11}^R = \left(1 + \frac{12y_0^2}{d^2}\right) \frac{r^2}{Ebd} \tag{4-13}$$

$$D_{12}^R = -\frac{12r^2y_0}{Ebd^3} \tag{4-14}$$

$$D_{22}^R = \frac{12r^2}{Ebd^3} \tag{4-15}$$

These values are now combined directly with the corresponding dome displacements so that

$$\begin{aligned}
D_{11} &= D_{11}^D + D_{11}^R \\
D_{12} &= D_{12}^D + D_{12}^R \\
D_{22} &= D_{22}^D + D_{22}^R
\end{aligned} \tag{4-16}$$

Compatibility The standard compatibility equations are written and solved for X_1 and X_2.

It should be noted that from the membrane theory the ring is in tension and the dome usually is entirely in compression (see Chap. 3). The membrane ring tension, $T = N'_\alpha r \cos \alpha$, forces the ring outward. Compatibility requires that the dome follow this movement. The ring will not move as far as the membrane theory predicts (Δ_H^R) because of the restraint of the dome. Thus the ring force is reduced (by X_1) and dome hoop tension occurs along with some bending moments. As the ring size (stiffness) increases, it "keeps" more of the tension, and in the limit case where $A_R = \infty$ the system becomes a fixed dome with the small correction forces and moments noted already. In the other direction as the ring size (stiffness) decreases, it "releases" more of the tension until in the limit case where $A_R = 0$, $X_1 = N'_\alpha \cos \alpha$, and $X_2 = 0$; the system becomes a hinged free-sliding dome on vertical supports and the dome hoop tension near the edge is very large.

4-4 RING-PRESTRESSING ANALYSIS

In the dome-ring system, the ring acts as a circular tension tie and may logically be prestressed. Wire-wrapping devices, as illustrated in Fig. 4-7, make ring prestressing practical. [4]

The analysis for ring prestressing follows the same line as in the preceding section. The problem is to consider the same dome-ring system but under a different loading.

Primary System (Fig. 4-6a) The only force on the system is the uniform horizontal component of the circumferential prestressing force F:

$$H_F = \frac{F}{r + b}$$

Fig. 4-7 Merry-go-round wire winder stressing 0.192 wire to 150,000 psi initial stress. (Courtesy of the Preload Company, Inc.)

Errors Of the four possible errors, only the two ring displacements occur. From Eqs. (4-4) and (4-5) they are, where H_F and e_F are taken positive as shown in Fig. 4-9a,

$$D^R_{10} = \left(\frac{1}{A_R} + \frac{y_0 e_F}{I_R} \right) \frac{r^2}{E} H_F \qquad (a)$$

$$D^R_{20} = -\frac{r^2 e_F}{E I_R} H_F \qquad (b)$$

Eccentricity to the prestressing force is unusual so Eq. (b) normally does not apply and Eq. (a) reduces to

$$D^R_{10} = \frac{r^2 H_F}{E A_R} \qquad (4\text{-}17)$$

These general equations are given because some eccentricity may occur during the prestressing operation. But for most analyses Eq. (4-17) is sufficient.

Corrections and Compatibility These are the same as in the previous section. The numerical values for all the displacements D_{11}, D_{12}, and D_{22} are the same; only the numerical values for D_{10} and D_{20} are different. The solution of the compatibility equations will result in new values for X_1 and X_2. The action of the dome-ring system will be the same as before. A very stiff ring "keeps" the compression caused by the prestressing, whereas a flexible ring releases compression to the dome edge.

4-5   DOME-WALL ANALYSIS

It is possible to resist the horizontal component of the membrane meridional thrust by building the dome into a cylindrical wall. All values for this analysis have been derived in Chaps. 2 and 3, and a summary will be given here.

Primary System Figure 4-8 illustrates the system in which the dome is considered supported fully by the wall. Under dome load the wall has a continuous edge force N'_α which is resolved into its two components:

$$H = N'_\alpha \cos \alpha$$
$$V = N'_\alpha \sin \alpha$$

Errors The dome edge will translate, D_{10}^D, and rotate, D_{20}^D. The vertical force $N'_\alpha \sin \alpha$ normally causes no errors in the wall, but the horizontal force $N'_\alpha \cos \alpha$ causes the wall to move outward and rotate, or from Chap. 2,

$$D_{11}^W = \frac{1}{2\beta^3 D}$$

$$D_{21}^W = -\frac{1}{2\beta^2 D}$$

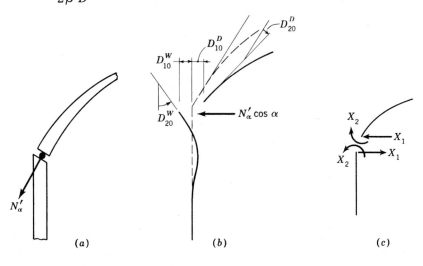

(a)   (b)   (c)

Fig. 4-8   Dome-wall analysis.

For the edge load $N'_\alpha \cos \alpha$ these values become

$$D^W_{10} = \frac{N'_\alpha \cos \alpha}{2\beta^3 D} \tag{4-18}$$

$$D^W_{20} = -\frac{N'_\alpha \cos \alpha}{2\beta^2 D} \tag{4-19}$$

Corrections   Displacements due to correction force $X_1 = 1$ and correction moment $X_2 = 1$ are now determined by adding the values already derived for the dome and wall.

$$\begin{aligned} D_{11} &= D^D_{11} + D^W_{11} \\ D_{12} &= D^D_{12} + D^W_{12} \\ D_{22} &= D^D_{22} + D^W_{22} \end{aligned} \tag{4-20}$$

Compatibility   The two compatibility equations are again written in the same form as before and the values of X_1 and X_2 are determined. It will be noted that the cylinder replaces the ring by absorbing a large tension force concentrated near the cylinder edge. It should also be noted that where the dome angle $\alpha = 90°$, the dome displacements D^D_{11}, D^D_{12}, and D^D_{22} will exactly equal the corresponding wall displacements, because our approximate bending theory of domes is based on the assumption that near the edge a dome of radius a acts like a cylinder of radius a.

The same procedure is applied where other loads are considered. The most common are wall loads due to internal or external pressure. In the case of water pressure, where the water level inside the cylinder is at the point of dome-wall junction, there will be no D^W_{10} error and only a rotation

$$D^W_{20} = -\frac{\gamma r^2}{E h_W}$$

The sign is reversed from that given by Eq. (2-23b) because the origin is taken at the top of the wall, and positive rotation corresponds to the positive direction of X_2 as shown in Fig. 4-8.

Likewise if the external pressure is caused by wall prestressing which equals zero at the dome-wall junction, only a rotation will exist.

$$D^W_{20} = \frac{\rho_F r^2}{E h_W}$$

where ρ_F is the external pressure due to prestressing. It is usually made somewhat larger than the water density, γ, to provide a slight residual wall compression under full water load.

As in Chap. 2, the analysis is based on the assumption that the cylinder is infinitely long and that therefore the effect of restraint at the cylinder bottom is not felt at the top. Where the cylinder is short, this assumption may lead to serious errors in the results, and the general solution given by Eq. (2-32) should be used.

4-6 DOME-RING-WALL ANALYSIS

In the preceding section the wall replaced the ring, and high tensile stresses were thereby developed in the edge region of the cylinder. This high tension often requires heavy reinforcement and an increase in wall thickness. In liquid-retaining structures it is not desirable to have large regions of high ring tension because of the possibility that cracks and leaks may develop.

Usually it is more desirable to "collect" the tension forces back up into a ring. As we have seen with the dome-ring problem, an increase in ring size—starting from the limiting case of no ring—increases the ring tension and correspondingly decreases the dome hoop tension. Figure 4-9, which includes the wall in the system, illustrates this action. Such a ring can be more easily prestressed or reinforced than the lengths of dome and wall which would otherwise carry the tension. The general method of analysis will now be given.

Primary System   Figure 4-9a illustrates the system in which all stress resultants for the dome and the wall are determined by the membrane theory.

Errors   There will now be eight errors (see Table 4-1); namely, translation and rotation of the dome, the top of the ring, the bottom of the ring, and the wall (D_{10}^D, D_{20}^D, D_{10}^R, D_{20}^R, D_{30}^R, D_{40}^R, D_{30}^W, and D_{40}^W, respectively). Figure 4-9 illustrates this system and shows that, in addition to all the previous displacements, there may be an additional ring rotation if the center line of the wall does not intersect the centroid of the ring cross section. Equation (4-8) must be modified:

$$\Delta_\alpha^R = -\frac{r^2 e}{EI_R} N_\alpha' - \frac{r^2 e_w}{EI_R} N_\alpha' \sin \alpha$$

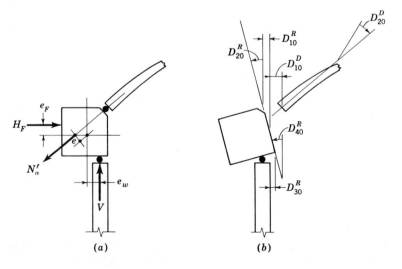

Fig. 4-9 Dome-ring-wall analysis.

Table 4-1 DISPLACEMENTS FOR DOME-RING-WALL ANALYSIS (ERRORS)
(Rectangular ring cross section)

Load		Load on dome	Ring prestressing	Pressure on wall
Dome	$D_{10}{}^D$	See Sec. 3-4	0	0
	$D_{20}{}^D$		0	0
Ring top	$D_{10}{}^R$	$\left(\cos\alpha + \dfrac{12y_0 e_t}{d^2}\right)\dfrac{r^2}{Ebd}N'_\alpha$	$\left(1 + \dfrac{12y_0 e_F}{d^2}\right)\dfrac{r^2}{Ebd}H_F$	0
	$D_{20}{}^R$	$-\dfrac{12r^2 e_t}{Ebd^3}N'_\alpha$	$-\dfrac{12r^2 e_F}{Ebd^3}H_F$	0
Ring bottom	$D_{30}{}^R$	$\left(-\cos\alpha + \dfrac{6e_t}{d}\right)\dfrac{r^2}{Ebd}N'_\alpha$	$-\left(1 - \dfrac{6e_F}{d}\right)\dfrac{r^2}{Ebd}H_F$	0
	$D_{40}{}^R$	$+\dfrac{12r^2 e_t}{Ebd^3}N'_\alpha$	$\dfrac{12r^2 e_F}{Ebd^3}H_F$	0
Wall	$D_{30}{}^W$	0	0	See Sec.
	$D_{40}{}^W$	0	0	2-5

or, as shown in Table 4-1, where $e_t = e + e_W \sin\alpha$,

$$\Delta_\alpha^R = -(e + e_w \sin\alpha)\frac{r^2 N'_\alpha}{EI_R} = D_{20}^R = -D_{40}^R \qquad (4\text{-}21)$$

Corrections   There will now be four corrections: a force X_1 and a moment X_2, which correspond to the required dome-ring values; and a force X_3 and a moment X_4, which correspond to a ring-wall analysis (which is the same as a dome-ring analysis in which the dome is considered to have the radius of the cylinder).

It is now a relatively simple matter to express all the correction displacements due to the forces shown in Fig. 4-10. Consider first the horizontal displacement at the junction of the dome and ring. Due to the force X_1 we obtain, as before,

$$D_{11} = D_{11}^D + D_{11}^R$$

From the force X_2 again comes

$$D_{12} = D_{12}^D + D_{12}^R$$

and from the ring-wall forces X_3 and X_4 come the displacements:

$$D_{13} = D_{13}^R$$

and

$$D_{14} = D_{14}^R$$

The displacements D_{13}^R and D_{14}^R are easily derived from the basic ring Eqs. (4-1) to (4-3). To solve for D_{13}^R, for example, refer to Fig. 4-11a and replace X_3 at the bottom

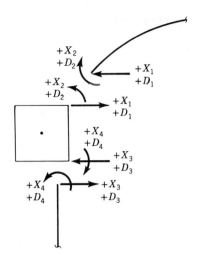

Fig. 4-10 Dome-ring-wall forces showing positive directions for the forces X and the displacements D.

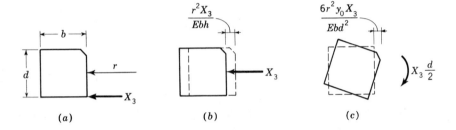

Fig. 4-11 Ring displacements.

edge by a force X_3 at the ring centroid and a moment $X_3 d/2$. The total horizontal displacement at the junction of the dome and ring will be composed of two factors (see Fig. 4-11b and c) given by Eqs. (4-1) and (4-2).

$$\Delta_H = -\frac{r^2 X_3}{E A_R} + \frac{r^2 y_0(d/2) X_3}{E I_R}$$

$$D_{13} = -\left(\frac{1}{A_R} - \frac{y_0 d/2}{I_R}\right) \frac{r^2}{E} \tag{4-22}$$

The horizontal displacement due to X_4 comes directly from Eq. (4-2) with $M_\alpha = X_4 = 1$.

$$D_{14} = \frac{y_0 r^2}{E I_R} \tag{4-23}$$

The rotation of the ring at this junction due to X_3 is derived from Eq. (4-3).

$$D_{23} = -\frac{r^2 d/2}{E I_R} \tag{4-24}$$

and the rotation due to X_4 will be

$$D_{24} = -\frac{r^2}{EI_R} \tag{4-25}$$

In a similar manner the displacements at the ring-wall junction can be derived.

$$D_{31} = D_{13} = -\left(\frac{1}{A_R} - \frac{y_0 d/2}{I_R}\right)\frac{r^2}{E}$$

$$D_{32} = D_{23} = -\frac{r^2 d/2}{EI_R}$$

$$D_{33} = D_{33}^R + D_{33}^W \tag{4-26}$$

where D_{33}^R and D_{34}^R come from Eqs. (4-9) and (4-10) with $y_0 = -d/2$, and D_{33}^W and D_{34}^W have been given previously as D_{11}^W and D_{12}^W in the dome-wall analysis.

$$D_{34} = D_{34}^R + D_{34}^W \tag{4-27}$$

$$D_{41} = D_{14} = \frac{y_0 r^2}{EI_R}$$

$$D_{42} = D_{24} = -\frac{r^2}{EI_R}$$

$$D_{43} = D_{34} \tag{4-28}$$

$$D_{44} = D_{44}^R + D_{44}^W \tag{4-29}$$

where $D_{44}^R = D_{22}^R$ and D_{44}^W was previously given as D_{22}^W in the dome-wall analysis.

Compatibility.   There will now be four simultaneous compatibility equations to solve for the four corrections X_1, X_2, X_3, and X_4. In order to avoid major mistakes, it is well to check the results by comparison with simpler analyses results (see Table 4-8).

4-7   DESIGN CONSIDERATIONS FOR PRESTRESSED CONCRETE TANKS

Table 4-2 presents typical dimensions for a wide range of prestressed concrete tanks.[4] Figure 4-12 gives, in more detail, the dimensions of a 1-million-gal prestressed concrete water tank to be built above the grade. This tank will serve as the basis for three separate examples: (1) It will be assumed that the ring is free to rotate and translate on the wall—the dome-ring problem, (2) the system will be considered just as it is shown in the figure—the dome-ring-wall problem, and (3) a finite-element analysis will be made for the dome-ring-wall problem. The idealized junctions of dome and ring and wall are shown in Fig. 4-13.

4-8   DOME-RING PROBLEM

In Fig. 4-12 the dome varies in thickness from 2½ to 6 in. over an 8-ft length at the edge. How shall we compute edge movements and forces in such a case? A solution of the basic dome equations in Chap. 3 can be obtained by numerical integra-

Table 4-2

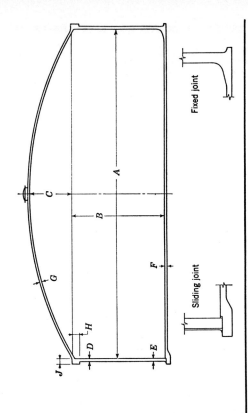

Fixed joint

Sliding joint

Typical dimensions

Tank capacity, U.S. gal	Interior diameter A	Liquid depth B	Rise of dome C	Wall thickness at top D	Wall thickness at bottom E	Floor thickness F	Dome-shell thickness G	Dome-ring height H	Dome-ring thickness J
100,000	41'0"	10'3"	5'1½"	4¾"	4¾"	2"	2"	8"	5¾"
250,000	55'6"	14'0"	6'11"	4¾"	4¾"	2"	2"	8½"	6"
500,000	70'0"	17'6"	8'9"	4¾"	4¾"	2"	2"	12"	6¾"
750,000	80'0"	20'0"	10'0"	4¾"	6"	2"	2"	14"	7½"
1,000,000	88'6"	22'0"	11'1½"	4¾"	7"	2"	2"	15"	8½"
1,500,000	101'0"	25'6"	12'8"	4¾"	9"	2"	2"	17"	9¾"
2,000,000	111'0"	28'0"	13'10½"	4¾"	9½"	2"	2½"	19"	10¾"
2,500,000	119'6"	30'0"	14'11"	8¾"	10¼"	2"	2½"	20"	12"
5,000,000	151'0"	37'6"	18'10½"	8¾"	17½"	2"	4"	27"	15¼"
10,000,000	190'0"	47'6"	23'9"	8¾"	29"	2"	4½"	35"	19¼"

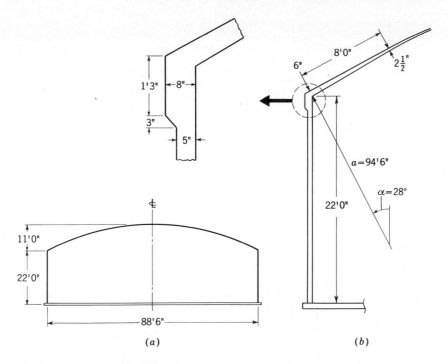

Fig. 4-12 Domed tank design.

tion, but a good approximation[5] can be obtained by assuming the thickness to be constant and equal to the actual thickness at a certain distance from the edge:

$$S_a = 0.5 \sqrt{ah_a} \qquad (4\text{-}30)$$

where S_a = arc distance from dome edge
$\quad a$ = radius of dome
$\quad h_a$ = average thickness of thickened region

This approximation includes the assumption that the thickened region of the dome extends a distance from the edge of

$$S = 2\sqrt{ah_a} \qquad (4\text{-}31)$$

For this example, with $h_a = 4.25$ in., Eqs. (4-30) and (4-31) would give $S_a = 2.9$ ft and $S = 11.6$ ft, and the thickness at S_a would be 5.12 in. In this design lower values were used: the actual thickened region was made only 8 ft, and an average thickness of 4 in. was used in the computation for edge effects.

The ACI Committee 344 report on the "Design and Construction of Circular Prestressed Concrete Structures" suggests wall thickness at $S_a = 0.4\sqrt{ah_e}$ be used as the equivalent edge thickness when h varies, where h_e is the actual edge thickness. Thus in our case, with $h_e = 6$ in., $S_a = 2.75$ ft. at which point $h =$

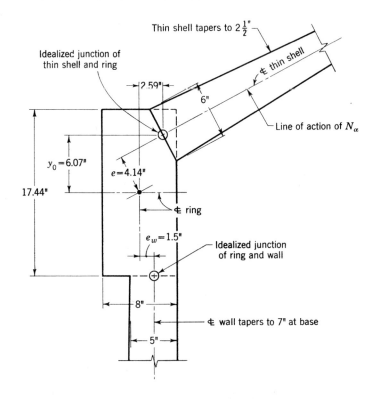

Fig. 4-13 Idealization of dome-ring-wall junctions.

2.5 + (5.25/8.00) 3.5 = 4.8 in.[6] The use of computer-based numerical analyses can provide a more complete answer to the question of how shell thickness variations influence behavior. This question is further discussed in Sec. 4-11.

   If the larger thickness had been assumed, the dome would have been "stiffer" and hence would have taken more of the horizontal thrust by hoop tension. The ring, being then less "stiff" by comparison, would have less tension. The comparison is shown in Table 4-3. Forces over the dome, plotted in Fig. 4-14, are obtained as in Table 3-4 by combining membrane values with those due to X_1 and X_2 given in Table 4-3. Note that as the ring size decreases, the dome hoop tension values $(N_\theta)_\alpha$ increase substantially. The values for ring displacements are based on $e = 4.26$ in., a slight discrepancy from the value shown in Fig. 4-13; the discrepancy has a negligible effect on the design.

   Table 4-4 gives the analysis for gravity loads and ring prestressing. Since there is some uncertainty concerning the value of h to choose, the ring-prestressing solution is given for both assumptions $h = 4$ in. and $h = 5.12$ in. The result is that a design based on 4 in. and zero hoop tension under full load gives a prestressing force ($H_F = 3.4$ kips/ft) 5 percent larger than the force required for a design based upon 5.12 in.

Table 4-3 DOME-RING ANALYSIS (Load uniformly distributed over the dome surface)

	Units	Fixed dome (ring stiff- ness ∞)	$h = 4$ in.	$h = 5.12$ in.	Free dome (ring stiff- ness = 0)
1. Primary system		Dome as a membrane			
		No ring	Ring free	Ring free	No ring
2. Errors					
$ED_{10}{}^D$		$-3,300q$	$-3,300q$	$-2,570q$	
$ED_{10}{}^R$		0	$-193,000q$	$-193,000q$	
ED_{10}		$-3,300q$	$-196,300q$	$-195,570q$	
$ED_{20}{}^D$		$+295q$	$+295q$	$+230q$	
$ED_{20}{}^R$		0	$+204,000q$	$+204,000q$	
ED_{20}		$+295q$	$+204,295q$	$+204,230q$	
3. Corrections					
$ED_{11}{}^D$		$+2,740$	$+2,740$	$+1,885$	
$ED_{11}{}^R$		0	$+4,961$	$+4,961$	
ED_{11}		$+2,740$	$+7,701$	$+6,846$	
$ED_{12}{}^D$		$+1,360$	$+1,360$	$+823$	
$ED_{12}{}^R$		0	$-5,812$	$-5,812$	
ED_{12}		$+1,360$	$-4,452$	$-4,989$	
$ED_{22}{}^D$		$+1,350$	$+1,350$	$+718$	
$ED_{22}{}^R$		0	$+11,489$	$+11,489$	
ED_{22}		$+1,350$	$+12,839$	$+12,207$	
4. Solution					
X_1	kips/ft	$+2.63q$	$+20.47q$	$+23.35q$	$+44.15q*$
X_2	ft-k/ft	$-2.87q$	$-8.68q$	$-7.2q$	0
$H_R = N_\alpha \cos \alpha + X_1$	kips/ft	$+41.52q$	$+23.68q$	$+20.80q$	0
$(N_\theta)_\alpha$ $(q = 0.09)$	kips/ft	-0.78	$+26.8$	$+33.4$	$+81.5$
$M_\alpha = X_2$ $(q = 0.09)$	ft-k/ft	-0.26	-0.78	-0.65	0

$* X_1 = -N'_\alpha \cos \alpha$

$[H_F = 3.4(33.4/35.2) = 3.22 \text{ kips/ft}]$, even though the full-load hoop tension for the 4-in. dome assumption is 25 percent less than that obtained from the 5.12-in. assumption. The conclusion is that if prestressing is used, the differences in assumptions tend to cancel out since prestressing is an opposite effect. Note that the bending moments are of the same sign. If there were no prestressing, it would be well to base the reinforcement on a conservative interpretation of these analyses. It should be remembered that the results are highly localized and that any increase in reinforcement is limited to a small area of the dome.

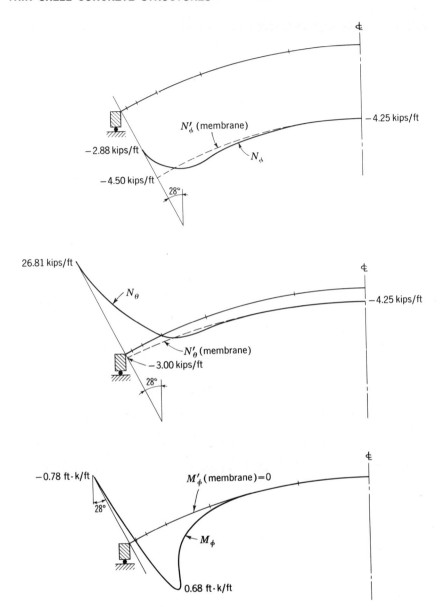

Fig. 4-14 Dome-ring system under dome loads.

Where prestressing is used, there is no object in designing a larger ring. The more flexible the ring, the more hoop tension in the dome under dome loads, but also the more counteracting hoop compression in the dome due to prestressing. The main criteria for the ring are that it not be overstressed in initial compression and that it provide sufficient vertical surface for the placement of the prestressing tendons.

Table 4-4 DOME-RING ANALYSIS

	Units	Dome D.L.	Dome T.L.	Ring pre-stressing	Ring pre-stressing
1. Primary system		$q = 1.0$ ksf $h = 4$ in.	$q = 1.0$ ksf $h = 4$ in.	$H_F = 1.0$ kips/ft $h = 4$ in.	$H_F = 1.0$ kips/ft $h = 5.12$ in.
2. Errors					
$ED_{10}{}^D$		$-3,300$	$-3,300$	0	0
$ED_{10}{}^K$		$-193,000$	$-193,000$	$+2,021$	$+2,021$
ED_{10}		$-196,300$	$-196,300$	$+2,021$	$+2,021$
$ED_{20}{}^D$		$+295$	$+295$	0	0
$ED_{20}{}^K$		$+204,000$	$+204,000$	0	0
ED_{20}		$+204,295$	$+204,295$	0	0
3. Corrections					
ED_{11}		$+7,701$	$+7,701$	$+7,701$	$+6,846$
ED_{12}		$-4,452$	$-4,452$	$-4,452$	$-4,989$
ED_{22}		$+12,839$	$+12,839$	$+12,839$	$+12,207$
4. Solution					
X_1	kips/ft	$+20.47$	$+20.47$	-0.326	-0.420
X_2	ft-k/ft	-8.68	-8.68	-0.114	-0.172
H_R	kips/ft	$+23.68$	$+23.68$	-0.674	-0.580
Actual load		$q = 0.04$ ksf	$q = 0.09$ ksf	$H_F = 3.4$ kips/ft	$H_F = 3.40$ kips/ft
X_1	kips/ft	$+0.82$	$+1.84$	-1.11	-1.43
$X_2 = M_\alpha$	ft-k/ft	-0.35	-0.78	-0.39	-0.59
H_R	kips/ft	$+0.95$	$+2.14$	-2.29	-1.97
$(N_\theta)_\alpha$	kips/ft	$+11.85$	$+26.8$	-26.8	-35.2
	psi	$+165$	$+372$	-372	-490

4-9 DOME-RING-WALL PROBLEM

The primary system is shown in Fig. 4-9b; the errors are given in Table 4-5 for three loadings: uniform surface load on the dome, temperature differential between dome and ring, and prestressing of ring. The wall varies in thickness from 5 in. at the top to 7 in. at the base to account for the increasing water pressure. As discussed in Sec. 2-6, little error arises by using the edge thickness of 5 in. in the analysis for bending. Table 4-6 includes the correction displacements, and Table 4-7 gives the results. The stress resultants and moments are plotted in Fig. 4-15.

This analysis is straightforward, complicated only by the necessity for the solution of four simultaneous equations.

It is interesting to see what differences would occur if we simply took no account of the ring. For this purpose some of the results are compared in Table 4-8 with those

Table 4-5 DOME-RING-WALL ANALYSIS

	Dome load, $q = 1$ kip/ft^2	Temperature differential, $+10°$ between dome and ring	Ring prestressing, $H_F = 1.0$ kip/ft
1. Primary system	Dome as a membrane, ring free, wall with a free top		
2. Errors			
$ED_{10}{}^D$	$-3,300$	$+1,145$	0
$ED_{10}{}^R$	$-209,780$	0	$+2,021$
$ED_{10}{}^W$	0	0	0
ED_{10}	$-213,080$	$+1,145$	$+2.021$
$ED_{20}{}^D$	$+295$		0
$ED_{20}{}^R$	$+237,000$		
$ED_{20}{}^W$	0		
ED_{20}	$+237,295$	0	0
$ED_{30}{}^D$	0		0
$ED_{30}{}^R$	$-83,400$		-2.021
$ED_{30}{}^W$	0		0
ED_{30}	$-83,400$	0	-2.021
$ED_{40}{}^D$	0		
$ED_{40}{}^R$	$-237,000$		
$ED_{40}{}^W$	0		
ED_{40}	$-237,000$	0	0

of a dome-wall analysis. The results are close in every case except wall moment, which is substantially reduced by the ring. The effect of the ring collecting tension from the dome and wall is observed. The maximum wall tension is reduced by 35 percent; maximum dome tension by 18 percent. This ring is, however, only 3 in. wider than the wall; and, as might be expected, it does not exert a dominating influence on the system. In fact, where 3.89 kips/ft of radial prestressing force is applied to the ring, only 1.02 kips stays in the ring, with the remainder being transferred to the dome and the wall. A bigger ring would, of course, collect more gravity-load tension and also keep a correspondingly greater prestressing compression.

If only the dome-wall analysis were made, the prestressing force required for zero tension would be 3.2 kips/ft. But if such a force were included in the more exact

Table 4-6 DOME-RING-WALL ANALYSIS CORRECTION DISPLACEMENTS

	ED_{11}	ED_{12}	ED_{13}	ED_{14}	ED_{22}	ED_{23}	ED_{24}	ED_{33}	ED_{34}	ED_{44}
Dome	+2.740	+1.360	0	0	+1.350	0	0	0	0	0
Ring	+4.961	−5.812	+2.201	+5.812	+11,489	−8,343	−11,489	+8,084	+8,343	+11,489
Wall	0	0	0	0	0	0	0	+2,860	−872	+530
Total	+7.701	−4.452	+2.201	+5.812	+12,839	−8,343	−11,489	+10,944	+7,471	+12,019

Table 4-7 DOME-RING-WALL ANALYSIS, SOLUTION, AND RESULTING FORCES AND MOMENTS

	Dome D.L., $q = 1.0$ ksf	Dome T.L., $q = 1.0$ ksf	Temperature variation, $\Delta t = +10^0$	Ring prestressing, $H_F = 1.0$ kips/ft
X_1	+23.19	+23.19	−0.2770	−0.425
X_2	−26.21	−26.21	+0.1440	+0.460
X_3	−9.93	−9.93	−0.0353	+0.313
X_4	−10.41	−10.41	+0.2940	+0.450
	$q = 0.04$ ksf	$q = 0.09$ ksf	$\pm 30^0$	$H_F = 3.89$
X_1	+0.925	+2.080	∓0.830	−1.65
$X_2 = M_\alpha$	−1.050	−2.360	±0.432	+1.79
X_3	−0.397	−0.894	∓0.106	+1.22
$X_4 = M_{y=0}$	−0.416	−0.936	±0.882	+1.75
Dome				
N'_θ	−1.33	−3.00	0	0
N_{θ_1} $(20.6X_1)$	+19.05	+42.86	∓17.10	−34.0
N_{θ_2} $(10.3X_2)$	−10.80	−24.30	±4.45	+18.45
N_θ	+6.92	+15.56	∓12.65	−15.55
Ring				
$H' = N'_\alpha \cos \alpha$	+1.77	+3.98	0	−3.89
H_1 $(-X_1)$	−0.93	−2.08	±0.43	+1.65
H_3 (X_3)	−0.40	−0.89	∓0.11	+1.22
H	+0.44	+1.01	±0.32	−1.02
Wall				
N'_θ	0	0	0	0
N_{θ_3} $(-26.7X_3)$	+10.60	+23.80	±28.30	−32.5
N_{θ_4} $(+8.18X_4)$	−3.40	−7.65	±7.21	+14.3
N_θ	+7.20	+16.15	±35.51	−18.2

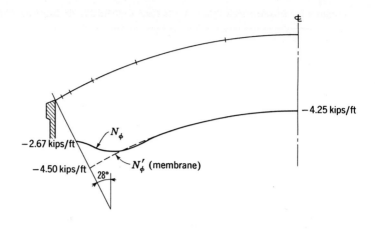

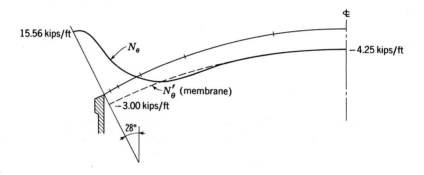

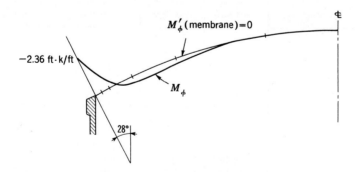

Fig. 4-15 (a) Dome-ring-wall system under dome loads.

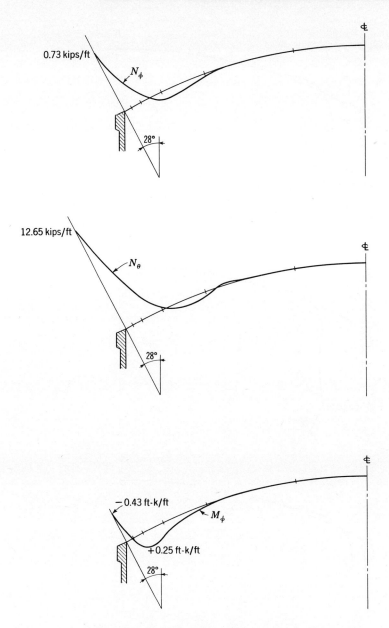

Fig. 4-15. (*b*) **Dome-ring-wall system under temperature variation between dome and wall of −30°.**

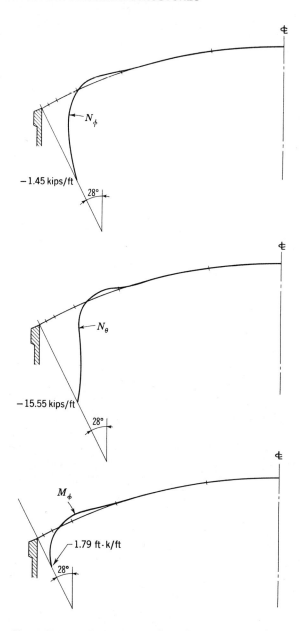

Fig. 4-15 (c) Dome-ring-wall system under ring prestressing.

Table 4-8 COMPARISON OF ANALYSES

	Dome ring	Dome wall*	Dome-ring wall
Dome load, $q = 0.09$ kip/ft			
Dome			
N_θ kips/ft	+26.81	+19.00	+15.56
M_α ft-k/ft	—0.78	−2.46	−2.36
Ring			
H kips/ft	+2.14		+1.01
Wall			
N_θ kips/ft		+24.30	+16.15
M_y ft-k/ft		−2.46	−0.94
Temperature, $+30°$			
Dome			
N_θ		−11.14	−12.65
M_α		+0.18	+0.43
Ring			
H			+0.32
Wall			
N_θ		+18.27	+35.51
M_y		+0.18	+0.88
Prestressing force, H_F kips/ft			
	3.4	3.2	3.89
Dome			
N_θ	−26.81	−19.5	−15.55
M_α	−0.39	+1.65	+1.79
Ring			
H_F	−2.29		−1.02
Wall			
N_θ		−24.30	−18.20
M_y		+1.65	+1.75

* Results from Table 4-5 in first edition.

analysis, as shown in Table 4-8, some tension would remain, since it takes a higher prestress to offset the shell tensions in the dome-ring-wall system.

Two other factors are important to observe in this complete system: (1) as in any prestressed structure, initial conditions must be investigated, and (2) volume changes will occur. Both of these conditions involve considerable uncertainties, and liberal assumptions are usually made. In this example a dead load of 40 psf is considered present when prestressing is applied. Consider the dome forces and moments first due to initial prestressing, 25 percent greater than the final, and dead load only

$$N_\theta = +6.92 - 1.25(15.6) = -12.58 \text{ kips/ft}$$

which for a 4-in. dome thickness corresponds to a stress of -262 psi. The thickness at the edge is really 6 in., so the stresses are less. The wall stress will be about the same.

The dome-edge moment initially will be

$$M_\alpha = +1.79(1.25) - 1.05 = +1.19 \text{ ft-k/ft}$$

It should be remembered that these values (with some allowance for prestressing loss) are the normal permanent forces and moments on the structure, while full dome load and volume changes will occur only at certain times. If a volume change corresponding to 30° temperature drop occurs between the dome and the ring (representing the top of the wall), then a ring compression will occur in the wall and a hoop tension will occur in the dome. Such forces may well exist because of shrinkage of the dome. Volume changes will also occur due to creep under prestressing, but the stresses from such strains will often be substantially reduced by plastic flow. Notice also that a volume change corresponding to a 30° rise in temperature causes substantial ring tension in the wall. Since it is hardly likely that this condition will occur during a live load (snow) on the dome, some of the prestressing compression will be available for reducing these tensions.

Often temperature differentials will occur between the dome and wall. The dome, for example, receives the main brunt of the sun's heat, while the wall temperature is stabilized by the liquid stored inside. Hence in summer the wall will be at a lower temperature than the dome; in winter the reverse is likely.

All of these observations should alert the designer to the needs of the structure and to a rational interpretation of the analysis. It is not intended that specific values given here be applied on such a system. It is, however, recommended that the three effects—dome surface load, volume change, and prestressing—be investigated by analysis. The designer must then determine what actual values to take for gravity loads, volume change strains, and prestressing forces.

4-10 REINFORCEMENT FOR THE DOME-RING-WALL SYSTEM

The dimensioning of the reinforcement may be done in the following steps:

1. *Dome hoop reinforcement* is provided to take all the hoop tension. For values of ψ between 0° and 1° (an arc length of 1.65 ft) the total tensile force (from Fig. 4-15) is
Dome load:

$$T_a = \frac{15.56 + 14.86}{2}(1.65) = 25.1 \text{ kips}$$

Temperature drop:

$$T_b = \frac{12.65 + 9.05}{2} \,(1.65) = 17.9 \text{ kips}$$

Final ring prestressing:

$$T_c = -\,\frac{15.55 + 14.56}{2} \,(1.65) = -24.8 \text{ kips}$$

$$T_a + T_b + T_c = 18.2 \text{ kips}$$

$$A_s = \frac{18.2}{20} = 0.91 \text{ in.}^2$$

for which we provide five No. 4 bars. Note that the prestress nearly cancels the dome-load stresses, and we supply reinforcement mainly for temperature (or shrinkage) effects. In a similar manner we obtain for the regions

$$1° \text{ to } 2° \qquad A_s = 0.56 \text{ in.}^2$$
$$2° \text{ to } 5° \qquad A_s = 0.35 \text{ in.}^2$$

for which we provide three No. 4 bars and two No. 4 bars, respectively (see Fig. 4-16). Past 5° there is no longer any appreciable tension due to the edge effects.

2. *Minimum dome reinforcement* will be supplied throughout the shell equal to at least 0.003bh as specified by ACI 344 (Ref. 6). For the 2½-in. shell, about 0.09 in.²/ft in each direction would be required, and we choose a welded wire fabric 4 × 4 − 6/6 which provides 0.087 in.²/ft in each of two directions.

3. *Dome meridional bending reinforcement* will be provided to resist the combined effects of N_ϕ and M_ϕ. Let us consider first the section at the dome edge where $\psi = 0°$. From Fig. 4-15 we find

Dome dead load (assumed to be 40 psf):

$$M_\phi = -1.05 \text{ ft-k/ft}$$
$$N_\phi = -1.19 \text{ kips/ft}$$

Dome live load (assumed to be 50 psf):

$$M_\phi = -1.31 \text{ ft-k/ft}$$
$$N_\phi = -1.48 \text{ kips/ft}$$

Temperature drop (assumed to be $-30°$F):

$$M_\phi = -0.43 \text{ ft-k/ft}$$
$$N_\phi = +0.73 \text{ kip/ft}$$

Final ring prestressing (assumed to be 3.89 kips/ft):

$$M_\phi = +1.79 \text{ ft-k/ft}$$
$$N_\phi = -1.45 \text{ kips/ft}$$

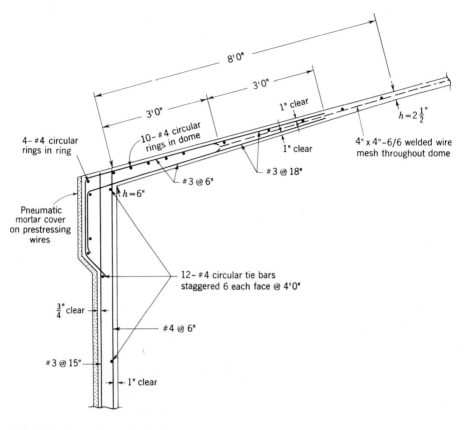

Fig. 4-16 Dome-ring-wall reinforcement.

If we neglect the temperature drop and consider initial prestressing, we obtain a maximum N_ϕ of -4.48 kips/ft, or an average stress of $-4480/(6 \times 12) = -62$ psi. We conservatively disregard this small compressive load in our computations for bending-moment reinforcement. *

We consider a 1-in. clearance for the steel and thus use

$$d = 6 - 1.5 = 4.5 \text{ in.}$$

Estimating $j = \frac{7}{8}$, we obtain for a full working-load moment of -1.00 ft-k/ft,

$$A_s = \frac{M}{f_s j d} = \frac{1.00 \times 12}{20 \times \frac{7}{8} \times 4.5} = 0.152 \text{ in.}^2/\text{ft}$$

Because the prestressing effects do not change significantly as overloads are

---

* This assumption was noted in Sec. 1901 (c) of ACI-318-63.

applied, we check for ultimate-load conditions.

$$M_D = -1.05 - 0.43 + 1.79 = +0.31 \text{ ft-k/ft}$$
$$M_L = -1.31 \text{ ft-k/ft}$$

and

$$M_u = +0.31(1.4) - 1.31(1.7) = -1.80 \text{ ft-k/ft}$$

With $f'_c = 4000$ psi and $f_y = 40{,}000$ psi,

$$M_u = \phi A_s f_y \left(d - \frac{a}{2} \right) = 1.90 \text{ ft-k/ft}$$

and

$$A_s f_y = 0.85 f'_c ba$$

from which

$$A_s = 0.14 \text{ in.}^2/\text{ft}$$

The initial conditions should also be checked where

Dome dead load:

$$M_\phi = -1.05 \text{ ft-k/ft}$$

Temperature rise:

$$M_\phi = +0.43 \text{ ft-k/ft}$$

Initial ring prestressing:

$$M_\phi = +1.25(1.79) = +2.24 \text{ ft-k/ft}$$

from which

$$M_\phi = +1.62 \text{ ft-k/ft}$$

By working-load analysis

$$A_s = \frac{1.62 \times 12}{20 \times \frac{7}{8} \times 4.5} = 0.246 \text{ in.}^2/\text{ft}$$

Since this total moment is temporary and most probably will not increase, the use of $f_s = 20{,}000$ psi is overly conservative. It is more logical to use ultimate-load analysis where

$$M_u = +1.62(1.4) = +2.37 \text{ ft-k/ft}$$

from which

$$a = 0.18 \text{ in.}$$

and

$$A_s = 0.19 \text{ in.}^2/\text{ft}$$

We may satisfy the bending requirements at $\psi = 0°$ by providing No. 3 bars at 8 in. at the top and No. 3 bars at 7 in. on the bottom of the shell.

At $\psi = 1°$ we find

Dome dead load:

$$M_\phi = -0.45 \text{ ft-k/ft}$$

Dome live load:

$$M_\phi = -0.57 \text{ ft-k/ft}$$

Temperature drop:

$$M_\phi = 0.04 \text{ ft-k/ft}$$

Final ring prestressing:

$$M_\phi = +0.73 \text{ ft-k/ft}$$

The depth of the section is

$$h = 6 - 3.5 \frac{1.65}{8.0} = 5.28 \text{ in.}$$

and we take $d = 3.75$ in.

The final condition gives

$$M_u = +0.28(1.4) - 0.57(1.7) = -0.60 \text{ ft-k/ft}$$

from which

$$A_s = 0.06 \text{ in.}^2/\text{ft}$$

and the initial condition

$$M_u = [+0.73(1.25) + 0.04 - 0.45] \ 1.4 = +0.71 \text{ ft-k/ft}$$

from which

$$A_s = 0.07 \text{ in.}^2/\text{ft}$$

We may use No. 3 at 18 in. in both cases.

At $\psi = 2°$ the moments are insignificant, and the minimum reinforcement would suffice.

We shall provide No. 3 bars at 6 in. top and bottom at $\psi = 0°$, stop two-thirds of the bars at 3 ft, and the remainder, No. 3 bars at 18 in., at 6 ft as shown in Fig. 4-16.

4. *Ring prestressing* is obtained by circular rings of tensioned steel with an assumed final stress of 120,000 psi. The total final tension force required is

$$T = H_F r = 3.89 \text{ kips/ft} \times 44.25 \text{ ft} = 172 \text{ kips}$$

and

$$A_s = 172/120 = 1.44 \text{ in.}^2$$

which may be supplied by thirty wires, each 0.25 in. in diameter. An initial stress of 150,000 psi is required to compensate for the assumed losses of 20 percent.

5. *Ring reinforcement* is needed to carry the tension resulting from temperature changes where

$$H = +0.32 \quad \text{and} \quad T = 0.32(44.25) = 14.2 \text{ kips}$$
$$A_s = 14.2/20 = 0.71 \text{ in.}^2$$

and we provide four No. 4 bars.

6. *Horizontal wall prestressing* is provided to counteract the tension due to liquid pressures.

The base of the wall is assumed to be hinged, and the analysis for water pressure follows the method illustrated in Sec. 2-8.

Primary System This is a free-sliding wall assumed to be free at the top.

Error From (2-23a)

$$D_{10} = -\frac{\gamma r^2 H}{Eh} = \frac{(0.0624)(44.25)^2(22)}{E(0.583)} = -\frac{4610}{E}$$

Correction From (2-36)

$$D_{11} = \frac{1}{2\beta^3 D}$$

where

$$\beta^4 = \frac{3(1 - \nu^2)}{r^2 h^2} = \frac{35/12}{(44.25)^2(0.583)^2} = 0.00439 \text{ ft}^{-4}$$

$$\beta^3 = 0.017$$

$$D = \frac{Eh^3}{12(1 - \nu^2)} = \frac{E(0.583)^3}{35/3} = 0.017E$$

so that

$$D_{11} = \frac{1}{2 \times 0.017 \times 0.017E} = \frac{1730}{E}$$

Compatibility

$$X_1 D_{11} + D_{10} = 0$$

$$X_1 = -\frac{D_{10}}{D_{11}} = \frac{4610}{1730} = 2.67 \text{ kips/ft}$$

The values for ring stress resultants and moments are now obtained by substituting $X_1 = -Q_0$ and $X_2 = M_0 = 0$ into expressions in Table 2-3 and thence into (2-35). The results are presented in Tables 4-9 and 4-10.

Table 4-9 WALL-RING STRESS RESULTANTS (In kips per foot)

Section, y/H	Wall loads Water (1)	Prestressing Initial* (2)	Prestressing Final (3)	Ring load prestressing Initial† (4)	Ring load prestressing Final (5)	Dome loads D.L. (6)	Dome loads T.L. (7)	Dome loads +30°‡ (8)	Max. tension Loadings§	Max. tension Tension	Wall prestressing, steel area in.²
0 (top)	0	0	0	−23.1	−18.5	7.3	16.4	10.1	13568	0	0.05
0.1	5.8	−7.9	−6.0	−15.0	−12.0	4.0	9.0	1.8	1357	0	0.11
0.2	12.6	−17.0	−13.0	−6.0	−4.8	1.3	2.9	−1.2	1357	0	0.16
0.3	19.5	−26.4	−20.2	−0.9	−0.7	0	0	−1.5	1356−8	+0.1	0.21
0.4	26.8	−36.2	−27.7	0.9	0.7	−0.3	−0.8	−0.8	1356−8	+0.3	0.27
0.5	34.2	−46.1	−35.3	0.9	0.7	−0.3	−0.6	−0.3	1356−8	0	0.32
0.6	40.5	−54.6	−41.9	0.5	0.4	−0.1	−0.3	0	1356	0	0.37
0.7	44.0	−59.5	−45.5	0.1	0.1	0	−0.1	0.1	13568	0	0.43
0.8	40.4	−54.5	−41.8	0	0	0	0	0.1	13568	0	0.48
0.9	25.6	−34.6	−26.4	0	0	0	0	0	1356	0	0.53
1.0 (base)	0	0	0	0	0	0	0	0			

* f_{si} = 150,000 psi, f_{sf} = 115,000 psi for wall prestressing.
† f_{si} = 150,000 psi, f_{sf} = 120,000 psi for ring prestressing.
‡ Dome +30° with respect to wall.
§ Values from the loadings in the column numbers indicated (−8 refers to a loading of −30°).

Table 4-10 WALL MOMENTS (In foot-kips per foot)

Section, y/H	Wall loads Water (1)	Prestressing Initial* (2)	Prestressing Final (3)	Ring load, prestressing Initial† (4)	Ring load, prestressing Final (5)	Dome loads D.L. (6)	Dome loads T.L. (7)	Dome loads +30°‡ (8)	Maximum moments Positive Loadings§	Maximum moments Positive Values	Maximum moments Negative Loadings§	Maximum moments Negative Values
0 (top)	0	0	0	+2.18	+1.75	-0.42	-0.94	+0.88	468	+2.64	57-8	-0.07
0.1	0	0	0	-0.03	-0.02	+0.12	+0.26	+0.75	568	+0.85	46-8	-0.66
0.2	0	0	0	-0.59	-0.47	+0.20	+0.45	+0.37	568	+0.10	46-8	-0.76
0.3	+0.13	-0.17	-0.13	-0.46	-0.37	+0.13	+0.30	+0.10	357	-	1246-8	-0.47
0.4	+0.06	-0.08	-0.06	-0.23	-0.18	+0.05	+0.12	-0.02	357	-	12468	-0.22
0.5	-0.27	+0.37	+0.28	-0.05	-0.04	+0.01	+0.02	-0.04	247-8	+0.38	13568	-0.06
0.6	-0.93	+1.26	+0.96	+0.01	+0.01	-0.01	-0.02	-0.03	246-8	+1.29	13568	0
0.7	-1.93	+2.61	+1.99	+0.03	+0.02	-0.01	-0.02	-0.01	246-8	+2.64	13568	
0.8	-3.14	+4.24	+3.24	+0.01	+0.01	0	-0.01	0	24	+4.25	1357	
0.9	-3.2	+4.33	+3.31	+0.01	+0.01	0	0	0	24	+4.34	135	
1.0 (base)	0	0	0	0	0	0	0	0		0		0

* f_{st} = 150,000 psi, f_{sf} = 115,000 psi for wall prestressing.

† f_{st} = 150,000 psi, f_{sf} = 120,000 psi for ring prestressing.

‡ Dome +30° with respect to wall.

§ Values from the loadings in the column numbers indicated (−8 refers to a loading of −30°).

The wall prestressing is obtained by wrapping wires around the shell. The intensity of the force is made to vary in a nearly linear manner from zero at the top to a maximum at the base. In this way, the effects on the wall of the prestressing may be considered to be the same as those of the water pressure, but of opposite sign.

The water pressure at the wall base is

$$\gamma H = 0.0624 \times 22 = 1.373 \text{ kips/ft}^2$$

which gives a ring tension of

$$\gamma Hr = 1.373 \times 44.25 = 60.9 \text{ kips/ft}$$

The initial prestressing force required will be

$$F_i = 60.9 \times 150/115 = 79.4 \text{ kips/ft}$$

for which we provide

$$A_s = 79.4/150 = 0.53 \text{ in.}^2/\text{ft}$$

Table 4-9 gives the steel area required over the entire wall in increments of $H/10$.

Note that a final stress of only 115,000 psi is used rather than the 120,000 psi used in the ring. The effect of this reduction is to provide a residual compression on the wall and thus to decrease the danger of leakage.

In practice the variation in prestressing force on the wall is usually made in 1-ft steps rather than in the 2.2-ft steps shown in Table 4-9.

During construction, care should be exercised to apply the prestressing force smoothly over the height of the wall. If all of the required force at one level were applied before any of the forces at adjacent levels, a concentrated ring load would result and temporary local bending moments would occur.

In practice the wall prestressing will be protected by a coating of pneumatically applied mortar which should be built up after the tank is filled to minimize tensile stresses.

7. *Wall bending reinforcement* must be provided to resist the values shown in Table 4-10. Again we neglect the axial load and compute reinforcement first for the top section of the wall by working-load analysis. We take a clearance of 1 in. and then $d = 5 - 1.3 = 3.7$ in.

$$A_s = \frac{2.64 \times 12}{20 \times \frac{7}{8} \times 3.7} = 0.48 \text{ in.}^2/\text{ft}$$

Since live loads reduce this moment, ultimate-load analysis will result in less reinforcement.

$$M_u = 1.4(2.64) = 3.70 \text{ ft-k/ft}$$

and with $f_c' = 4000$ psi and $f_y = 40,000$ psi,

$$M_u = 0.9 \times 40 \times A_s \left(3.7 - \frac{a}{2}\right) = 3.70 \times 12 \text{ in.-k/ft}$$

from which

$$A_s = 0.35 \text{ in.}^2/\text{ft}$$

for which we may provide No. 4 bars at 6 in.

At the section of maximum moment, 2.2 ft from the base, we obtain

$$M_y = 1.4(4.34) = 6.06 \text{ ft-k/ft}$$
$$d = 5 + 0.9(2.0) - 1.3 = 5.5 \text{ in.}$$

$$M_u = 0.9 \times 40 \times A_s \left(5.5 - \frac{a}{2}\right) = 6.06 \times 12 \text{ in.-k/ft}$$

from which

$$A_s = 0.38 \text{ in.}^2/\text{ft}$$

for which we choose No. 4 bars at 6 in.

As with the dome bending, these computations are slightly conservative because we have neglected the compressive axial load.

On the outside of the wall, we obtain, from Table 4-10, a maximum moment of -0.76 ft-k/ft at 4.4 ft from the wall top. For this moment, where the clearance is ¾ in.,

$$M_u = 1.4(0.76) = 1.07 \text{ ft-k/ft}$$
$$d = 5 + 0.2(2.0) - 1.0 = 4.4 \text{ in.}$$

$$M_u = 36A_s \left(4.4 - \frac{a}{2}\right) = 1.07 \times 12 \text{ in.-k/ft}$$

$$0.98A_s = a$$

from which

$$A_s = 0.08 \text{ in.}^2/\text{ft}$$

for which we provide No. 3 bars at 15 in.

Although no ring tension reinforcement is required because of the prestressing, to position the vertical bars we shall supply twelve No. 4 rings, six on each face and staggered. The reinforcement is shown in Fig. 4-16.

8. *Hinged-base reinforcement* is provided because the hinge must transfer both a vertical load and a horizontal load from wall to base slab. The total vertical load is

Dome weight = 0.945 kip/ft
Dome live load = 2.12 kips/ft
Ring weight = 0.052 kip/ft
Wall weight = 1.65 kips/ft
V_{TL} = 4.767 kips/ft V_{DL} = 2.647 kips/ft

The horizontal load due to initial prestressing, from step 6 is

$$H_F = \left(\frac{150}{115}\right)(2.67) = 3.48 \text{ kips/ft}$$

As an approximation we consider all the loads on the hinge to be taken by the crossed bars. We can write two equations of equilibrium (see Fig. 4-17a) in which C is the bar with maximum compression.

$$\Sigma V = 0 \qquad C \frac{5}{5.39} - T \frac{5}{5.39} = 4.767 \text{ kips/ft}$$

$$\Sigma H = 0 \qquad C \frac{2}{5.39} + T \frac{2}{5.39} = 3.48 \text{ kips/ft}$$

from which

$$C = 7.25 \text{ kips/ft}$$
$$T = 2.10 \text{ kips/ft}$$

The tension value will increase to 3.25 kips/ft where dome live load is not considered. We provide 7.25/20 = 0.37 in.^{2}/ft or No. 5 bars at 10 in. for both C and T for simplicity and No. 3 bar horizontal ties between each two sets of crossed bars.

Figure 4-17b shows the reinforcement at the hinged base. The dimensions and reinforcement of the base are schematic, and proper details depend upon foundation conditions. Also a water stop would normally be included between the wall and the base. A completed prestressed concrete tank is shown in Fig. 4-18.

Use of the hinged base does lead to large vertical bending moments and costly

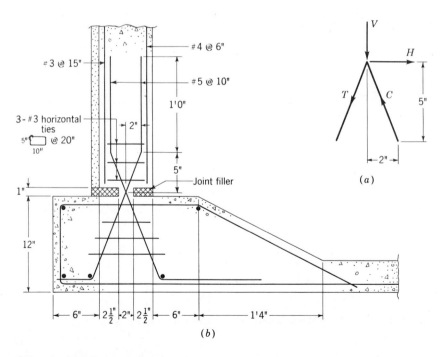

Fig. 4-17 Wall-base reinforcement.

Fig. 4-18   Prestressed concrete reservoir, Akron, Ohio. Capacity 1.5 million gal. (Courtesy of the Preload Company, Inc.)

reinforcing details at the hinges. To overcome these difficulties, partially sliding bases have been devised for which a slightly different wall analysis is required. Further design and construction details are given in the ACI 344 report[6] and in writings on completed tanks.[7-10]

4-11   FINITE-ELEMENT ANALYSES FOR THE DOME RING WALL

An idealization comparable to that in Fig. 4-13 appears in Fig. 4-19 from which a finite-element analysis was carried out for both dead and live loads. Again, as with the fixed-dome example in Chap. 3, the finite-element study uses the correct shell thicknesses for dead loading and hence gives N_ϕ values less than those in Fig. 4-15a. Table 4-11 summarizes these values and shows that the classical theory gives results close to those found from the numerical analysis. The differences are not significant for design and the reinforcement found in Sec. 4-10 will not be affected appreciably by the finite-element results.

Table 4-12 summarizes separately the dead-load and live-load results from the finite-element analysis to demonstrate the refinement possible and to emphasize structural behavior. Element 29 is near the crown and shows values for N_ϕ and N_θ very close to those from membrane theory: for dead load with $q = 31.5$ psf (2½-in. shell) and for live load with $q = 50$ psf, $N_\phi' = N_\theta' = -aq/2 = -1.47$ kips/ft and -2.36 kips/ft, respectively. The values of N_ϕ at elements 17 and 15 (Table 4-12) are not quite correct because of the distorted shapes of those elements.

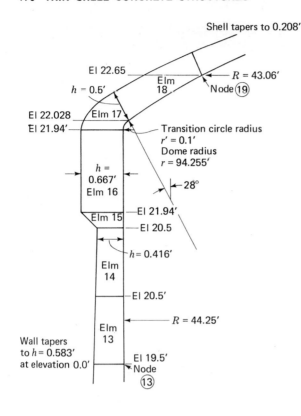

Fig. 4-19 Idealization of dome-ring wall junction for finite-element analysis.

Table 4-11 DOME EDGE VALUES

| | Membrane theory | Fixed $h = 4$* | Dome h variest† | Dome ring wall | |
				$h = 4$*	h variest†
N_ϕ‡	−4.50	−4.29	−3.59	−2.67	−2.65
N_θ‡	−3.00	−0.78	−0.59	15.56	19.32
M_ϕ§	0	−0.26	−0.63	−2.36	−2.87

* In inches.
† Finite-element solution (element 18 in Table 4-12).
‡ In kips per foot.
§ In kips.

Table 4-12 gives considerable insight into the behavior by showing how a more refined idealization of the shell-ring junctions leads to a substantially greater moment at the ring top (-2.93 kips in element 17) compared to -2.36 kips from the simplified solution. The increase of 24 percent is due largely to the haunching effect of the varying thickness, an effect neglected in the simplified approach.

Table 4-12

Element number	Dead load				Live load ($q = 50$ psf)			
	N_ϕ^*	N_θ^*	M_ϕ†	Q_ϕ^*	N_ϕ^*	N_θ^*	M_ϕ†	Q_ϕ^*
29	−1.44	−1.46	0	0	−2.40	−2.30	0.00	0.00
20	−1.28	5.23	−0.47	−0.23	−1.88	8.53	−0.72	−0.29
19	−1.12	6.61	−0.90	−0.34	−1.60	10.30	−1.25	−0.40
18	−1.11	7.62	−1.22	−0.09	−1.54	11.70	−1.65	−0.05
17	−1.22	8.77	−1.25	0.41	−1.61	13.45	−1.68	0.64
16	−1.01	9.16	−0.89	0.49	−1.19	13.70	−1.15	0.70
15	−0.97	7.02	−0.53	−0.08	−1.03	10.30	−0.66	−0.13
14	−1.11	4.95	−0.26	0.28	−1.18	7.16	−0.31	0.39
13	−1.17	4.01	−0.04	0.17	−1.18	5.73	−0.01	0.23

* In kips per foot.
† In kips.

Table 4-12 also shows that the shears Q_ϕ carry a significant part of the vertical dome load thus resulting in N_ϕ values near the ring substantially lower than N_ϕ'. Also, the value for N_θ in the ring (element 16) is somewhat less at 10.97 kips/ft (13.70 × 40/50) than that found from the simplified results for $q = 40$ psf where $H = 0.44$ kip/ft, $T = 0.44 \times 44.25 = 19.5$ kips, and $N_\theta = T/(17.44/12) = 13.5$ kips/ft. This is so because some of the ring tension is carried in the transition elements 15 and 17.

The main conclusion from this study is that the approximate solution provides a reasonable basis for design even though the more rigorous solution does give somewhat different results. Where the finite-element program is readily available, it can be used as a basis for design, but it is always important to study such results carefully in the light of more approximate solutions.

4-12 DESIGN FACTORS FOR A HYPERBOLIC COOLING TOWER SHELL

The overall dimensions for a natural draft cooling tower depend upon the thermal criteria, the construction economy, and structural questions. The thermal conditions usually set the diameter in the base region where the heat exchange takes place, in the air inlet areas where the diagonal columns support the shell, and in the overall height which produces the draft. Construction economy can modify these dimensions, for example, where wind loads are high, by requiring a lower but wider tower to reduce the foundation costs.

Structural requirements influence the meridional shell profile, the size of columns, the shell thickness and its variation, the need for top edge rings, and the shell reinforcement. To study these requirements in some detail, we shall consider the Trojan Tower described in Chap. 2.

A report of the ACI-ASCE committee on concrete shell construction defined the principal design factors as analysis; loadings for gravity, for wind, for earthquake, and for ice and thermal loadings; stability, strength, and serviceability; and reinforcement.[11] In Chap. 2, we have already discussed analysis for various loadings; here, we shall outline the factors related to reinforcement, first in the shell, then at the juncture of the shell and columns. The question of stability is presented in Chap. 9.

The committee recommended that the load factors used for the shell be $U = 0.9D + 1.3L$ (where D and L are in opposite directions), where L is the wind load based upon a 100-year-return fastest mile including a gust factor. For Portland, Oregon, the 100-year-return wind is between 100 and 110 mph, which from the ANSI Code would give for exposure C wind pressures from 90 ft up to 580 ft of from 46 psf to 79 psf.[12] The actual design was based on an earlier report which led to values from 46.6 psf to 92.5 psf. The static computations for dead load and wind gave at the tower base (Fig. 2-10),

$$N_{\phi D} = -76 \text{ kips/ft}$$
$$N_{\phi W} = 116 \text{ kips/ft}$$

for a tension steel requirement of

$$A_s = \frac{1}{\phi f_y} [-0.9 (76) + 1.3 (116)]$$

$$= \frac{1}{(0.9)(60)} (-68.5 + 151.5)$$

$$= \frac{83}{54} = 1.54 \text{ in.}^2/\text{ft}$$

The minimum required is $0.0035A_c$ in each direction so that at the lintel where $h = 3.5$ ft,

$$A_{s,\min} = 0.0035 \times 12 \times 42 = 1.76 \text{ in.}^2/\text{ft}$$

more than the amount gotten from the loads. At a distance about 180 ft above the lintel where the shell thickness becomes a minimum of 10 in., the comparable values are

$$N_{\phi D} = -30.0 \text{ kips/ft}$$
$$N_{\phi W} = +72.8 \text{ kips/ft}$$

from which

$$A_s = 1.39 \text{ in.}^2$$

compared to a minimum required of

$$A_s = 10 \times 12 \times 0.0035 = 0.42 \text{ in.}^2$$

which shows that the critical section can occur well above the base. In the circumferential direction, the stresses do not normally control and minimum steel will be used throughout. As discussed in Chap. 9, this circumferential steel is essential to the integrity of the tower, especially since thermal gradients will cause meridional cracking.

In this tower, the earthquake loads did not control the design and neither did dynamic influences.[13] An important question is the design of reinforcement near the base to account for the concentrated reactions of the columns onto the lintel. This theoretically complex problem is normally solved by using a deep-beam approximation to get tensile steel requirements at the free edge between columns and up in the shell over the supports.

Figure 4-20 gives some idea of the behavior of the shell in its lower support region where it is supported by columns of total width C spaced a distance L apart. The diagrams shown are based upon analytic solutions given by Franz Dischinger[14] in 1932 and applied to concrete deep beams some years later by the Portland Cement Association.[15] As shown in Ref. 15, the loading for the analytic solution is taken at the free edge and the beam depth H is taken equal to the column spacing L, whereas in reality the load is N_ϕ distributed uniformly at some distance above the edge. Thus the solutions shown are approximate and serve mainly to indicate the type of special reinforcement needed: along the lower edge near midspan and higher up over the supports. Also, these solutions neglect the shell curvature by assuming a planar deep beam.

For the Trojan Tower, Fig. 4-21 shows the distribution of N_ϕ at the lintel level, where the maximum tension is about 116 kips/ft and it occurs at $\theta = 0°$. The maximum compression comes at about 69° and is about −84 kips/ft. This compression combines with the dead load of about −76 kips/ft to give −160 kips/ft. The appropriate load factors for dead load and wind are

$$N_{\phi v} = 0.75(1.4N_{\phi D} + 1.7N_{\phi W})$$
$$= 1.05N_{\phi D} + 1.28N_{\phi W}$$
$$= -80 - 107 = -187 \text{ kips/ft}$$

This load is taken as constant over the span between columns even though it will be somewhat less away from $\theta = 69°$ as seen in Fig. 4-21. Because there are 44 supports around the lintel of a radius of about 182 ft, the span will be $2\pi(182)/44 =$

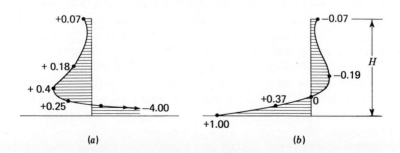

(a) (b)

Fig. 4-20 Deep-beam approximation for the behavior of the base of a cooling tower. (a) Stress coefficients at support. (b) Stress coefficients at midspan.

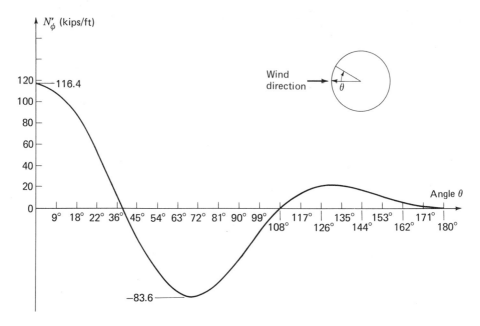

Fig. 4-21 N'_ϕ **at tower base for wind load.**

26 ft. The lintel width b varies from 40 in. at the base to 30 in. where $H = L = 26$ ft. Thus the maximum stresses are, from Fig. 4-20, $f_s = -4.0(187/12)/40 = -1560$ psi at the supports and $f_m = 1.0(187/12)/40 = 390$ psi at midspan. The maximum tension over the supports would be only $(0.4)(15.6)/40 = 156$ psi. Even though these low tensions would not normally cause cracking, it is good practice in thin shells to consider that all tension be taken by reinforcement. The horizontal reinforcement can be proportioned on the basis of the diagrams in Fig. 4-20, where, for example, over the support from a distance of $0.1H$ from the base up to $0.5H$ the amount of tension would be about $0.3(0.4H)q = 0.12qH$ or about $0.12(187)26 = 585$ kips, requiring $A_s = 585/0.9(60) = 10.8$ in². Additional steel would be required from $0.5H$ to H above the base as well. At midspan the total tension would be about $0.37(0.1H)q/2 + 0.37(0.1H)q + 0.63(0.1H)q/2 = 0.088qH = 0.09(187)26 = 440$ kips, requiring $A_s = 440/54 = 8.2$ in².

There is already a minimum shell reinforcement of $0.0035(40)(12) = 1.68$ in²/ft in the lintel area, and because of the approximate nature of this analysis, it is good practice to add the steel found from the deep-beam analysis to the minimum. To provide shear reinforcement some of the midspan horizontal steel should be bent up near the supports.

The diagonal columns need special care where there are significant earthquake loads; it is especially important that the column steel be well anchored into the shell and confined by spirals to prevent pulling out by combined vertical and horizontal vibrations.

REFERENCES

1. Timoshenko, S. P., and S. Woinowsky-Krieger, *Theory of Plates and Shells,* 2d ed., McGraw-Hill Book Company, New York, 1959.
2. Timoshenko, S. P., *Strength of Materials,* 3d ed., vol. 2, D. Van Nostrand Company, Inc., Princeton, N.J., 1956.
3. Girkmann, K., *Flächentragwerke,* 5th ed., Springer-Verlag OHG, Vienna, 1959.
4. Dobell, C., "Prestressed Concrete Tanks," *Proceedings, First U.S. Conference on Prestressed Concrete,* MIT Press, Cambridge, 1951.
5. Hanna, M. M., "Thin Spherical Shells under Rim Loading," *Fifth Congress of the International Association for Bridge and Structural Engineering,* Lisbon, 1956.
6. "Design and Construction of Circular Prestressed Concrete Structures," *J. ACI,* Rep. of ACI Committee 344, no. 9, proc. v. 67, September 1970, pp. 657–672.
7. Crom, J. M., "Design of Prestressed Tanks," *Proceedings ASCE,* vol. 76, October 1950.
8. Doanides, P. J., "Some Notes on Precast Prestressed Tanks," *S. African Institute of Civil Engineers,* July 1955, p. 207.
9. Creasy, L. R., *Prestressed Concrete Cylindrical Tanks,* John Wiley & Sons, Inc., New York, 1961.
10. "Preload Prestressed Concrete Tank Design Concepts," Bull. T-25, The Preload Company, Inc., Westbury, N.Y.
11. "Reinforced Concrete Cooling Tower Shells—Practice and Commentary," *J. ACI,* January 1977, pp. 22–31.
12. "Building Code Requirements for Minimum Design Loads in Buildings and Other Structures," *American National Standard,* ANSI, New York, 1972, pp. 14–16.
13. Abel, J. F., and D. P. Billington, "Effect of Shell Cracking on Dynamic Response of Concrete Cooling Towers," *Proceedings, IASS World Congr. Space Enclosures,* Montreal, 1976, pp. 895–904.
14. Dischinger, F., "Contribution to the Theory of Wall-Like Girders," *Publ.,* vol. 1, Int. Assoc. Bridge and Struct. Eng., Zurich, 1932.
15. "Design of Deep Girders," Portland Cem. Assoc., Skokie, Ill. (undated).

5

ANALYSIS OF CIRCULAR
CYLINDRICAL SHELLS

5-1   INTRODUCTION

The type of structural system treated in this chapter (Fig. 5-1) consists of circular cylindrical shells, longitudinal edge beams, and transverse frames. Cylindrical shells are often referred to as *barrel shells*.

A cylindrical shell may be defined as a curved slab which has been cut from a full cylinder (Fig. 5-2*a*). The slab is bounded by two straight "longitudinal" edges parallel to the axis of the cylinder and by two curved "transverse" edges in planes perpendicular to the axis. The slab is, therefore, curved in only one direction. When the curvature is constant, the cylindrical shell is circular.

The longitudinal edge beams stiffen the shell edges and act together with the shell in carrying loads to the transverse frames. When the edge beams are fully supported by continuous foundations, the curved slab carries loads like a barrel arch and may be so analyzed; such systems are not considered here.

The transverse frames fully support the shell and edge beams and are usually composed of ribs, columns, and footings. The ribs are normally arches built monolithically with the shell.

The analysis can be carried out in two phases: first for a simply supported shell, in which the boundary conditions along the straight longitudinal edges are satisfied; and second, for a shell built together with the transverse frames, in which the boundary conditions along the curved transverse edges are studied. The first phase has been studied extensively (see Sec. 5-6) and many methods are available for its solution. A complete formulation and a method of solution will be developed in some detail in Secs. 5-4 and 5-5. Tables derived from such solutions are included in the Appendix. For the second phase, few methods are available, and only the results from a relatively simple approximate method are included here (Sec. 5-10).

These two phases can be considered separately because the restraint of the transverse frames usually affects only a narrow zone of the shell adjacent to the curved edge.

For the case of a shell without longitudinal edge beams we will divide the first phase again into our usual four-step procedure:

Primary System   The shell is assumed to carry the surface loads solely by stress resultants derived from the membrane theory.

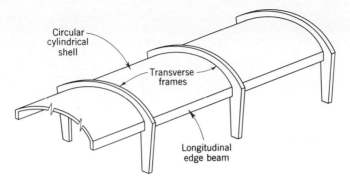

Fig. 5-1 Definitions of barrel-shell system.

Errors These are the forces required by the membrane theory at the free edges.

Corrections Forces (called *line loads*) are applied along the free edges.

Equilibrium This can be achieved by setting the line loads equal to the membrane stress resultants along the edge. No equations need be solved to determine the correction forces.

The stress resultants from the membrane theory are given in the next section, and the effect on the shell of the line loads is described in Sec. 5-5. For this method of analysis, it is necessary to have in some convenient form influence coefficients for line loads; tables in the Appendix, taken from ASCE Manual 31,[1] are included for that purpose. This method is presented in this chapter and illustrated in Chap. 6.

An alternative procedure presented in the first edition only is to solve the basic differential equation directly, obtaining a particular integral and eight independent solutions. Each independent solution includes an arbitrary constant which is dependent upon the boundary conditions. With eight boundary conditions specified, the eight arbitrary constants are obtained from the solution of eight simultaneous equations.

Cylindrical shells are usually described as either long, intermediate, or short, depending upon the ratio of transverse radius r to longitudinal length L. The tables from Manual 31 do not include the intermediate category and are based upon a division point of $r/L = 0.6$.

A different division, presented by Gibson,[2] specifies:

1. Long shells where $r/L < 0.4$
2. Intermediate shells where $0.4 < r/L < 2.0$
3. Short shells where $r/L > 2.0$

In long shells the line loads usually produce internal forces of significant magnitude throughout the entire surface of the shell. These internal forces are usually so large that the membrane values become negligible by comparison. The primary system,

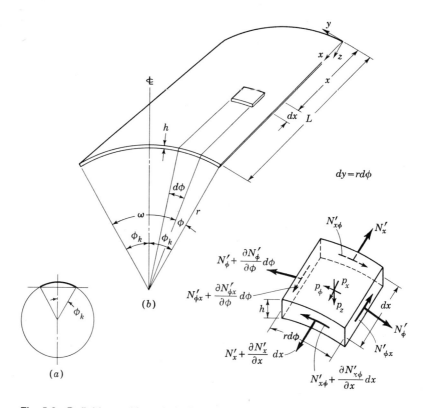

Fig. 5-2 Definitions of barrel-shell geometry and forces.

therefore, does not correspond closely to the real system. In fact, as r/L decreases, the real system approaches the behavior of a beam of curved cross section (see Fig. 5-2). An approximate analysis on this basis is described in Sec. 6-2.

In short shells the line loads produce internal forces in the shell which are usually restricted to the region near the longitudinal edges. In fact, as r/L increases, the greater part of the shell behaves as assumed in the primary system, i.e., with only membrane values. In such cases, short shells are like thin-barrel arches supported near the longitudinal edges by slightly curved diaphragms. These diaphragms carry the arch thrust to the transverse frames by in-plane stresses like deep beams.

5-2 STRESS RESULTANTS FROM THE MEMBRANE THEORY

Figure 5-2b defines the dimensions of the shell. The expressions for the membrane stress resultants, already derived in Chap. 2 as (2-3), are

$$N'_\phi = -p_z r$$

$$N'_{x\phi} = -\frac{1}{r} \int \frac{\partial N'_\phi}{\partial \phi}\, dx - \int p_\phi\, dx + f_1(\phi)$$

$$N'_x = -\frac{1}{r} \int \frac{\partial N'_{\phi x}}{\partial \phi}\, dx - \int p_x\, dx + f_2(\phi)$$

(5-1)

Note that the subscripts used in (5-1) are different from those in (2-3) even though the equations are the same. This change is introduced to carry the general sense of x as the horizontal axis, y as the vertical or tangential axis, and z as the radial axis. Note that ϕ is measured from the longitudinal edge.

The three general equations contain only three unknown stress resultants N'_ϕ, N'_x, and $N'_{x\phi} = N'_{\phi x}$. These stress resultants, therefore, can easily be computed for distributed loads. As an example, the stress resultants due to the shell dead load will be derived. The results for some other loading conditions are summarized in Table 5-1.

For a uniform shell dead load p_d the loading components are

$$p_z = p_d \cos(\phi_k - \phi)$$
$$p_\phi = -p_d \sin(\phi_k - \phi)$$
$$p_x = 0$$

N'_ϕ is obtained directly from the first of (5-1),

$$N'_\phi = -p_d r \cos(\phi_k - \phi) \tag{5-2}$$

When this value for N'_ϕ is substituted into the second of Eqs. (5-1),

$$N'_{x\phi} = \frac{1}{r} \int \frac{\partial[p_d r \cos(\phi_k - \phi)]}{\partial \phi} dx + \int p_d \sin(\phi_k - \phi) dx + f_1(\phi)$$

Differentiating and integrating, we obtain

$$N'_{x\phi} = p_d \sin(\phi_k - \phi)x + p_d \sin(\phi_k - \phi)x + f_1(\phi)$$

In a simple beam, $N'_{x\phi} = 0$ at $x = L/2$, from which

$$f_1(\phi) = -p_d \sin(\phi_k - \phi)L$$

so that

$$N'_{x\phi} = 2p_d \sin(\phi_k - \phi)x - p_d \sin(\phi_k - \phi)L$$

or

$$N'_{x\phi} = -p_d \sin(\phi_k - \phi)L \left(1 - \frac{2x}{L}\right) \tag{5-3}$$

Substituting this value for $N'_{x\phi}$ into the third of (5-1), we have

$$N'_x = -\frac{1}{r} \int \frac{\partial[-p_d \sin(\phi_k - \phi)L(1 - 2x/L)]}{\partial \phi} dx + f_2(\phi)$$

from which

$$N'_x = -p_d \frac{L}{r} \cos(\phi_k - \phi) \int \left(1 - \frac{2x}{L}\right) dx + f_2(\phi)$$

which gives

$$N'_x = -p_d \frac{L}{r} \cos(\phi_k - \phi)x \left(1 - \frac{x}{L}\right) + f_2(\phi)$$

Table 5-1   COMPARISON OF MEMBRANE STRESS RESULTANT DISTRIBUTIONS UNDER DIFFERENT LOADINGS (ALL STRESS RESULTANTS HAVE A NEGATIVE SIGN)

Membrane stress resultant, kips/ft	Load, ksf	Shell constants, ft	Transverse distribution	Longitudinal distribution
N'_ϕ	p_u	r	$\cos^2(\phi_k - \phi)$	1
$N'_{\phi x}$	p_u	$r(L/r)$	$\cos(\phi_k - \phi)\sin(\phi_k - \phi)$	$(3/2)(1 - 2x/L)$
N'_x	p_u	$r(L/r)^2$	$\cos^2(\phi_k - \phi) - \sin^2(\phi_k - \phi)$	$(3x/2L)(1 - x/L)$
N'_ϕ	p_d	r	$\cos(\phi_k - \phi)$	1
$N'_{\phi x}$	p_d	$r(L/r)$	$\sin(\phi_k - \phi)$	$1 - 2x/L$
N'_x	p_d	$r(L/r)^2$	$\cos(\phi_k - \phi)$	$(x/L)(1 - x/L)$
N'_ϕ	p_u	r	$\cos^2(\phi_k - \phi)$	$(1/n)\sin(n\pi x/L)$
$N'_{\phi x}$	p_u	$r(L/r)$	$\cos(\phi_k - \phi)\sin(\phi_k - \phi)$	$(3/n^2\pi)\cos(n\pi x/L)$
N'_x	p_u	$r(L/r)^2$	$\cos^2(\phi_k - \phi) - \sin^2(\phi_k - \phi)$	$(3/n^3\pi^2)\sin(n\pi x/L)$
N'_ϕ	p_d	r	$\cos(\phi_k - \phi)$	$(1/n)\sin(n\pi x/L)$
$N'_{\phi x}$	p_d	$r(L/r)$	$\sin(\phi_k - \phi)$	$(2/n^2\pi)\cos(n\pi x/L)$
N'_x	p_d	$r(L/r)^2$	$\cos(\phi_k - \phi)$	$(2/n^3\pi^2)\sin(n\pi x/L)$

In a simple beam the longitudinal force N'_x must be zero at $x = 0$ and $x = L$, so that $f_2(\phi) = 0$ and

$$N'_x = -p_d \frac{L}{r} \cos(\phi_k - \phi)x \left(1 - \frac{x}{L}\right) \tag{5-4}$$

In Table 5-1 these equations are separated into three parts to emphasize the distribution of membrane stress resultants in a cylindrical shell. The first part of each equation contains terms which depend only upon the magnitude of the load and the dimensions of the shell. The second and third parts of each equation give, respectively, the transverse and the longitudinal distributions of stress resultants. The transverse and longitudinal distributions are not interrelated, so that where two types of surface loads have the same transverse distribution but different longitudinal distributions, the transverse distributions of stress resultants will be the same.

The results of Eqs. (5-2) to (5-4) are plotted in Fig. 5-3 for a full circular cylinder simply supported.

The same force distribution may also be obtained by considering the shell as a hollow circular beam, spanning in the direction of its axis of rotation.

In fact, Eqs. (5-2) to (5-4) give values of N'_ϕ, $N'_{x\phi}$, and N'_x which can be obtained from ordinary beam formulas as done for shell walls by Eqs. (2-6):

$$f_\phi = \frac{N'_\phi}{h} = -\frac{p_z r}{h} \tag{a}$$

$$v = \frac{N'_{x\phi}}{h} = -\frac{VQ}{2Ih} \tag{b}$$

$$f_x = \frac{N'_x}{h} = -\frac{My}{I} \tag{c}$$

where for a ring cross section

$$M = p_d 2\pi r \left(\frac{Lx}{2} - \frac{x^2}{2}\right)$$

$$y = r \cos (\phi_k - \phi)$$
$$I = \pi h r^3$$

$$V = p_d 2\pi r \left(\frac{L}{2} - x\right)$$

$$Q = 2hr^2 \sin (\phi_k - \phi)$$

so that (a), (b), and (c) may be rewritten as

$$N'_\phi = -p_d r \cos (\phi_k - \phi) \tag{5-5}$$

$$\frac{N'_{x\phi}}{h} = -\frac{p_d 2\pi r[(L/2) - x]2hr^2 \sin (\phi_k - \phi)}{\pi h r^3 (2h)}$$

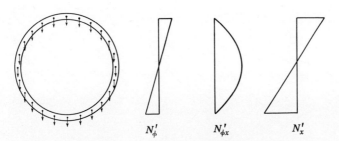

$$N'_\phi \qquad\qquad N'_{\phi x} \qquad\qquad N'_x$$

Fig. 5-3 Stress resultants in a full cylindrical shell.

which gives

$$N'_{x\phi} = -p_d \sin (\phi_k - \phi) L \left(1 - \frac{2x}{L}\right) \tag{5-6}$$

and

$$\frac{N'_x}{h} = -\frac{p_d 2\pi r[(Lx/2) - x^2/2]r \cos (\phi_k - \phi)}{\pi h r^3}$$

which gives

$$N'_x = -p_d \frac{L}{r} \cos (\phi_k - \phi) x \left(1 - \frac{x}{L}\right) \tag{5-7}$$

Equations (5-6) and (5-7) are the same as (5-3) and (5-4).

The normal cylindrical shell is only a small slice of a full cylinder. Therefore, for the membrane theory to give a correct picture of the internal stress resultants there must be reactions at the longitudinal edges equal to the membrane values N'_ϕ and $N'_{\phi x}$. The distribution of membrane values in a cylindrical shell is shown in Fig. 5-4, and the required edge reactions in Fig. 5-5. Where the shell has free edges, these reactions are zero. Therefore, correction forces or line loads $T_L = -N'_\phi$ and $S_L = -N'_{\phi x}$ must be applied to give the true resultant reactions. The displacements and internal forces due to these line loads are considered in the bending theory in Secs. 5-4 and 5-5.

To overcome mathematical difficulties in the bending theory, it is desirable to express the uniformly distributed line loads as sums of partial loads. A uniform line load T_L (Fig. 5-6a) may be represented by a Fourier series:

$$(T_L)_x = \frac{4}{\pi} T_L \sum_{n=1,3,5...}^{\infty} \frac{1}{n} \sin \frac{n\pi x}{L}$$

The surface loads must also be expressed by a Fourier series to provide compatible forces and deformations along the length of the free edges. For example, the Fourier series for the uniform shell dead load is

$$(p_d)_x = \frac{4}{\pi} p_d \sum_{n=1,3,5...}^{\infty} \frac{1}{n} \sin \frac{n\pi x}{L}$$

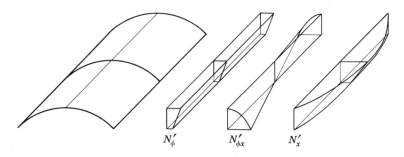

N'_ϕ $N'_{\phi x}$ N'_x

Fig. 5-4 Membrane values in a free-edged barrel shell.

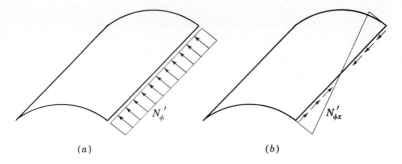

(a)   (b)

Fig. 5-5   Edge errors in a free-edged barrel shell.

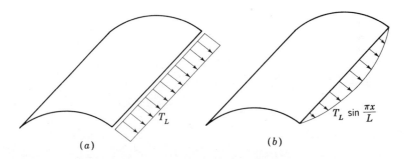

(a)   (b)

Fig. 5-6   Edge corrections in a free-edged barrel shell.

The first three terms of a sine series are plotted in Fig. 5-7 and show a close correspondence to the uniform load. For long shells it is usually sufficient to use just the first term of the series (Fig. 5-6b), although for short shells it is sometimes advisable to include the second term as well.

5-3   DISPLACEMENTS FROM THE MEMBRANE THEORY

Figure 5-8 defines the displacements of a differential shell element from which the change in the length of dx and $r\,d\phi$ are

$$\epsilon_x = \frac{\partial u}{\partial x} \qquad (a)$$

$$\epsilon_\phi = \frac{v + (\partial v/\partial\phi)\,d\phi - v}{r\,d\phi} + \frac{(r - w)\,d\phi - r\,d\phi}{r\,d\phi} = \frac{\partial v}{r\,d\phi} - \frac{w}{r} \qquad (b)$$

similar to (a) and (b) of Sec. 2-5 for cylinders except for the term in v.

The shearing strain, defined as the change in the angle between the two sides at the local origin,

$$\gamma_{x\phi} = \frac{(\partial u/\partial\phi)\,d\phi}{r\,d\phi} + \frac{(\partial v/\partial x)\,dx}{dx} = \frac{\partial u}{r\,\partial\phi} + \frac{\partial v}{\partial x} \qquad (c)$$

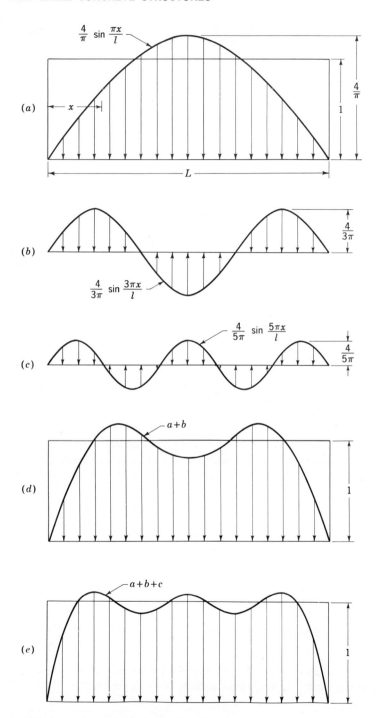

Fig. 5-7 Plots of terms in a Fourier series.

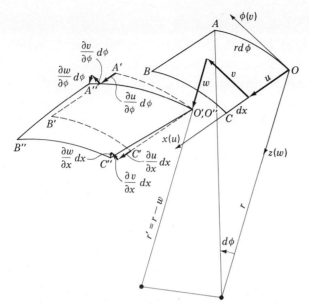

Fig. 5-8 Displacements in barrel shells.

The rotation of the dx side in the xz plane

$$\phi_x = \frac{(\partial w/\partial x)\, dx}{dx} = \frac{\partial w}{\partial x} \qquad (d)$$

and of the $r\, d\phi$ side in the ϕz plane

$$\phi_\phi = \frac{v}{r} + \frac{\partial w}{r\, \partial \phi} \qquad (e)$$

where the term in v arises because of the curvature.

The strains may now be written in terms of the membrane resultants following Eqs. (2-20):

$$\epsilon_x = \frac{\partial u}{\partial x} = \frac{1}{Eh}\,(N'_x - \nu N'_\phi)$$

$$\epsilon_\phi = \frac{\partial v}{r\, \partial \phi} - \frac{w}{r} = \frac{1}{Eh}\,(N'_\phi - \nu N'_x) \qquad (5\text{-}8)$$

and for shearing strain where the modulus of rigidity $G = E/2(1 + \nu)$,

$$\gamma_{x\phi} = \frac{\partial u}{r\, \partial \phi} + \frac{\partial v}{\partial x} = \frac{1}{Gh}\, N'_{x\phi} = \frac{2(1 + \nu)}{Eh}\, N'_{x\phi} \qquad (5\text{-}9)$$

For concrete ν is usually about 1/6 and has only a small influence on deformations. It seems reasonable, therefore, to neglect ν so that simpler expressions may be obtained. The first of Eqs. (5-8) is then solved directly for the longitudinal displacement

where

$$u = \frac{1}{Eh} \int N'_x \, dx + f_3(\phi) \tag{5-10}$$

The tangential displacement may be obtained from Eq. (5-9):

$$v = -\frac{1}{r} \int \frac{\partial u}{\partial \phi} \, dx + \frac{2}{Eh} \int N'_{x\phi} \, dx + f_4(\phi) \tag{5-11}$$

and the radial displacement from the second of Eqs. (5-8):

$$w = \frac{\partial v}{\partial \phi} - \frac{rN'_\phi}{Eh} \tag{5-12}$$

These displacements can be evaluated for any type of load. As an example, consider the case of the shell under a loading uniformly distributed over the shell surface. The uniform longitudinal distribution is approximated by the first term of a Fourier series from which, with $k = n\pi/L$, $p = (4/\pi)p_a$, and $n = 1$,

$$p_z = p \cos (\phi_k - \phi) \sin kx$$
$$p_\phi = -p \sin (\phi_k - \phi) \sin kx$$
$$p_x = 0$$

The stress resultants derived from (5-1) are

$$N'_\phi = -pr \cos (\phi_k - \phi) \sin kx$$

$$N'_{x\phi} = -\frac{2p}{k} \sin (\phi_k - \phi) \cos kx + f_1(\phi) \tag{5-13}$$

$$N'_x = -\frac{2p}{rk^2} \cos (\phi_k - \phi) \sin kx - \frac{x}{r} \frac{\partial f_1(\phi)}{\partial \phi} + f_2(\phi)$$

When (5-13) are introduced into Eqs. (5-10) to (5-12),

$$u = \frac{1}{Eh} \left[\frac{2p}{rk^3} \cos (\phi_k - \phi) \cos kx - \frac{x^2}{2r} \frac{\partial f_1(\phi)}{\partial \phi} + xf_2(\phi) \right] + f_3(\phi) \tag{5-14}$$

$$v = -\frac{1}{Ehr} \left[\frac{2p}{rk^4} \sin (\phi_k - \phi) \sin kx - \frac{x^3}{6r} \frac{\partial^2 f_1(\phi)}{\partial \phi^2} + \frac{x^2}{2} \frac{\partial f_2(\phi)}{\partial \phi} \right]$$
$$- \frac{x}{r} \frac{\partial f_3(\phi)}{\partial \phi} - \frac{2}{Eh} \left[\frac{2p}{k^2} \sin (\phi_k - \phi) \sin kx - xf_1(\phi) \right] + f_4(\phi) \tag{5-15}$$

$$w = \frac{1}{Ehr} \left[\frac{2p}{rk^4} \cos (\phi_k - \phi) \sin kx + \frac{x^3}{6r} \frac{\partial^3 f_1(\phi)}{\partial \phi^3} - \frac{x^2}{2} \frac{\partial^2 f_2(\phi)}{\partial \phi^2} \right]$$
$$- \frac{x}{r} \frac{\partial^2 f_3(\phi)}{\partial \phi^2} + \frac{2}{Eh} \left[\frac{2p}{k^2} \cos (\phi_k - \phi) \sin kx + x \frac{\partial f_1(\phi)}{\partial \phi} \right] + \frac{\partial f_4(\phi)}{\partial \phi}$$
$$+ \frac{pr^2}{Eh} \cos (\phi_k - \phi) \sin kx \tag{5-16}$$

In the case of a simply supported shell with uniformly distributed load,

$$f_1(\phi) = f_2(\phi) = f_3(\phi) = f_4(\phi) = 0$$

and Eqs. (5-14) to (5-16) reduce to

$$u = \frac{2p}{Ehr}\frac{1}{k^3}\cos(\phi_k - \phi)\cos kx \qquad (5\text{-}17)$$

$$v = -\frac{2p}{Eh}\left(\frac{1}{r^2}\frac{1}{k^4} + 2\frac{1}{k^2}\right)\sin(\phi_k - \phi)\sin kx \qquad (5\text{-}18)$$

$$w = \frac{2p}{Ehr^2}\frac{1}{k^4}\left(1 + 2r^2k^2 + \frac{r^4k^4}{2}\right)\cos(\phi_k - \phi)\sin kx \qquad (5\text{-}19)$$

It is useful to have the vertical and horizontal displacements and these are easily obtained from (5-18) and (5-19) as follows:

$$\begin{aligned}
\Delta_V &= -v\sin(\phi_k - \phi) + w\cos(\phi_k - \phi) \\
&= \frac{2p}{Ehr^2}\frac{1}{k^4}\Big[(1 + 2r^2k^2)\sin^2(\phi_k - \phi) \\
&\qquad\qquad + \left(1 + 2r^2k^2 + \frac{r^4}{2}k^4\right)\cos^2(\phi_k - \phi)\Big]\sin kx \\
&= \frac{2p}{Ehr^2}\frac{1}{k^4}\left[1 + 2r^2k^2 + \frac{r^4}{2}k^4\cos^2(\phi_k - \phi)\right]\sin kx \qquad (5\text{-}20)
\end{aligned}$$

and

$$\begin{aligned}
\Delta_H &= v\cos(\phi_k - \phi) + w\sin(\phi_k - \phi) \\
&= \frac{2p}{Ehr^2}\left(-\frac{1}{k^4} - 2r^2\frac{1}{k^2} + \frac{1}{k^4} + 2r^2\frac{1}{k^2} + \frac{r^4}{2}\right)\sin(\phi_k - \phi) \\
&\qquad\qquad\qquad\qquad\qquad\qquad \times \cos(\phi_k - \phi)\sin kx \\
&= \frac{pr^2}{Eh}\sin(\phi_k - \phi)\cos(\phi_k - \phi)\sin kx \qquad (5\text{-}21)
\end{aligned}$$

where Δ_V is positive for downward displacement and Δ_H is positive for inward displacement.

5-4 COMPLETE FORMULATION USING THE THEORY OF SHALLOW SHELLS

Because of the importance of this theory and its wide applicability, we shall present the derivations fully (and relate them to the general formulations of Chap. 2). Figure 5-9 shows a differential element acted upon by the bending moments and radial shears which were neglected in the membrane theory. From the definitions

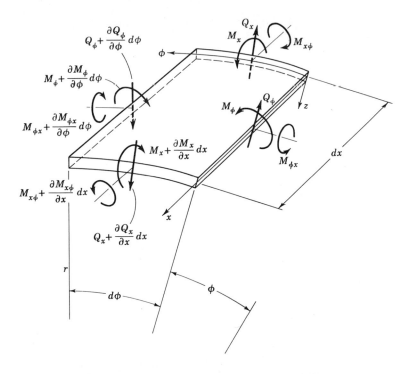

Fig. 5-9   Bending moments and shear forces in barrel shells.

given in Sec. 5-2, the general equilibrium equations (2-25) can now be rewritten as

$$\frac{\partial N_x}{\partial x} r + \frac{\partial N_{\phi x}}{\partial \phi} + p_x r = 0$$

$$\frac{\partial N_\phi}{\partial \phi} + \frac{\partial N_{x\phi}}{\partial x} r - Q_\phi + p_\phi r = 0$$

$$\frac{\partial Q_x}{\partial x} r + \frac{\partial Q_\phi}{\partial \phi} + N_\phi + p_z r = 0 \qquad (5\text{-}22)$$

$$-\frac{\partial M_\phi}{\partial \phi} + \frac{\partial M_{x\phi}}{\partial x} r + Q_\phi r = 0$$

$$+\frac{\partial M_x}{\partial x} r + \frac{\partial M_{\phi x}}{\partial \phi} - Q_x r = 0$$

These equations are the same as those derived in Chap. 2 except that the coordinate system is different. The first three equations of (5-22) can be obtained from the comparable equations of (2-25) by replacing θ by ϕ and y by x. For the last two, again the same coordinate substitution is made, but in addition the sign of all the terms except the twisting moments must change because in the two right-hand coordinate systems the x coordinate is θ (the curved coordinate) for Eqs. (2-25) and is x (the straight coordinate) for Eqs. (5-22).

It is now necessary to express the stress resultants and stress couples in terms of displacements, as done in Sec. 2-6 for axisymmetrical cases. Using Eqs. (5-8) and (5-9), we can write (where $K = Eh/(1 - \nu^2)$)

$$N_x = \frac{Eh}{1 - \nu^2}(\epsilon_x + \nu\epsilon_\phi) = K\left[\frac{\partial u}{\partial x} + \nu\left(\frac{\partial v}{r\,\partial\phi} - \frac{w}{r}\right)\right]$$

$$N_\phi = \frac{Eh}{1 - \nu^2}(\epsilon_\phi + \nu\epsilon_x) = K\left[\frac{\partial v}{r\,\partial\phi} - \frac{w}{r} + \nu\frac{\partial u}{\partial x}\right] \qquad (5\text{-}23)$$

$$N_{x\phi} = N_{\phi x} = \frac{Eh}{2(1 + \nu)}\gamma_{x\phi} = K(1 - \nu)\left(\frac{\partial u}{r\,\partial\phi} + \frac{\partial v}{\partial x}\right)$$

For the stress couples, we need first to express the change of curvature and the change of twist of a surface in terms of its displacements. The rotation at the local origin in the xz plane, $\phi_x = \partial w/\partial x$ given in (d) of Sec. 5-3, changes over dx to $\phi_x + (\partial\phi_x/\partial x)\,dx$ and the same for ϕ_ϕ in the ϕz plane. Thus the changes in rotation over dx and $r\,d\phi$, respectively, are

$$\chi_x = \frac{\partial\phi_x}{\partial x} = \frac{\partial^2 w}{\partial x^2} \qquad (a)$$

$$\chi_\phi = \frac{\partial\phi_\phi}{r\,\partial\phi} = \frac{1}{r^2}\frac{\partial v}{\partial\phi} + \frac{1}{r^2}\frac{\partial^2 w}{\partial\phi^2} \qquad (b)$$

These changes in rotation define the changes in curvature. The change in twist is defined as the average of the changes in rotation ϕ_x in the ϕ direction plus the changes in rotation ϕ_ϕ in the x direction, i.e.,

$$\chi_{x\phi} = \tfrac{1}{2}\left(\frac{\partial\phi_x}{r\,\partial\phi} + \frac{\partial\phi_\phi}{\partial x}\right)$$

$$= \tfrac{1}{2}\left(\frac{\partial^2 w}{r\,\partial x\,\partial\phi} + \frac{1}{r}\frac{\partial v}{\partial x} + \frac{1}{r}\frac{\partial^2 w}{\partial x\,\partial\phi}\right)$$

$$\chi_{x\phi} = \frac{\partial^2 w}{r\,\partial x\,\partial\phi} + \frac{1}{2r}\frac{\partial v}{\partial x} \qquad (c)$$

The stress couples are defined by the changes in curvature and twist as

$$M_x = -D(\chi_x + \nu\chi_\phi)$$
$$M_\phi = -D(\chi_\phi + \nu\chi_x)$$
$$M_{x\phi} = -M_{\phi x} = D(1 - \nu)\chi_{x\phi}$$

which, after substituting in (a) to (c), give

$$M_x = -D\left[\frac{\partial^2 w}{\partial x^2} + \frac{\nu}{r^2}\left(\frac{\partial v}{\partial\phi} + \frac{\partial^2 w}{\partial\phi^2}\right)\right]$$

$$M_\phi = -D\left(\frac{\partial v}{r^2\,\partial\phi} + \frac{\partial^2 w}{r^2\,\partial\phi^2} + \nu\frac{\partial^2 w}{\partial x^2}\right) \qquad (5\text{-}24)$$

$$M_{x\phi} = -M_{\phi x} = D(1 - \nu)\left(\frac{\partial^2 w}{r\,\partial x\,\partial\phi} + \frac{1}{2r}\frac{\partial v}{\partial x}\right)$$

These eleven equations, (5-22) to (5-24), contain eleven unknowns: three in-plane stress resultants, three stress couples, two out-of-plane stress resultants, and three displacements. Thus a solution is possible.

A straightforward combination leads to an equation, the solution for which is extremely involved. A much simpler solution is possible if the theory of shallow shells is used. Such a solution has been derived independently by Donnell,[3] Vlasov,[4] and Jenkins[5] and has been presented by Gibson.[2]

In deriving the differential equation used in shallow-shell theory, it will be assumed that the radius r, the thickness h, and the material properties E and ν are everywhere constant. The derivation consists of two similar parts, one dealing with a plate-type equation and the other with a deep-beam-type equation.

The biharmonic plate-bending equation comes from expressing the two bending equations of (5-22) in terms of the radial displacement w and substituting the results into the force equilibrium equation, the third of (5-22), just as was done in Chap. 2 to get Eq. (2-28). Differentiating the fourth of (5-22) with respect to ϕ and the fifth with respect to x gives

$$\frac{\partial Q_\phi}{r \, \partial \phi} = \frac{\partial^2 M_\phi}{r^2 \, \partial \phi^2} - \frac{\partial^2 M_{x\phi}}{r \, \partial x \, \partial \phi} \tag{d}$$

$$\frac{\partial Q_x}{\partial x} = \frac{\partial^2 M_x}{\partial x^2} + \frac{\partial^2 M_{\phi x}}{r \, \partial x \, \partial \phi} \tag{e}$$

We next substitute Eqs. (5-24) into (d) and (e) after making the *first approximation for shallow shells*, i.e., that the in-plane displacements are negligible in the terms for changes of curvature. Thus

$$\frac{\partial Q_\phi}{r \, \partial \phi} = -D \left[\frac{\partial^4 w}{r^4 \, \partial \phi^4} + \nu \frac{\partial^4 w}{r^2 \, \partial \phi^2 \, \partial x^2} + (1 - \nu) \frac{\partial^4 w}{r^2 \, \partial \phi^2 \, \partial x^2} \right] \tag{f}$$

$$\frac{\partial Q_x}{\partial x} = -D \left[\frac{\partial^4 w}{\partial x^4} + \nu \frac{\partial^4 w}{r^2 \, \partial x^2 \, \partial \phi^2} + (1 - \nu) \frac{\partial^4 w}{r^2 \, \partial x^2 \, \partial \phi^2} \right] \tag{g}$$

We note that all terms with ν cancel out in each of (f) and (g). When these two expressions are added together, we find

$$\frac{\partial Q_\phi}{r \, \partial \phi} + \frac{\partial Q_x}{\partial x} = -D \nabla^4 w \tag{h}$$

where

$$\nabla^4 w = \frac{\partial^4 w}{\partial x^4} + 2 \frac{\partial^4 w}{r^2 \, \partial x^2 \, \partial \phi^2} + \frac{\partial^4 w}{r^4 \, \partial \phi^4} \tag{i}$$

and ∇^4 is the biharmonic operator. The equation for equilibrium of forces in the z direction, the third of (5-22), now becomes

$$D \nabla^4 w - \frac{N_\phi}{r} = p_z \tag{j}$$

This equation is the same one that is used for plate bending except for the addition of the in-plane term N_ϕ / r. We have one equation with two unknowns w and N_ϕ as derived from six of the original eleven equations. Using the remaining five equations, we next derive a second equation in the same unknowns.

The biharmonic deep-beam equation is that second equation, and it comes from the first two of Eqs. (5-22) and from Eqs. (5-23). In the plate-bending equation, we found a potential function w by which to express all the stress couples. In the same way we define a potential function F by means of which to express all of the in-plane stress resultants:

$$N_x = \frac{\partial^2 F}{r^2\, \partial\phi^2} - \int p_x\, dx$$

$$N_\phi = \frac{\partial^2 F}{\partial x^2} - \int p_\phi r\, d\phi \qquad (5\text{-}25)$$

$$N_{x\phi} = -\frac{\partial^2 F}{r\, \partial x\, \partial\phi}$$

These functions are chosen such that they satisfy the first two equilibrium equations (5-22) except for the Q_ϕ term. Dropping Q_ϕ in the second of (5-22) thus is the *second approximation for shallow shells*. Making this approximation allows Eqs. (5-25) to satisfy the in-plane equilibrium equations.

In shell theory these two approximations are directly coupled so that once one is made, the other must be made as well. When the shell equations are derived by a principle of virtual work, the dropping of in-plane displacements in the rotation term ϕ_ϕ leads automatically to the elimination of Q_ϕ in the in-plane force equilibrium equation. Physically this means that the shell is flat enough so that the rotations are mainly due to w and that Q_ϕ plays negligible role in carrying the loads. Where the shell is not flat, these approximations must be questioned. For the circular cylindrical shell roofs considered here, Scordelis has found that this theory is useful down to ratios of $r/l \geq 0.2$ for single shells free at the longitudinal edges and under uniform load. For typical interior shells (in multibarrels) fixed at those edges also under uniform loads, he has found the theory applicable down to $r/l > 0.4$.[6]

For the complete derivation, we next need to find an equation similar to the third of (5-22) from which Eq. (j) came. This new equation needs to use the remaining unused expressions of (5-23).

These three can be rewritten directly in the form of (5-8) and (5-9):

$$Eh\epsilon_x = N_x - \nu N_\phi = Eh\,\frac{\partial u}{\partial x} \qquad (k)$$

$$Eh\epsilon_\phi = N_\phi - \nu N_x = Eh\left(\frac{\partial v}{r\, \partial\phi} - \frac{w}{r}\right) \qquad (l)$$

$$Eh\gamma_{x\phi} = 2(1+\nu)N_{x\phi} = Eh\left(\frac{\partial u}{r\, \partial\phi} + \frac{\partial v}{\partial x}\right) \qquad (m)$$

Since the goal is to eliminate u and v, we can differentiate (k), (l), and (m) as follows:

$$\frac{\partial^2(N_x - \nu N_\phi)}{r^2\, \partial\phi^2} = Eh\,\frac{\partial^3 u}{r^2\, \partial x\, \partial\phi^2} \qquad (n)$$

$$\frac{\partial^2(N_\phi - \nu N_x)}{\partial x^2} = Eh\left(\frac{\partial^3 v}{r\,\partial x^2\,\partial\phi} - \frac{1}{r}\frac{\partial^2 w}{\partial x^2}\right) \tag{o}$$

$$2(1+\nu)\frac{\partial^2 N_{x\phi}}{r\,\partial x\,\partial\phi} = Eh\left(\frac{\partial^3 u}{r^2\,\partial x\,\partial\phi^2} + \frac{\partial^3 v}{r\,\partial x^2\,\partial\phi}\right) \tag{p}$$

Now when (n) and (o) are added and (p) is subtracted from the result, all terms in u and v are eliminated leaving

$$\frac{\partial^2(N_x - \nu N_\phi)}{r^2\,\partial\phi^2} + \frac{\partial^2(N_\phi - \nu N_x)}{\partial x^2} - 2(1+\nu)\frac{\partial^2 N_{x\phi}}{r\,\partial x\,\partial\phi} = -Eh\frac{1}{r}\frac{\partial^2 w}{\partial x^2} \tag{q}$$

Finally when the expressions of (5-25) are put into (q), we find

$$\nabla^4 F + \frac{Eh}{r}\frac{\partial^2 w}{\partial x^2} = f(p) \tag{r}$$

where

$$\nabla^4 F = \frac{\partial^4 F}{\partial x^4} + 2\frac{\partial^4 F}{r^2\,\partial x^2\,\partial\phi^2} + \frac{\partial^4 F}{r^4\,\partial\phi^4}$$

and

$$f(p) = \int\frac{\partial^2 p_x}{r^2\,\partial\phi^2}\,dx + \int\frac{\partial^2 p_\phi}{\partial x^2}r\,\partial\phi - \nu\frac{\partial p_\phi}{r\,\partial\phi} - \nu\frac{\partial p_x}{\partial x}$$

Equation (r) is the biharmonic equation used for studying deep beams loaded in their plane except for the addition of the term in $\partial^2 w/\partial x^2$. Eq. (j) needs now to be modified by substituting in the second of Eqs. (5-25), and the result is two equations in two unknowns:

$$\nabla^4 w - \frac{1}{Dr}\frac{\partial^2 F}{\partial x^2} = f'(p) \tag{5-26}$$

$$\nabla^4 F + \frac{Eh}{r}\frac{\partial^2 w}{\partial x^2} = f(p) \tag{5-27}$$

where

$$f'(p) = \frac{p_z}{D} - \frac{1}{rD}\int p_\phi r\,d\phi \tag{s}$$

These two 4th-order equations can be combined to give one 8th-order equation in w by operating on (5-26) with ∇^4 and on (5-27) by $(1/Dr)\partial^2/\partial x^2$, thus giving

$$\nabla^4\nabla^4 w - \frac{1}{Dr}\nabla^4\frac{\partial^2 F}{\partial x^2} = \nabla^4 f'(p) \tag{t}$$

$$\frac{\partial^2}{Dr\,\partial x^2}\nabla^4 F + \frac{Eh}{Dr^2}\frac{\partial^4 w}{\partial x^4} = \frac{\partial^2}{Dr\,\partial x^2}f(p) \tag{u}$$

Adding (u) and (t) eliminates F and gives

$$\nabla^8 w + \frac{Eh}{Dr^2}\frac{\partial^4 w}{\partial x^4} = f''(p) \tag{5-28}$$

where

$$\nabla^8 w = \left(\frac{\partial^2}{\partial x^2} + \frac{\partial^2}{r^2 \partial \phi^2}\right)^4 w \qquad (v)$$

and

$$f''(p) = \nabla^4 f'(p) + \frac{1}{Dr} \frac{\partial^2}{\partial x^2} f(p)$$

$$= \frac{1}{D}\left(\nabla^4 p_z - 2\frac{\partial^3 p_\phi}{r^2 \partial x^2 \partial \phi} - \frac{1}{r^4}\frac{\partial^3 p_\phi}{\partial \phi^3} + \frac{1}{r^3}\frac{\partial^3 p_x}{\partial x \partial \phi^2}\right.$$

$$\left. - \nu \frac{\partial^3 p_\phi}{r^2 \partial x^2 \partial \phi} - \nu \frac{\partial^3 p_x}{r \partial x^3}\right) \qquad (w)$$

The use of (5-28) can be simplified further. The membrane theory solution can be used in place of the particular solution to (5-28) which considers the complex loading expression (w). A study of the accuracy of the membrane theory approximations leads to the conclusion that even with short shells, the membrane theory results are not noticeably modified by including bending (see the first edition of this book).

The primary system and resulting errors can be obtained directly from either the membrane theory or the particular integral. The determination of the corrections is not so easy because the difficult problem of solving the homogeneous part of (5-28) must be tackled. Since the circular cylindrical shell is one of the few shell systems for which a solution has been well worked out, a derivation is given here.

5-5   SOLUTION OF THE SHALLOW-SHELL EQUATIONS

We further simplify Eq. (5-28) assuming $\nu = 0$, so that homogeneous equation can be written as

$$r^2 \left(\frac{\partial^2}{\partial x^2} + \frac{\partial^2}{r^2 \partial \phi^2}\right)^4 w + \frac{12}{h^2}\frac{\partial^4 w}{\partial x^4} = 0 \qquad (5\text{-}29)$$

for which the solution has the form

$$w = \sum_{n=1,3\ldots}^{\infty} A_m e^{M\phi} \sin kx \qquad (5\text{-}30)$$

where A_m represents eight arbitrary constants which depend upon the longitudinal boundary conditions and M in $e^{M\phi}$ represents eight roots; these depend only upon the dimensions of the shell. The longitudinal variation is described by the Fourier series; only the first term will be retained in this discussion.

The roots M are determined by substituting (5-30) into (5-29):

$$r^2 \left(-k^2 + \frac{M^2}{r^2}\right)^4 + \frac{12}{h^2}k^4 = 0$$

or

$$\left[\frac{M^2 - (kr)^2}{Q^2 \sqrt{2}}\right]^4 + 1 = 0 \qquad (a)$$

where

$$Q^8 = 3(kr)^4 \left(\frac{r}{h}\right)^2 \tag{b}$$

and with

$$\gamma = \left(\frac{kr}{Q}\right)^2$$

(a) becomes

$$\left(\frac{M^2}{Q^2} - \gamma\right)^4 + 4 = 0 \tag{5-31}$$

The resulting eight roots are complex and conjugate:

$$\begin{aligned} M_1 &= \pm (\alpha_1 \pm i\beta_1) \\ M_2 &= \pm (\alpha_1' \pm i\beta_1') \end{aligned} \tag{5-32}$$

where

$$\alpha_1 = Q \sqrt{\frac{\sqrt{(1+\gamma)^2 + 1} + (1+\gamma)}{2}} = Q m_1$$

$$\beta_1 = Q \sqrt{\frac{\sqrt{(1+\gamma)^2 + 1} - (1+\gamma)}{2}} = Q n_1$$

$$\alpha_1' = Q \sqrt{\frac{\sqrt{(1-\gamma)^2 + 1} - (1-\gamma)}{2}} = Q m_2 \tag{5-33}$$

$$\beta_1' = Q \sqrt{\frac{\sqrt{(1-\gamma)^2 + 1} + (1-\gamma)}{2}} = Q n_2$$

The only remaining problem is to determine the eight arbitrary constants A_m, four for each longitudinal edge. Gibson gives such solutions, but the availability of computer programs has rendered those analytic results of less practical value.

With the A_m known, the problem now is to obtain corrections, i.e., edge effects due to edge loads (line loads) as described in Sec. 5-2. For a symmetrical shell, for example, these effects are obtained by setting three boundary conditions equal to zero and the fourth equal to one; the resulting four simultaneous equations are solved for the four arbitrary constants.

Thus Table 1A of Manual 31 (see Appendix) was prepared. Manual 31 also included a table for short-barrel shells.

The solution given by (5-33) carries the additional mathematical restriction that all functions containing $\sin kx$ vanish along the boundaries $x = 0$ and $x = L$. Hence, values containing $\sin kx$ such as w, v, M_x, N_ϕ, and N_x must be taken as zero at both ends of the shell.

The principal physical restrictions thus imposed are that the transverse frames, which fully support the shell at $x = 0$ and $x = L$, provide (1) complete rigidity in their vertical planes, i.e., $v_0 = w_0 = v_L = w_L = 0$; and (2) complete flexibility in planes parallel to the shell middle surface, i.e., $M_{x_0} = N_{x_0} = M_{x_L} = N_{x_L} = 0$. Ideally, then, the shell is assumed to be supported by a wall both so deep that no in-plane displacement can occur and so thin that no out-of-plane shear or bending can be developed. Even though actual shell supports deviate greatly from this ideal (see Figs. 5-1 and 6-3), the theory presented here has been widely used in the design of many successfully built roof structures.

Equation (5-30) was used primarily to separate the variables and thus treat the solution like that of an ordinary differential equation, in the same manner as a Lévy-type solution for plates. The restriction on two of the four boundaries is the price paid for this mathematical convenience. Notice that the formulations for the shells in Chaps. 2 and 3 were reduced to ordinary differential equations, by assuming axisymmetrical geometries and loadings, again to avoid dealing with a partial differential equation.

There are five forces at the free edge, N_ϕ, M_ϕ, Q_ϕ, $M_{\phi x}$, and $N_{\phi x}$, but only four constants of integration. Therefore it is necessary to combine two of the forces into one, $Q'_\phi = Q_\phi - \partial M_{\phi x}/\partial x$. The justification for this combination originally introduced for plate analysis can be found in Ref. 7.

5-6 COMPLETE SOLUTION BY OTHER THEORIES

A number of formulations similar to that described in the preceding sections have been proposed, and several valuable papers have compared these various theories numerically. The purpose of this section is briefly to review some of these theories and to compare them with the formulation for shallow shells just presented.

Flügge[8] first presented the rigorous theory concisely, and Dischinger[9] brought it into a form suitable for design. Because of the complicated nature of this rigorous theory, simplifications were proposed by Finsterwalder,[10] Schorer,[11] and Donnell.[3] The Donnell theory was apparently derived independently by Vlasov[4] and by Jenkins,[5] because Donnell was not concerned with roof structures and his paper was apparently unnoticed by designers for some time. A new formulation was made by Parme for the ASCE Manual 31[1], and a "new approximate" theory has been developed by Holand.[12] MacNamee,[13] Holand,[12] and others have compared these various theories, and the following discussion is based largely on Holand's book.

If we begin with Eq. (5-31):

$$\left(\frac{M^2}{Q^2} - \gamma\right)^4 + 4 = 0$$

then we may write the formulations of Flügge, Holand, Donnell, and Schorer, respectively, as

$$\left[\frac{M^2}{Q^2} - \left(\gamma - \frac{1}{2Q^2}\right)\right]^4 + 4 - \delta = 0 \tag{5-34}$$

$$\left[\frac{M^2}{Q^2} - \left(\gamma - \frac{1}{2Q^2}\right)\right]^4 + 4 = 0 \tag{5-35}$$

$$\left(\frac{M^2}{Q^2} - \gamma\right)^4 + 4 = 0 \tag{5-36}$$

$$\left(\frac{M^2}{Q^2}\right)^4 + 4 = 0 \tag{5-37}$$

where δ represents a complex quantity (see page 31 of Ref. 12) which is normally small and which is neglected by Holand, who shows that for the normal range of shell roofs the resulting error is below 1 percent. Holand compares the values of the roots which are obtained from each of the Eqs. (5-34) to (5-37) (page 53 of Ref. 12) and shows that the Donnell equation gives roots with an error of less than 1 percent for all except very long shells where the error rapidly increases. Two independent dimensionless parameters are used by Holand: first,

$$Q = \sqrt[8]{\left(\frac{n\pi r}{L}\right)^4 3(1 - \nu^2) \left(\frac{r}{h}\right)^2} \tag{5-38}$$

which has already been defined by (b) of Sec. 5-5 for the case where $\nu = 0$; and second,

$$\kappa = \frac{(n\pi r/L)^2}{Q^2} = \frac{n\pi r}{L} \sqrt{\frac{h}{r}} \sqrt[4]{\frac{1}{3(1 - \nu^2)}} \tag{5-39}$$

which is the same as the quantity γ defined in Sec. 5-5.

A third but dependent parameter is also used:

$$\gamma_H = \kappa - \frac{1}{2Q^2} \tag{5-40}$$

which is the same as γ except for the term $1/2Q^2$. Thus we see that the only difference between the formulation of Holand and that of Donnell (by the theory of shallow shells) is the term

$$\frac{1}{2Q^2} = \frac{L}{n\pi r} \sqrt{\frac{h}{r}} \sqrt[4]{\frac{1}{48(1 - \nu^2)}}$$

compared to κ.

If we rewrite (5-40) and use (5-39), we find that

$$\gamma_H = \frac{2(n\pi r/L)^2 - 1}{2Q^2}$$

We see that γ_H diverges from γ as r/L decreases, an indication of the range of applicability of the theory of shallow shells for cylindrical shells.

Although γ_H changes substantially for an r/L ratio of 0.4 (given by Gibson as the highest value for long shells), the effect upon the roots m_1, n_1, m_2, and n_2 is very small. For example, in the shell analyzed in Sec. 6-4,

$$\gamma = 0.09257$$
$$Q = 4.1303$$
$$\frac{1}{2Q^3} = 0.0293$$
$$\gamma_H = 0.0633$$

and

$$m_1 = 1.131 \quad \text{as compared to } 1.1344$$
$$n_1 = 0.445 \quad \text{as compared to } 0.4408$$

Table 5-2

Method	Equation	m_1	n_1
Flügge	(5-34)	1.1294	0.4420
Holand	(5-35)	1.1288	0.4429
Donnell	(5-36)	1.1399	0.4386
Schorer	(5-37)	1.0986	0.4551

Holand (page 53 of Ref. 12) gives for $Q = 4.183$ the values, shown here in Table 5-2, of the roots resulting from (5-34) to (5-37), which indicate an error of less than 1 percent when (5-36) is used. Holand points out that even though the error in the roots is negligible, the errors which finally appear in some of the stress resultants and stress couples may be considerably higher. For example, in the case of $Q = 4.183$, the errors in the coefficients for $N_{\phi x}$ are about 3.2 percent. The other values given by Holand show less error.

Although exact limits will not be set upon the range of shallow-shell theory, it appears reasonable to use (5-36) for short and for intermediate shells; where shells are long, the more rigorous methods of Holand or the ASCE Table 1 can be used. In many cases, the design of long shells can be based on the beam-arch theory of Lundgren (Sec. 6-2).

5-7 EDGE BEAMS

Normally there will be two types of barrel shells with edge beams: vertical beams (Fig. 5-10) and horizontal beams (Fig. 5-11). Vertical beams are usually employed for long shells, where the principal structural action is longitudinal bending. Horizontal beams are commonly used with short shells, where the principal structural action is transverse arching.

Consider first the vertical edge beam of Fig. 5-10. In the case of the free-edge analysis, line loads T_L and S_L are added so that the resultant edge forces will be zero. If we start from this point, i.e., assume that the shell has been correctly analyzed for the free-edge case, then the effect of a vertical edge beam may be thought of as restricting vertical movement (Δ_v) and longitudinal edge strain (N_x/Eh). If the

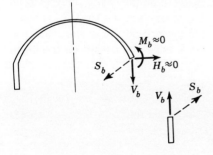

Fig. 5-10 Long-barrel shell with edge beams.

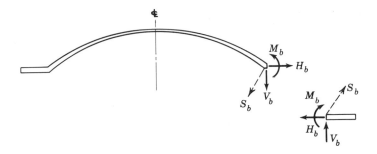

Fig. 5-11   Short-barrel shells and edge beams.

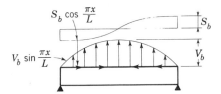

Fig. 5-12   Edge-beam loadings.

beam is slender, it is reasonable to assume that it offers negligible resistance to rotation and horizontal translation, and the unknown edge forces are V_b and S_b (see Fig. 5-10). The determination of V_b and S_b now depends upon the amount of restraint the edge beam provides to vertical deflection and longitudinal edge strain (or stress, since we assume the validity of Hooke's law).

The analysis is divided into the usual four steps:

Primary System   The shell, supporting surface loads, has free edges; and the edge member acts as a simple beam carrying its own dead weight.

Errors   Primary-system errors are (1) the difference in the vertical deflections of the shell edge and the edge beam, $D_{10}^S + D_{10}^B$, and (2) the difference in the longitudinal stresses in the shell edge and the top of the edge beam, $f_{20}^S + f_{20}^B$.

Corrections   As shown in Fig. 5-12, (1) a vertical force $V_b = X_1$, and (2) a shearing force $S_b = X_2$. V_b generally acts upward on the shell and downward on the beam, thereby reducing the vertical deflection and the longitudinal stresses in the shell. S_b generally acts inward on the shell and outward on the beam, thereby again reducing the vertical deflection and the longitudinal stresses in the shell. V_b is taken positive when it acts downward on the shell and upward on the beam, and S_b is assumed positive when it acts outward on the shell edge and inward on the beam (Fig. 5-12).

Compatibility   This is restored by

$$X_1(D_{11}^S + D_{11}^B) + X_2(D_{12}^S + D_{12}^B) + D_{10}^S + D_{10}^B = 0 \qquad (5\text{-}41)$$
$$X_1(f_{21}^S + f_{21}^B) + X_2(f_{22}^S + f_{22}^B) + f_{20}^S + f_{20}^B = 0 \qquad (5\text{-}42)$$

where (5-41) represents vertical displacements and (5-42) represents longitudinal stresses.

Vertical displacements are positive in the direction of positive V_b. For stresses, tension is taken as positive in the shell and compression as positive in the beam.

It is necessary to compute the values for these displacements and stresses at the shell edge and in the edge beam. Values for the shell are available from the derivations discussed in Sec. 5-5 and presented in the tables of the Appendix. The corresponding values for the beam come from the ordinary flexural theory as long as the beam does not behave like a deep beam. Since the shell line loads are represented by successive terms of a Fourier series, it is necessary that the same representation be used for the edge-beam loading in order to have compatibility with the shell over the full length of the longitudinal edges.

The ordinary flexural theory is based on the following relationships:

$$EI \frac{d^4 y}{dx^4} = w_x = \text{loading} \tag{5-43a}$$

$$EI \frac{d^3 y}{dx^3} = -S_x = \text{shear} \tag{5-43b}$$

$$EI \frac{d^2 y}{dx^2} = -M_x = \text{moment} \tag{5-43c}$$

$$\frac{dy}{dx} = -\int_0^L \frac{M_x}{EI} dx + f_1(x) = \theta = \text{slope} \tag{5-43d}$$

$$y = -\iint_0^L \frac{M_x}{EI} dx + \int_0^L f_1(x) dx + f_2(x) = \text{deflection} \tag{5-43e}$$

Let us consider the effect of a vertical load varying as a sine function so that Eq. (5-43a) becomes

$$EI \frac{d^4 y}{dx^4} = w_x = V_b \sin kx \tag{a}$$

Then Eqs. (5-43c) and (5-43e) become

$$EI \frac{d^2 y}{dx^2} = -M = -V_b \frac{1}{k^2} \sin kx \tag{b}$$

$$y = \frac{V_b}{EI} \frac{1}{k^4} \sin kx \tag{5-44}$$

The longitudinal stress at the top of the beam can be obtained from (b):

$$f_t = \frac{M}{Z_t} = \frac{V_b}{Z_t} \frac{1}{k^2} \sin kx \tag{5-45}$$

Next we consider the effect of a longitudinal shearing force at the top edge of the beam, which varies as a cosine function

$$S = S_b \cos kx \tag{c}$$

At any distance x from the beam support, this shear will produce a resultant thrust at the edge of the beam,

$$T = -\int_0^x S_b \cos kx\, dx = -S_b \frac{1}{k} \sin kx \qquad (d)$$

This thrust gives a bending moment Te and an axial load T on the beam. The vertical deflection, due to the bending moment, is easily computed from Eq. (5-43e), with $f_1(x) = f_2(x) = 0$:

$$y = -\left(\frac{S_b}{EI} \frac{1}{k^3} \sin kx\right) e \qquad (5\text{-}46)$$

The longitudinal stress at the top of the beam is

$$f_t = \frac{T}{A} + \frac{Te}{Z_t} \qquad (5\text{-}47)$$

For an edge beam of rectangular cross section, where $I = bd^3/12$, $Z = bd^2/6$, and $e = d/2$, these formulas can be summarized as

For $V_b = 1$:

$$D_{11}^B = \frac{L^4}{Ebd^3} \frac{0.12319}{n^4} \sin kx$$

$$f_{21}^B = -\frac{L^2}{bd^2} \frac{0.60793}{n^2} \sin kx$$

For $S_b = 1$: $\qquad\qquad\qquad\qquad\qquad\qquad\qquad\qquad (5\text{-}48)$

$$D_{12}^B = -\frac{L^3}{Ebd^2} \frac{0.19351}{n^3} \sin kx$$

$$f_{22}^B = \frac{L}{bd} \frac{1.2732}{n} \sin kx$$

The values for beam displacement and stress due to full loads on the beam alone can be obtained directly from Eqs. (5-48). The vertical uniform load may be represented by the Fourier series:

$$(w_{B+L})_x = -\frac{4}{\pi} w_{B+L} \sum_{n=1,3,5\ldots}^{\infty} \frac{1}{n} \sin kx \qquad (e)$$

where w_{B+L} is the load due to the beam weight and to the live load in kips per foot. Therefore, from Eqs. (5-48),

$$D_{10}^B = -\frac{4}{\pi} w_{B+L} \frac{L^4}{Ebd^3} \frac{0.12319}{n^4} \sin kx$$

$$\qquad\qquad\qquad\qquad\qquad\qquad\qquad\qquad\qquad (5\text{-}49)$$

$$f_{20}^B = \frac{4}{\pi} w_{B+L} \frac{L^2}{bd^2} \frac{0.60793}{n^2} \sin kx$$

For the case of a horizontal edge beam, H_b replaces V_b in Eqs. (5-48), and d and b are the beam width and depth, respectively (Fig. 5-11). The beam is again assumed to have negligible rotational and transverse (vertical in this case) stiffness. However, there will be both a vertical load V_{B+L} and a bending moment M_{B+L} at the shell edge.

$$V_{B+L} = w_{B+L} \qquad\qquad (f)$$

$$M_{B+L} = w_{B+L}\,\frac{d}{2} \qquad\qquad (g)$$

Thus, the horizontal edge beam is assumed to carry its own load as a cantilever of span d, not as a beam of span L. Ratio of depth b to span L for all edge beams will usually not exceed $\frac{1}{3}$, so the ordinary flexural theory may be applied.

5-8 PRESTRESSING WITHIN EDGE BEAMS

Edge beams normally serve two purposes: to stiffen the shell edge and to act together with the shell in carrying flexural stresses, usually tension. Often the edge beams may be substantially reduced in size if prestressing is used. Smaller beams will, however, provide less stiffening.

The analysis for edge-beam prestressing is similar to that described in Sec. 5-7.

Primary System   The shell has free edges under no load, and the edge beam acts as a simple beam under the prestressing load.

Errors   These are (1) the differences in the vertical deflections of the shell edge (zero in this case) and the edge beam, and (2) the difference in the longitudinal stresses in the shell edge (zero in this case) and the top of the edge beam.

Corrections   These are the same as those derived in Sec. 5-7.

Compatibility.   This can be expressed by Eqs. (5-41) and (5-42). The only terms which differ from those derived in Sec. 5-7 are those which represent the errors.

These errors may be easily derived from Eqs. (5-48). With the edge beam considered as a simple beam, a parabolic tendon profile will cause (1) a vertical upward deflection, and (2) a longitudinal stress at the top of the edge beam:

$$D_{10}^B = -\int\!\!\int_0^L \frac{M_F}{EI}\,dx\,dx + \int_0^L f_1(x)\,dx + f_2(x) \qquad\qquad (a)$$

and

$$f_{20}^B = \frac{F}{A} - \frac{M_F}{Z_t} \qquad\qquad (b)$$

The bending moment M_F due to the eccentricity of the tendon profile is assumed to be zero at the supports. The deflections and stresses caused by this bending moment may be computed by introducing the equivalent uniform load:

$$w_e = \frac{8Fe_c}{L^2} \qquad (c)$$

which may be represented by the Fourier series:

$$(w_e)_x = \frac{4}{\pi} w_e \sum_{n=1,3,5...}^{\infty} \frac{1}{n} \sin kx \qquad (d)$$

For an edge beam of rectangular cross section, Eq. (a) may be expressed in the same form as the first of Eqs. (5-49):

$$D_{10}^B = \frac{4}{\pi} \frac{8Fe_c}{L^2} \frac{L^4}{Ebd^3} \frac{0.12319}{n^4} \sin kx \qquad (5\text{-}50)$$

and the second term of Eq. (b) as the second of Eq. (5-49):

$$\frac{M_F}{Z_t} = \frac{4}{\pi} \frac{8Fe_c}{L^2} \frac{L^2}{bd^2} \frac{0.60793}{n^2} \sin kx \qquad (e)$$

For the first term of Eq. (b) the uniform compression is given by the Fourier series:

$$\left(\frac{F}{A}\right)_x = \frac{4}{\pi} \frac{F}{A} \sum_{n=1,3,5}^{\infty} \frac{1}{n} \sin kx \qquad (f)$$

Substituting Eqs. (e) and (f) into Eq. (b) and using only the first term of the Fourier series, we obtain

$$f_{20}^B = \frac{4}{\pi} \frac{F}{bd} \sin kx - \frac{4}{\pi} \frac{8Fe_c}{L^2} \frac{L^2}{bd^2} (0.60793) \sin kx \qquad (5\text{-}51)$$

With the values from Eqs. (5-50) and (5-51) used in Eqs. (5-41) and (5-42), V_b^F and S_b^F are determined.

5-9 MULTIPLE-BARREL SHELLS

Figure 5-13 shows a cross section through a system of multiple-barrel shells. A completely general analysis, which includes edge beams, would require the solution of four compatibility equations at each longitudinal edge of the shell. For the system in Fig. 5-13, twenty simultaneous equations would have to be solved, four each at the interior valleys and two each at the exterior edges.

It is frequently assumed that all interior shells are in the center of a symmetrical system with an infinite number of shells. On this basis, only one interior valley need be considered. For long shells, disturbances at one edge may significantly affect the opposite edge. Thus the first interior valley may be greatly affected by the exterior edge. Nevertheless, in Manual 31[1] it is recommended to compute the line loads as if all interior valleys were symmetrical. This simplification probably does not affect the design significantly, even though the designer may have difficulty in estimating the

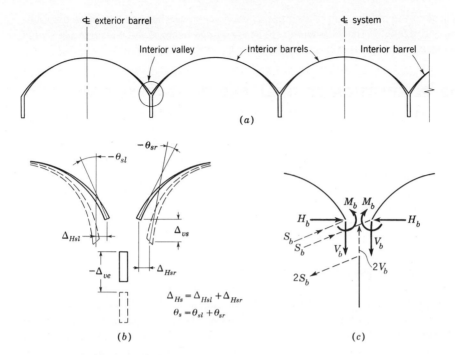

Fig. 5-13   Multiple-barrel shells.

effect on the analysis. For short shells, disturbances at one edge have negligible effects at the opposite edge.

The analysis of an interior valley will be outlined.

Primary System   This is a series of shells with free edges under surface loads and edge beams as simple beams under their own loads.

Errors   These are (1) the difference in vertical deflections of the shell edges and edge beam, (2) the difference in longitudinal stresses in the shell edges and the top of the edge beam, (3) the horizontal displacement of the shells, and (4) the rotation of the shells.

Corrections   These are expressed as in Eq. (5-48) except that all values V_b, S_b, H_b, and M_b are required. The displacements and stresses due to these corrections are obtained from the tables in the Appendix for the shell and from Sec. 5-7 for the edge beam.

Compatibility   This is restored by

$$\Sigma \Delta_{vs} + \Sigma \Delta_{ve} = 0$$
$$\Sigma f_{xs} + \Sigma f_{xe} = 0$$
$$\Sigma \Delta_{hs} = 0 \qquad\qquad\qquad (5\text{-}52)$$
$$\Sigma \theta_s = 0$$

Because of the assumption of symmetry, each adjoining shell has the same displacement and stress. In the case where there is no edge beam at the interior valley, only the last two of Eqs. (5-52) are used, since $V_b = S_b = 0$.

5-10 INTERACTION OF SHELL AND TRANSVERSE RIBS

Perhaps the most uncertain part of the analysis of cylindrical shells is the determination of internal forces and displacements at the junction of a shell with the transverse rib. This type of problem was already considered when we analyzed the dome and circular ring junction and the cylindrical wall and circular ring junction. In fact, it has been demonstrated that these two problems are usually solved by the same formulas. We shall now see that even a part of a cylinder is analyzed by an approximation which assumes a full cylinder.

The theory has already been presented in Sec. 2-6. The analysis can be divided into two parts: (1) the analysis for restraint of the shell by a stiff rib, and (2) the analysis for the restraint of the rib by the shell. This is exactly the same way we approached dome analysis. The analytic example in Chap. 3 considered the dome fully fixed, i.e., restrained by a ring which would neither translate nor rotate; then the design example of Chap. 4 considered the ring as flexible. It was noted that the moments and the hoop stresses due to fixity were very small compared with those which occurred with the flexible ring. In a like manner, the moments and ring forces due to a cylindrical shell fixed into an immovable rib will usually be small compared to the moments and ring forces which occur when the rib deflects.

In the case of the dome and ring we computed the effect of the restraint of the shell into the ring and the effect of the restraint of the ring into the shell in one operation. This was possible because all the loads and displacements were axisymmetrical. In the case of the part cylinder, this cannot be done. It would be extremely complicated to compute displacements of part cylinders (D_{11}, D_{12}, and D_{22}) and even harder to achieve compatibility between shell and rib over the arc length. The rib is no longer a ring but usually an arch or haunched frame in which the displacements are not at all symmetrical about the axis of rotation of the cylinder. For this reason an approximate method of analysis is used. The rib and the shell are assumed to be a part of a closed ring and a full cylinder. This approximation may be valid near the crown for long shells and over a substantial arc length for short shells. It must be regarded only as a method to arrive at a reasonable concrete thickness and reinforcement where a shell and rib join. Near the longitudinal edges this analysis is not valid. For a more accurate solution, a detailed finite-element analysis can be used. Experienced designers, nevertheless, completed numerous shell roofs successfully by using the simplified approach described here.

The analytic procedure is to compute the loads on the transverse rib and to analyze it as a T beam in which an effective width of the shell b_e is taken as the flange. The normal stresses in the rib are then computed, based on a homogeneous cross section. From these stresses the strain ϵ_{ϕ_0} at the junction of rib and shell is computed. The value ϵ_{ϕ_0} then leads to a solution for the shell forces and moments as follows, where, from (2-30) with $\nu = 0$, $\beta = \sqrt[4]{3}/\sqrt{rh} = 1/0.76\sqrt{rh}$,

$$Q_0 = \frac{Eh}{\beta r} \epsilon_{\phi 0}$$

$$M_0 = -0.29 E h^2 \epsilon_{\phi 0}$$

$$N_\phi = E h \epsilon_{\phi 0} \tag{5-53}$$

$$b_e = 0.76 \sqrt{rh} = 1/\beta$$

$$N_{x\phi} = \frac{Eh}{r} b_e \frac{d\epsilon_{\phi 0}}{d\phi}$$

and the longitudinal variation in moment along the shell will be

$$M_x = -0.29 E h^2 \epsilon_{\phi 0} \psi(\beta x) \tag{5-54}$$

Where the rib provides no stiffness to rotation,

$$Q_0 = \frac{Eh}{2\beta r} \epsilon_{\phi 0}$$

$$M_0 = 0$$

$$N_\phi = E h \epsilon_{\phi 0} \tag{5-55}$$

$$b_e = 0.38 \sqrt{rh}$$

$$N_{x\phi} = \frac{Eh}{r} b_e \frac{d\epsilon_{\phi 0}}{d\phi}$$

and the longitudinal variation in moment along the shell will be

$$M_x = 0.29 E h^2 \epsilon_{\phi 0} \zeta(\beta x) \tag{5-56}$$

The full derivation of these formulas comes from Ref. 1 and appears in the first edition of this book.

A conservative design would provide for both limiting cases at an exterior rib or at an unsymmetrically loaded interior rib, where the interaction would not be given by either set of expressions. Since the whole approach is an approximation and since the forces are usually small, it is probably wise to be conservative. A series of tests has shown that this approach is reasonable on short barrels.[14] These tests also have shown that where a short length of barrel is included as an overhang outside of the exterior rib, the interaction approaches that of an interior arch. The report states that "the influence of the torsional stiffness (of the transverse rib) is insignificant for an overhang >4.5 in. The portions of the shell on both sides of the rib are then very nearly self-balancing." The 4.5-in. overhang refers to a model[14] in which

$$b_e = 0.76 \sqrt{108 \times 0.118} = 2.7 \text{ in.}$$

and therefore

$$\beta = \frac{1}{2.7 \text{ in.}}$$

From Table 2-3 we observe that the function $\phi(\beta x)$, and hence N_ϕ (see Eq. (5-64) in the first edition) is zero at $\beta x = 2.35$, so that in this case

$$x = 2.35 \times 2.7 = 6.3 \text{ in.}$$

The effect of the overhang will be to increase the arch stiffness, in this case actually to give a total flange 4 times that of the flange in the exterior rib with no overhang. This increased stiffness, as stated in the report, "may be important in increasing the buckling resistance of the rib and shell combination."

The short shell which was tested was a 1/30 model of a pair of reinforced concrete hangars actually constructed, [15] and the effective width computed for these shells was approximately that given by the fourth of Eqs. (5-53) and (5-55).

In connection with the rib analysis, it should be pointed out that where N_ϕ values in the shell are zero—such as at the longitudinal free edge of a shell—then there can be no effective width. Therefore, the moment of inertia of the rib at the springing line will usually be less than it is at the crown. The stability of the rib may be most critical at the springing line.

5-11 TRANSVERSE-FRAME ANALYSIS

Figure 5-14 shows an arch rib loaded by the in-plane shearing stress resultants $N_{x\phi}$ from the shell. There will also be some load transferred by the radial shearing stress resultants Q_x, but usually these later values are so small that they may be neglected in the rib analysis.

Thus the analysis of the rib is carried out directly by the standard methods of arch analysis (see, for example, Ref. 16), provided that the magnitude and distribution of $N_{x\phi}$ have been determined.

Sometimes the transverse frames are analyzed for the vertical loads corresponding to the surface loads on the shell. Such a loading on the ribs would be correct if the shell were made up of independent slab strips, which transferred their load by bending and shear (M_x and Q_x).

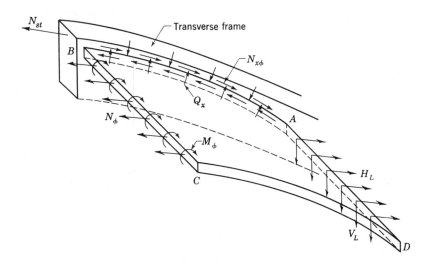

Fig. 5-14 Equilibrium of shell-arch system.

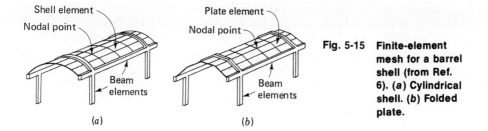

Fig. 5-15 **Finite-element mesh for a barrel shell (from Ref. 6). (a) Cylindrical shell. (b) Folded plate.**

When the slabs are connected, very little load is transferred by the bending and shear (M_x and Q_x), and the effect on the rib is to introduce, in addition to the total vertical load from the shell, a substantial tensile thrust, N_{st}, which can be computed approximately from

$$N_{st} + \int_B^C N_\phi \, dx = 0 \tag{5-57}$$

Eq. (5-57) expresses the equilibrium along a longitudinal line BC (Fig. 5-14) where C is taken at a section CD where the shear ($N_{x\phi}$) is zero. The load tends to be carried into the rib near the springing line as the shell becomes shorter (as r/L increases).

Many engineers analyze the rib for loads distributed as in the shell and then introduce a correction which consists of a tension thrust and a moment (if the shell centroid is eccentric to the rib centroid). This method is described in Ref. 1. It is not developed here, because it appears to offer only small computational advantage over the direct approach. The difficulty of analyzing arch frames for the $N_{x\phi}$ loads has been largely eliminated when standard computer programs are used. A series of tables in Ref. 17 can be conveniently applied as well.

5-12 ANALYSIS BY FINITE ELEMENTS

Fig. 5-15 shows a cylindrical barrel shell divided into shell elements and beam elements. Scordelis discusses some of the computer programs available to solve such problems,[6] and in Chap. 6 we shall describe the readily available general purpose program SAP (Structural Analysis Program) developed by Wilson and Bathe at Berkeley. The following brief introductory discussion derives from the paper by Clough and Johnson.[18]

The analysis procedure simply consists of dividing the shell system into elements, computing the stiffnesses of the elements, assembling them into a stiffness matrix for the entire shell system, and solving that matrix for any applied load. The solution will give nodal displacements which from element stiffnesses can be used to compute nodal forces and element stresses.

The great virtue of the method is that there is no restriction in principle on shell geometry, boundary conditions, and loading. In practice, complex shell systems can take considerable time in data preparation, can be costly to run on large computers, and may give questionable results since errors or sensitive assumptions are sometimes difficult to identify. In all cases, the designer needs to have a clear idea of structural behavior before attempting to use large programs.

The main issues in analysis for thin shell concrete structures are:

1. The element shape
2. The number of degrees of freedom at the nodes of the element
3. The linkages between shell and beam elements

As for element shape, it is easier to use flat-plate elements because in-plane and out-of-plane behavior are not coupled within the element. It would be more accurate geometrically to use curved elements, but it is more difficult to define their stiffness properties. For this reason, Clough and Johnson concluded that "it is not likely that the inclusion of curvature in the elements will greatly improve the finite-element approximation."

Some of the earliest element configurations were triangles; but for shells, quadrilaterals are regularly used because they provide better membrane strain distributions for lower degrees of freedom. Clough and Johnson show excellent results for a relatively coarse mesh when quadrilaterals are used with only 5 degrees of freedom for a cylindrical shell with free longitudinal edges, where $r/h = 100$, $r/L = 1.0$, $\phi_k = 40°$, and $h = 3$ in.

Scordelis described the method of including edge beams where the plate or shell elements can join the beam elements concentrically or eccentrically. The latter is an essential feature for shell systems, as we have already seen for dome rings.

REFERENCES

1. "Design of Cylindrical Concrete Shell Roofs," *Man. Eng. Pr.*, ASCE, no. 31, New York, 1952.
2. Gibson, J. E., *The Design of Cylindrical Shell Roofs*, 2d ed., D. Van Nostrand Company, Inc., Princeton, N.J., 1961.
3. Donnell, L. H., "Stability of Thin-Walled Tubes under Torsion," *NACA Rep.* 479, 1934.
4. Vlasov, V. Z., "Some New Problems on Shells and Thin Structures," *Izv. Akad. Nauk*, no. 1, 1947, pp. 27–53. (English translation in NACA TM1204.)
5. Jenkins, R. S., *Theory and Design of Cylindrical Shell Structures*, O. N. Arup Group of Consulting Engineers, London, 1947.
6. Scordelis, A. C., "Analysis of Cylindrical Shells and Folded Plates", *Concr. Thin Shells*, ACI, no. SP-28, 1971, pp. 207–236.
7. Timoshenko, S., and S. Woinowsky-Krieger, *Theory of Plates and Shells*, McGraw-Hill Book Company, 2d ed., 1959, p. 84.
8. Flügge, W., *Stresses in Shells*, Springer-Verlag OHG, Berlin, 1960.
9. Dischinger, F., "Die strenge Theorie der Kreiszylinderschale in ihrer Anwendung auf die Zeiss-Dywidag-schalen," *Beton u. Eisen*, vol. 34, pp. 257–264 and 283–294, 1935.
10. Finsterwalder, U., "Die Theorie der zylindrischen Schalengewölbe System Zeiss-Dywidag und ihre Anwendung auf die Grossmarkthalle in Budapest," Int. Assoc. Bridges and Struct. Eng., vol. 1, pp. 127–152, 1932.
11. Schorer, H., "Line Load Action on Thin Cylindrical Shells," *Proceedings ASCE*, vol. 61, pp. 281–316, 1935.
12. Holand, I., "Design of Circular Cylindrical Shells," *Norges Tekniske Vitenskapsakademie*, ser. 2, no. 3, 1957.

13. MacNamee, J. J., "Existing Methods for the Analysis of Concrete Shell Roofs," *Symposium on Concrete Shell Roof Construction,* London, 1952.
14. Thürlimann, B., and B. G. Johnston, "Analysis and Tests of a Cylindrical Shell Roof Model," *Proceedings ASCE,* vol. 80, April 1954.
15. Allen, J. E., "Construction of Long-Span Concrete Arch Hangar at Limestone Airforce Base," *J. ACI,* vol. 21, 1950, p. 405.
16. Michalos, James, *Theory of Structural Analysis and Design,* The Ronald Press Company, New York, 1958.
17. "Design Constants for Circular Arch Bents," Portland Cem. Assoc., Adv. Eng. Bull., no. 7, 1963.
18. Clough, R. W., and C. P. Johnson, "Finite Element Analysis of Arbitrary Thin Shells," *Concr. Thin Shells,* ACI, Publ. SP-28, 1971, pp. 333–363.

6

DESIGN OF CYLINDRICAL
SHELL ROOFS

6-1   INTRODUCTION

The analysis of cylindrical shells has been presented in Chap. 5. The purpose of this chapter is to demonstrate the application of this analysis within the context of a design. The design example used in this chapter is similar in its general dimensions to a large warehouse complex constructed at Middletown, Pennsylvania, and completed in 1959.[1]

Before describing the design of this structure, we shall discuss a simplified analysis for long shells, the beam-arch approximation, often used as a basis for design.

6-2   BEAM-ARCH APPROXIMATION

This method, developed by H. Lundgren[2] of Denmark, consists of separate analyses in which the shell is considered first as a beam and second as an arch.

Beam Analysis

Longitudinal stresses at any cross section of the long barrel are computed on the basis of the simple flexural theory (Fig. 6-1a):

$$f_x = \frac{N_x}{h} = \frac{M_x y}{I} \tag{6-1}$$

and shear stresses are computed by the corresponding simple expression (with $N_{x\phi} = N_{\phi x} = S$):

$$v = \frac{S}{h} = \frac{VQ}{Ib} \tag{6-2}$$

Arch Analysis

Transverse-ring stresses and bending moments are computed based on the consideration of any transverse slice of shell dx as an arch. Figure 6-1b shows such a slice and the forces acting on the free body. For an interior barrel, there will be an M and H at the edge. The vertical surface loads are held in equilibrium

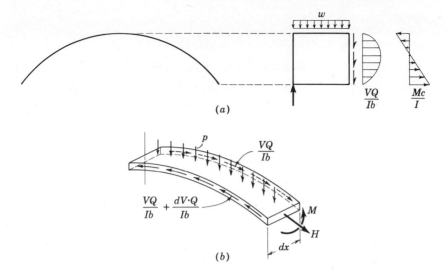

Fig. 6-1 Beam-arch analysis.

by the vertical component of the shear forces. Since the shear force in a beam under uniform load varies linearly from end to end, there will be a difference in the shear stress from one side of the shell slice to the other (for a prismatic member) of

$$\frac{dv}{dx} = \frac{dV}{dx}\frac{Q}{Ib} = w\frac{Q}{Ib}$$

which will have a vertical component of

$$\frac{wQ \sin (\phi_k - \phi)}{Ib}$$

The vertical force system, which consists of the surface load and the shear difference "support," will cause moments M_ϕ and thrusts N_ϕ in the "arch." These may be determined by the elastic center method described in Ref. 3. Some insight into the "arch" action may be gained by reference to Fig. 6-1b where the various forces acting on the arch are shown.

The beam analysis is extremely simple, but the arch analysis involves some lengthy computations.[4] For symmetrically loaded interior circular shells, Table 6-1 has been prepared,[5] which gives the shell values directly. For exterior shells with edge beams, the N_x and $N_{x\phi}$ values can often be closely approximated by the beam step, but the arch step involves more uncertainty. Methods for solving such problems have been presented by Lundgren.[2]

For initial layout and final check of the design example presented here, Table 6-1 will be used for interior barrels and to illustrate its limitations for exterior barrels without edge beams (Fig. 6-2).

The beam-arch method of analysis is based on the same assumptions as the more rigorous theory given in Chap. 5, except that the following factors are neglected:

Table 6-1*

$$N_x = \frac{L^2}{r}[p_u \text{ Col. (1)} + p_d \text{ Col. (5)}]$$

$$N_\phi = r[p_u \text{ Col. (2)} + p_d \text{ Col. (6)}]$$

$$S = -L[p_u \text{ Col. (3)} + p_d \text{ Col. (7)}]$$

$$M_\phi = r^2[p_u \text{ Col. (4)} + p_d \text{ Col. (8)}]$$

ϕ_k, deg	ϕ	Uniform transverse load				Dead-weight load			
		N_x (1)	N_ϕ (2)	S (3)	M_ϕ (4)	N_x (5)	N_ϕ (6)	S (7)	M_ϕ (8)
	ϕ_k	−6.010	−1.411	0.000	−0.00292	−6.167	−1.433	0.000	−0.00309
	$0.75\phi_k$	−4.875	−1.189	2.211	−0.00112	−5.003	−1.205	2.269	−0.00118
22.5	$0.50\phi_k$	−1.482	−0.614	3.533	0.00232	−1.521	−0.615	3.626	0.00245
	$0.25\phi_k$	4.137	0.049	3.084	0.00235	4.245	0.065	3.165	0.00249
	0	11.927	0.361	0.000	−0.00662	12.239	0.384	0.000	−0.00702
	ϕ_k	−4.855	−1.402	0.000	−0.00353	−5.012	−1.430	0.000	−0.00378
	$0.75\phi_k$	−3.937	−1.182	1.985	−0.00135	−4.064	−1.202	2.049	−0.00145
25.0	$0.50\phi_k$	−1.193	−0.612	3.170	0.00280	−1.232	−0.613	3.273	0.00300
	$0.25\phi_k$	3.342	0.044	2.765	0.00282	3.451	0.064	2.855	0.00304
	0	9.617	0.347	0.000	−0.00797	9.929	0.374	0.000	−0.00857
	ϕ_k	−4.000	−1.393	0.000	−0.00417	−4.158	−1.426	0.000	−0.00453
	$0.75\phi_k$	−3.242	−1.175	1.799	−0.00159	−3.370	−1.199	1.869	−0.00173
27.5	$0.50\phi_k$	−0.980	−0.609	2.871	0.00331	−1.018	−0.610	2.985	0.00360
	$0.25\phi_k$	2.755	0.038	2.503	0.00332	2.863	0.063	2.602	0.00363
	0	7.908	0.331	0.000	−0.00938	8.220	0.363	0.000	−0.01025
	ϕ_k	−3.350	−1.383	0.000	−0.00482	−3.508	−1.422	0.000	−0.00533
	$0.75\phi_k$	−2.714	−1.166	1.643	−0.00183	−2.842	−1.195	1.720	−0.00203
30.0	$0.50\phi_k$	−0.817	−0.606	2.622	0.00384	−0.856	−0.607	2.746	0.00424
	$0.25\phi_k$	2.308	0.032	2.284	0.00383	2.417	0.061	2.392	0.00426
	0	6.608	0.314	0.000	−0.01082	6.920	0.352	0.000	−0.01204
	ϕ_k	−2.844	−1.372	0.000	−0.00548	−3.002	−1.418	0.000	−0.00618
	$0.75\phi_k$	−2.303	−1.158	1.511	−0.00207	−2.431	−1.191	1.595	−0.00235
32.5	$0.50\phi_k$	−0.691	−0.603	2.410	0.00438	−0.729	−0.603	2.544	0.00492
	$0.25\phi_k$	1.960	0.026	2.098	0.00434	2.069	0.060	2.215	0.00492
	0	5.596	0.297	0.000	−0.01227	5.908	0.339	0.000	−0.01393

Table 6-1 (Continued)

ϕ_k, deg	ϕ	Uniform transverse load				Dead-weight load			
		N_x (1)	N_ϕ (2)	S (3)	M_ϕ (4)	N_x (5)	N_ϕ (6)	S (7)	M_ϕ (8)
35.0	ϕ_k	-2.442	-1.361	0.000	-0.00615	-2.601	-1.414	0.000	-0.00707
	$0.75\phi_k$	-1.977	-1.148	1.397	-0.00232	-2.105	-1.186	1.488	-0.00268
	$0.50\phi_k$	-0.591	-0.599	2.227	0.00491	-0.629	-0.600	2.372	0.00565
	$0.25\phi_k$	1.684	0.019	1.938	0.00484	1.793	0.058	2.064	0.00561
	0	4.794	0.278	0.000	-0.01370	5.105	0.326	0.000	-0.01591
37.5	ϕ_k	-2.118	-1.349	0.000	-0.00679	-2.278	-1.409	0.000	-0.00800
	$0.75\phi_k$	-1.174	-1.138	1.298	-0.00255	-1.842	-1.181	1.396	-0.00302
	$0.50\phi_k$	-0.510	-0.596	2.069	0.00544	-0.548	-0.596	2.224	0.00640
	$0.25\phi_k$	1.461	0.012	1.798	0.00532	1.571	0.057	1.933	0.00632
	0	4.146	0.260	0.000	-0.01509	4.458	0.312	0.000	-0.01796
40.0	ϕ_k	-1.853	-1.335	0.000	-0.00742	-2.013	-1.404	0.000	-0.00897
	$0.75\phi_k$	-1.498	-1.127	1.211	-0.00277	-1.627	-1.176	1.315	-0.00337
	$0.50\phi_k$	-0.443	-0.592	1.929	0.00595	-0.482	-0.592	2.095	0.00719
	$0.25\phi_k$	1.279	0.005	1.675	0.00578	1.389	0.055	1.819	0.00705
	0	3.616	0.241	0.000	-0.01641	3.928	0.297	0.000	-0.02006
45.0	ϕ_k	-1.449	-1.307	0.000	-0.00853	-1.610	-1.393	0.000	-0.01096
	$0.75\phi_k$	-1.170	-1.104	1.065	-0.00316	-1.299	-1.165	1.183	-0.00408
	$0.50\phi_k$	-0.343	-0.585	1.694	0.00688	-0.381	-0.583	1.882	0.00883
	$0.25\phi_k$	1.001	-0.011	1.468	0.00657	1.112	0.052	1.630	0.00854
	0	2.809	0.202	0.000	0.01872	3.120	0.266	0.000	-0.02437
50.0	ϕ_k	-1.160	-1.276	0.000	-0.00939	-1.322	-1.380	0.000	-0.01301
	$0.75\phi_k$	-0.935	-11079	0.947	-0.00344	-1.065	-1.152	1.079	-0.00480
	$0.50\phi_k$	-0.271	-0.578	1.504	0.00762	-0.308	-0.574	1.713	0.01054
	$0.25\phi_k$	0.802	-0.029	1.300	0.00714	0.914	0.049	1.481	0.01002
	0	2.232	0.164	0.000	-0.02052	2.543	0.234	0.000	-0.02871
55.0	ϕ_k	-0.946	-1.242	0.000	-0.00989	-1.109	-1.367	0.000	-0.01506
	$0.75\phi_k$	-0.761	-1.053	0.849	-0.00358	-0.892	-1.139	0.995	-0.00549
	$0.50\phi_k$	-0.217	-0.572	1.347	0.00807	-0.255	-0.563	1.578	0.01227
	$0.25\phi_k$	0.655	-0.048	1.161	0.00742	0.767	0.045	1.360	0.01144
	0	1.805	0.128	0.000	-0.02130	2.115	0.201	0.000	-0.03293
60.0	ϕ_k	-0.783	-1.205	0.000	-0.00992	-0.947	-1.352	0.000	-0.01705
	$0.75\phi_k$	-0.629	-1.025	0.766	-0.00355	-0.761	-1.124	0.927	-0.00613
	$0.50\phi_k$	-0.177	-0.566	1.213	0.00815	-0.214	-0.552	1.467	0.01398
	$0.25\phi_k$	0.543	-0.068	1.043	0.00734	0.656	0.042	1.261	0.01275
	0	1.481	0.095	0.000	-0.02118	1.790	0.167	0.000	-0.03688

* From Ref. 5.

1. Relative displacements within each transverse cross section
2. Longitudinal bending moments M_x and radial shearing forces Q_x
3. Torsional moments $M_{x\phi}$
4. Strain from the in-plane shearing forces

Assumptions 2 and 3 seem reasonable, since the forces neglected are usually small. It must be recalled, however, from Sec. 5-4, that the effects of these forces are included in the derivations for the other forces and moments, so that no completely general statement about their importance can be made here.

Assumption 4 is dangerous because it appears reasonable. In fact, it is just the assumption we make in the ordinary flexural theory for beams, frames, and arches. But the errors in displacements caused by this assumption can be important. It appears that the errors in forces and moments due to this assumption are not so great.

Finally we come to assumption 1, which is the heart of the beam-arch approximation and which theoretically is never correct. To demonstrate the meaning of this assumption, let us consider the single shell shown in Fig. 6-2 with the N_x and $N_{x\phi}$ values resulting both from the more rigorous analysis and from a beam analysis. There is little correspondence. Now let us investigate the vertical and horizontal deflection at the center of the span. From Table 1B in the Appendix (the membrane displacements are negligible in this case) we obtain

$$\Delta_V = \frac{L^4}{r^3 hE} [96.1(1.354) - 64.69(1.354) + 2.5(3.050)]$$

The membrane stress resultants, such as $N_{x\phi} = 3.050$ kips/ft, are derived in Sec. 6-4.

Table 6-2 MOMENT OF INERTIA*

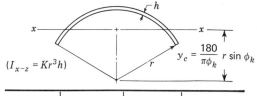

$(I_{x-z} = Kr^3 h)$

$$y_c = \frac{180}{\pi \phi_k} r \sin \phi_k$$

ϕ_k, deg	K	ϕ_k, deg	K
22.5	0.00041	37.5	0.00502
25.0	0.00068	40.0	0.00687
27.5	0.00110	45.0	0.01216
30.0	0.00168	50.0	0.02017
32.5	0.00249	55.0	0.03174
35.0	0.00358	60.0	0.04782

* From Ref. 5.

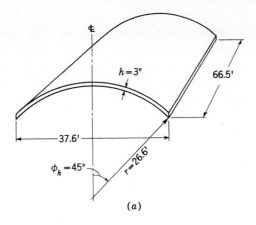

(a)

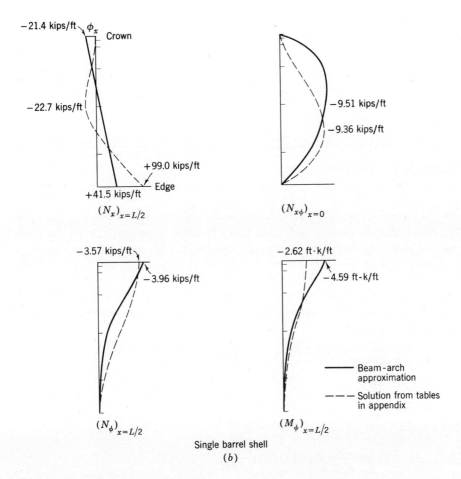

Single barrel shell
(b)

Fig. 6-2 Single-barrel shell results.

With $L^2/r^3h = 0.94$,

$$\Delta_V = 47.40 \frac{L^2}{E}$$

and

$$\Delta_H = \frac{L^4}{r^3hE} [+64.69(1.354) - 44.95(1.354) + 1.372(3.050)]$$

which gives

$$\Delta_H = 31.0 \frac{L^4}{r^3hE}$$

or

$$\Delta_H = 29.1 \frac{L^2}{E}$$

Now let us compute the deflection based on a beam analysis:

$$\Delta_V = \frac{ML^2}{\pi^2 EI}$$

$$M = 2r\phi_k \frac{4}{\pi} (0.08) \frac{66.5^2}{\pi^2} = 1915 \text{ ft-kips}$$

where ϕ_k is in radians and the load is 0.08 kip/ft^2.

$$I = r^3h(0.01216) = 57 \text{ ft}^4$$

from Table 6-2, and thus

$$\Delta_V = \frac{1}{\pi^2} \frac{1915}{57} \frac{L^2}{E} = 3.41 \frac{L^2}{E}$$

The beam analysis gives a deflection which is about 7 percent that of the more rigorous analysis, obviously a very poor correlation. We can realize the limits of the beam analysis when we study now the effect of assumptions 1 and 4 in the light of these computations. One of the principal effects of assumption 1 is that no horizontal movement may occur in the cross section. For this to be true at the edge, for example, Δ_H must equal zero. The fictitious force needed to achieve this condition would be

$$\Delta_H = 31.0 \frac{L^4}{r^3hE} - 44.95 \frac{L^4}{r^3hE} H = 0$$

$$H = \frac{31.0}{44.95} = 0.691 \text{ kip/ft}$$

The resulting vertical deflection would be

$$\Delta_V = 47.40 \frac{L^2}{E} - 64.69(0.691) \frac{L^4}{r^3hE}$$

$$= [47.4 - 44.6(0.94)] \frac{L^2}{E} = 5.50 \frac{L^2}{E}$$

which is now much closer to the deflection obtained from the beam analysis. The result of assumption 1 is seen to be, in this case, an error of 47.40/5.50 or about 870 percent.

If the deflections due to shear are neglected, as in assumption 4, then

$$\Delta_H = [31.0 - 1.372(3.050)]\,\frac{L^4}{r^3hE}$$

or

$$\Delta_H = 26.8\,\frac{L^4}{r^3hE} = 25.2\,\frac{L^2}{E}$$

and

$$H = \frac{26.8}{44.95} = 0.599 \text{ kip}$$

from which

$$\Delta_V = \frac{L^4}{r^3hE}\,[96.1(1.354) - 64.69(1.354 + 0.599)]$$

$$= \frac{L^4}{r^3hE}\,(4.00) = 3.74\,\frac{L^2}{E}$$

which is now only 10 percent greater than the deflection obtained from the beam analysis.

The beam-arch approximation is helpful when it emphasizes the essential behavior of the shell system, but its value in design is limited. Where either tables or a computer program is available for a more rigorous analysis, the beam-arch method will probably not be used.

6-3 LAYOUT FOR A SIMPLY SUPPORTED SINGLE BARREL

Figure 6-3 gives the layout of a roof unit made up of fifteen individual barrels covering a total area of 40,000 ft². The warehouse complex in Pennsylvania comprised sixteen such units covering 640,000 ft². There are a total of 240 barrels (Fig. 6-4). First we shall consider one barrel completely free of edge restraint. Actually these barrels are restrained (1) longitudinally by edge beams and by adjacent shells, and (2) transversely by the arch ribs.

The column spacing and column height are normally controlled by the owner; they will be considered as fixed in this case. The barrels are oriented in the direction of the long span primarily to permit a continuous movement of forms in the long direction of the building. A shell thickness of 3 in. is chosen for practical considerations. A minimum of three layers of reinforcement is normally used: top and bottom wire fabric with main bars in between. The cover over the fabric should be at least ⅜ in., preferably ½ in. Main bars may be as large as ¾ in. in diameter and the fabric will probably be about ¼ in. thick (two intersecting bars). This requires a minimum of 2 in. Decreasing the amount of concrete usually means one must use more careful (and hence more expensive) field labor; 3 in. of concrete will give ample cover and permit reasonably rapid construction. The shell is thickened both at the ribs and at the valleys to reduce flexural stresses (see Fig. 6-18). The local variation in thickness is neglected in the analyses that follow.[6]

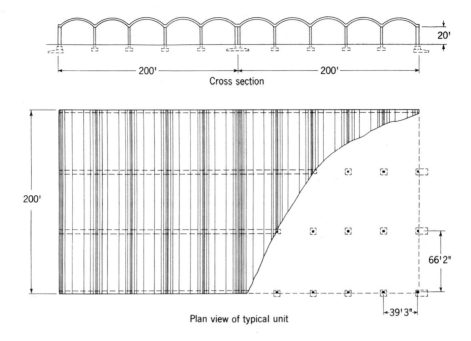

Cross section

Plan view of typical unit

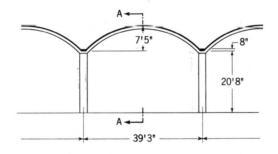

Section A-A

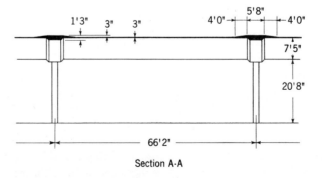

Fig. 6-3 Layout of a barrel-shell warehouse.

Fig. 6-4 Barrel-shell warehouse, Middletown, Pennsylvania. (Courtesy of Roberts & Schaefer Company, Inc.)

The choice of curvature remains to be made. The two limits are $r = 18.8$ ft (semicircle) and $r = \infty$. The first leads to a roof with low stresses but with a surface area over 1.5 times the covered area; furthermore, concrete placement would be extremely difficult on the nearly vertical slopes. The second limit, of course, leads to a slab. Here it would be helpful to use the beam analysis as a guide, except, as shown in Sec. 6-2, it is hardly applicable in this case.

It is not desirable to design shell slopes greater than 45° because of the difficulty in casting. On steep slopes, top forms are usually required, which often result in extra expense and poor results. Shells with slopes less than 45° will be easier to build, but the stresses will be higher. If a deep edge member were provided, it is likely that a solution near $\phi_k = 30°$ would be used. In this case, to avoid the deep edge beam, we choose $\phi_k = 45°$ and $r = 26.6$ ft (Fig. 6-2). It must be emphasized that this choice is not easy to set by rules of thumb; but it is important to remember that the slope should be neither too steep (for construction economy) nor too shallow (for material economy).

The shell loading is given as

	psf
1. Shell dead load	40
2. Roofing and mechanical equipment	10
3. Live load (snow)	30
Total	80

Loads 1 and 2 are mainly distributed uniformly over the surface of the shell. The distribution of 3 is not normally specified in codes, so it must be determined by the designer. Live load has often been considered to be uniformly distributed over the horizontal projection of the surface; but for simplicity of computation we shall assume that the snow is distributed evenly over the surface of the shell itself. The entire surface load is therefore distributed in the same manner as the shell weight. The layout is now complete, subject only to revisions required by the results of analysis.

6-4 ANALYSIS OF A SIMPLY SUPPORTED SINGLE BARREL BY TABLES

The analysis will be divided, as in the case of other systems, into four steps.

Primary System The stress resultants are determined from the membrane theory. The equations are given in Table 5-1. The membrane values for total load ($p = 80$ psf) are given in the first line of Table 6-3. Although we can easily compute the forces with the correct longitudinal distribution, a Fourier series will be used, with the first term considered as sufficiently accurate (as described in Sec. 5-2).

Errors The longitudinal edges are considered free in this example so that displacements along these edges are not errors, just as the free end of a cantilever cannot have displacement errors. For this part of the analysis the discontinuities along the curved transverse supports are not considered, so that the errors are solely the edge values $N'_{\phi k}$ and $N'_{\phi k x}$ which result from the membrane analysis, but which cannot exist. These have been computed from Table 5-1 as

$$N'_{\phi k} = - \frac{4}{\pi} pr \cos(\phi_k - \phi) \sin kx = -1915 \sin kx \text{ lb/ft}$$

$$N'_{\phi k x} = - \frac{4}{\pi} pr \, \frac{2}{rk} \sin(\phi_k - \phi) \cos kx$$

$$= -3050 \cos kx \text{ lb/ft}$$

$N'_{\phi k}$ is constant all along the edge and is approximated by a sine curve. $N'_{\phi k x}$ varies linearly with x and is approximated by a cosine curve. At $x = L/2$

$$N'_{\phi k} = -1915 \text{ lb/ft}$$

and at $x = 0$ or $x = L$

$$N'_{\phi k x} = -3050 \text{ lb/ft}$$

Corrections Line loads are added to the primary system to remove the errors. These line loads will be $-N'_{\phi k}$ and $-N'_{\phi k x}$. The shell constants are

$$\frac{r}{h} = \frac{26.6}{3/12} = 106 \approx 100$$

$$\frac{r}{L} = \frac{26.6}{66.5} = 0.4$$

$$\phi_k = 45°$$

Since $r/L < 0.6$ and $\phi_k \leq 45°$, this shell may be classified as a long barrel, and Table 1A in the Appendix may be used in its analysis. Since this table is set up on the basis of vertical and horizontal line loads, $-N'_{\phi_k}$ is resolved into

$$V_L = -N'_{\phi_k} \sin \phi_k$$

and

$$H_L = -N'_{\phi_k} \cos \phi_k$$

Compatibility   No solution is needed to compute the correction forces, since by inspection we know that

$$V_L = -N'_{\phi_k} \sin \phi_k = +1354 \text{ lb/ft}$$
$$H_L = -N'_{\phi_k} \cos \phi_k = +1354 \text{ lb/ft}$$

and

$$S_L = -N'_{\phi_k x} = +3050 \text{ lb/ft}$$

Of course, the real problem, that of determining the effect of these symmetrical line loads on the shell, involves the solution of four simultaneous equations, but these solutions are tabulated and the numerical work can, therefore, be avoided. Table 1A gives the required values which have been compiled into Table 6-3 and combined with the membrane forces to give the final results. These results have been plotted in Fig. 6-2 and compared with values obtained from a beam-arch analysis. A solution following shallow-shell theory as developed in Sec. 5-4 shows results that are close to those in Table 6-3. Either method will provide a reasonable basis for design in this case.

We may make some important observations at this point:

1. The beam analysis is not valid in this case. The center of compression is lower; hence the moment arm is reduced from the beam approximation. The result is a greatly increased tension at the edge.
2. The membrane theory is not even remotely helpful in depicting the action of this shell. It is helpful, however, as a primary system for analysis since it is a simple matter to compute the errors and, most important, the required corrections have been tabulated.
3. Large tension stresses due to transverse bending moments occur at the crown of the shell. These moments are large primarily because of the lack of edge restraint and the resulting large deflection of the edge relative to the crown.
4. These results, being close to those from shallow-shell theory, add evidence to Scordelis's observation that the latter theory is valid down to $r/l = 0.2$.

This example is not considered the best layout for a single-barrel roof. It is recommended that edge beams be added to reduce longitudinal tension and to control the edge displacements, thereby reducing transverse bending. If edge displacements are not critical, the tension stresses may be effectively controlled by prestressing. Detailed reinforcement will be given only for the case of the shell with edge beams.

Table 6-3 STRESS RESULTANTS AND COUPLES IN A SIMPLY SUPPORTED SINGLE-BARREL CYLINDRICAL SHELL

Loading	Multipliers	Angles ϕ (measured from edge)					
		45°	40°	30°	20°	10°	0°
		$N_x = T_x$ at $x = L/2$, kips/ft					
Load	Coefficient (C_m)* $4/\pi \times 0.080 \times 26.6 \times 6.25 \times C_m$	-0.2026 -3.430	-0.2019 -3.419	-0.1957 -3.319	-0.1837 -3.110	-0.1660 -2.811	-0.1433 -2.427
V_L	Coefficient (C)† $1.354 \times 6.25 \times C$	$+4.329$ $+36.651$	$+3.663$ $+31.011$	-0.918 -7.771	-6.034 -51.086	-3.355 -28.405	$+19.64$ $+166.272$
H_L	Coefficient (C) $1.354 \times 6.25 \times C$	-3.817 -32.315	-3.361 -28.454	-0.150 -1.269	$+3.796$ $+32.139$	$+3.104$ $+26.278$	-10.78 -91.264
S_L	Coefficient (C) $3.050 \times 6.25 \times C$	-0.0653 -1.243	-0.0723 -1.378	-0.1030 -1.963	-0.0344 -0.656	$+0.3651$ $+6.974$	$+1.388$ $+26.448$
Total		-0.337	-2.24	-14.32	-22.71	$+2.04$	$+99.03$
		$N_{\phi x} = S$ at $x = 0$, kips/ft					
Load	Coefficient (C_m) $4/\pi \times 0.08 \times 26.6 \times 2.5 \times C_m$	0 0	-0.0555 -0.375	-0.1648 -1.115	-0.2690 -1.821	-0.3652 -2.474	-0.4502 -3.050
V_L	Coefficient (C) $1.354 \times 2.5 \times C$	0 0	$+1.126$ $+3.813$	$+2.029$ $+6.872$	-0.008 -0.027	-3.194 -10.816	0 0
H_L	Coefficient (C) $1.354 \times 2.5 \times C$	0 0	-1.005 -3.403	-2.079 -7.042	-1.015 -3.437	$+1.260$ $+4.267$	0 0
S_L	Coefficient (C) $3.050 \times 2.5 \times C$	0 0	-0.0186 -0.142	-0.0673 -0.514	-0.1142 -0.870	-0.0449 -0.342	$+0.4000$ $+3.050$
Total		0	-0.11	-1.80	-6.16	-9.36	0

* Coefficients C_m taken from Table 5-1.

† Coefficients C taken from Table 1A in Appendix.

Table 6-3 (Continued)

Loading	Multipliers	Angles φ (measured from edge)					
		45°	40°	30°	20°	10°	0°
		$N_\phi = T_\phi$ at $x = L/2$, kips/ft					
Load	Coefficient (C_m)	−1.000	−0.9962	−0.9659	−0.9063	−0.8191	−0.7071
	$4/\pi \times 0.08 \times 26.6 \times C_m$	−2.709	−2.699	−2.618	−2.456	−2.219	−1.915
V_L	Coefficient (C)	−0.119	−0.277	−1.249	−1.889	−0.854	+0.7071
	$1.354 \times C$	−0.161	−0.374	−1.691	−2.558	−1.157	+0.957
H_L	Coefficient (C)	−0.372	−0.230	+0.703	+1.638	+1.497	+0.7071
	$1.354 \times C$	−0.504	−0.312	+0.952	+2.219	+2.027	+0.957
S_L	Coefficient (C)	−0.0657	−0.0630	−0.0384	+0.0158	+0.0716	0
	$3.050 \times C$	−0.200	−0.192	−0.117	+0.048	+0.219	0
Total		−3.57	−3.58	−3.47	−2.75	−1.13	0
		M_ϕ at $x = L/2$, kips					
V_L	Coefficient (C)	−0.1513	−0.1526	−0.1564	−0.1377	−0.0780	0
	$1.354 \times 26.6 \times C$	−5.451	−5.498	−5.635	−4.962	−2.811	0
H_L	Coefficient (C)	+0.0899	+0.0923	+0.1060	+0.1083	+0.0731	0
	$1.354 \times 26.6 \times C$	+3.240	+3.325	+3.819	+3.902	+2.635	0
S_L	Coefficient (C)	−0.0051	−0.0049	−0.0035	−0.0013	−0.0003	0
	$3.050 \times 26.6 \times C$	−0.413	−0.397	−0.283	−0.106	+0.024	0
Total		−2.62	−2.57	−2.10	−1.17	−0.15	0

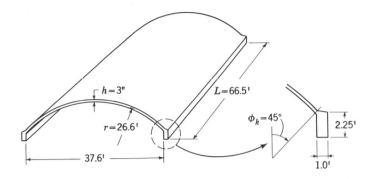

Fig. 6-5 Barrel shell with edge beams.

6-5 ANALYSIS OF A SIMPLY SUPPORTED SINGLE BARREL WITH EDGE BEAMS

The same shell considered in Sec. 6-4 will be analyzed with a vertical edge beam (Fig. 6-5).

Primary System In this case we shall take the final results of the analysis in Sec. 6-4 as the primary system. It may be reemphasized that a primary system need not be statically determinate; it is only necessary that all its internal forces and displacements be known. We now have the forces and can easily obtain the displacements from Tables 1A and 1B in the Appendix. In addition, the edge beams are considered as simple beams disconnected from the shell.

Errors As described in Sec. 5-7, there will be errors in this system because the vertical deflections and longitudinal stresses of the shell edge and beam top are generally not the same. The torsional rigidity and horizontal stiffness of the beam are neglected as recommended in Ref. 6. The effects of this assumption are discussed at the end of this section. The vertical deflection of the shell edge under total load $p = 80$ psf is (from Eqs. 5-20 and Table 1B in Appendix)

$$\Delta v = \frac{L^4}{r^3 hE} \left\{ pr \left[\left(\frac{2r}{\pi L} \right)^2 + \frac{2}{\pi^4} + \left(\frac{r}{L} \right)^4 0.500 \right] \right.$$

$$\left. + 96.10 V_L - 64.69 H_L + 2.500 S_L \right\} \sin kx$$

$$= \frac{L^4}{r^3 hE} [2.128(0.0649 + 0.0206 + 0.0128)$$

$$+ 130.12 - 87.59 + 7.62] \sin kx$$

$$= \frac{L^4}{r^3 hE} (50.41) \sin kx = \frac{L^2}{E} (47.401) \sin kx = D_{10}^S$$

The longitudinal stress resultant at ϕ_k from Table 6-3 is 99.03 kips/ft or a stress of $99.03/0.25 = 396.12$ kips/ft² $= f_{20}^S$ at $x = L/2$.

The vertical deflection and the corresponding top fiber stress of the edge beam under its own dead load are obtained from Eqs. (5-49).

$$D_{10}^B = -\frac{4}{\pi}\,(0.338)\,\frac{L^4}{Ebd^3}\,(0.123)\sin kx = -20.52\,\frac{L^2}{E}\,\sin kx$$

$$f_{20}^B = \frac{4}{\pi}\,(0.338)\,\frac{L^2}{bd^2}\,(0.608)\sin kx$$

which equals 227.7 kips/ft^2 at $x = L/2$. (Compression taken here as positive.)

Corrections The correction forces applied are V_b and S_b. The resulting vertical displacements and longitudinal stresses on the shell are

$$\Delta v = \frac{L^4}{r^3 hE}\,(96.10V_b + 2.50S_b)\sin kx$$

$$= \frac{L^2}{E}\,(90.32V_b + 2.35S_b)\sin kx = X_1 D_{11}^S + X_2 D_{12}^S$$

and

$$f^S = \left(\frac{L}{r}\right)^2 \frac{1}{h}\,(19.64V_b + 1.388S_b)\sin kx$$

$$= (491V_b + 34.7S_b)\sin kx = X_1 f_{21}^S + X_2 f_{22}^S$$

and the resulting values for the edge beam (from Eqs. 5-48) are

$$\Delta_V = \frac{L^4}{bd^3 E}\left[0.123V_b - \frac{d}{L}\,(0.1935)S_b\right]\sin kx$$

$$= \frac{L^2}{E}\,(47.83V_b - 2.542S_b)\sin kx = X_1 D_{11}^B + X_2 D_{12}^B$$

and

$$f_t^B = \frac{L}{bd}\left(-0.608\,\frac{L}{d}\,V_b + 1.273S_b\right)\sin kx$$

$$= (-530.84V_b + 37.63S_b)\sin kx = X_1 f_{21}^B + X_2 f_{22}^B$$

Compatibility The two simultaneous equations (5-41) and (5-42)

$$(90.32 + 47.83)\,V_b + (2.35 - 2.542)\,S_b + (47.40 - 20.52) = 0$$
$$(491 - 530.84)\,V_b + (34.7 + 37.63)\,S_b + (396.12 + 227.7) = 0$$

are solved to give

$$V_b = -0.207 \text{ kip/ft}$$
$$S_b = -8.725 \text{ kips/ft}$$

The resulting stress resultants and couples in the shell are given in Table 6-4 and plotted in Figs. 6-6 and 6-7.

In complicated problems of analysis it is highly desirable to check the results for static equilibrium. Such a check does not ensure that the analysis is correct, but

Table 6-4 SHELL WITH EDGE BEAMS

Loading	Multipliers	Angles ϕ (measured from edge)					
		15°	40°	30°	20°	10°	0°
$N_x = T_x$ at $x = L/2$, kips/ft							
Simple shell		−0.34	−2.24	−14.32	−22.71	+2.04	+99.03
V_b	C	+4.329	+3.663	−0.918	−6.034	−3.355	+19.64
	−1.294C	−5.603	−4.742	+1.187	+7.811	+4.341	−25.409
S_b	C	−0.0653	−0.0723	−0.1030	−0.0344	+0.3651	+1.388
	−54.542C	+3.560	+3.943	+5.617	+1.876	−19.906	−75.704
Total		−2.38	−3.04	−7.52	−13.02	−13.53	−2.08
$N_{\phi x} = S$ at $x = 0$, kips/ft							
Simple shell		0	−0.11	−1.80	−6.16	−9.36	0
V_b	C	0	+1.126	+2.029	−0.008	−3.194	0
	−0.516C		−0.580	−1.046	+0.003	+1.648	0
S_b	C	0	−0.0186	−0.0673	−0.1142	−0.0449	+0.4000
	−21.813C	0	+0.405	+1.468	+2.491	+0.979	−8.725
Total		0	−0.29	−1.38	−3.67	−6.73	−8.725
$N_\phi = T_\phi$ at $x = L/2$, kips/ft							
Simple shell		−3.57	−3.58	−3.47	−2.75	−1.13	0
V_b	C	−0.119	−0.277	−1.249	−1.889	−0.854	+0.707
	−0.207C	+0.025	+0.057	+0.259	+0.392	+0.177	−0.146
S_b	C	−0.0657	−0.0630	−0.0384	+0.0158	+0.0716	0
	−8.725C	+0.573	+0.549	+0.335	−0.137	−0.624	0
Total		−2.97	−2.97	−2.88	−2.49	−1.58	−0.146
M_ϕ at $x = L/2$, kips/ft							
Simple shell		−2.62	−2.57	−2.10	−1.17	−0.15	0
V_b	C	−0.1513	−0.1526	−0.1564	−0.1377	−0.0780	0
	−5.506C	+0.833	+0.839	+0.861	+0.759	+0.430	0
S_b	C	−0.0051	−0.0049	−0.0035	−0.0013	+0.0003	0
	−232.10C	+1.184	+1.137	+0.812	+0.302	−0.070	0
Total		−0.60	−0.60	−0.42	−0.11	+0.21	0

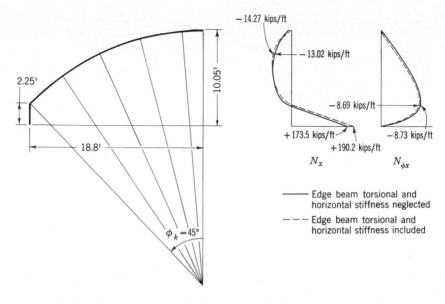

Fig. 6-6 N_x and $N_{\phi x}$ values for barrel shell with edge beams.

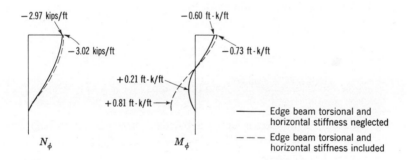

Fig. 6-7 N_ϕ and M_ϕ values for barrel shell with edge beams.

often the static computations reveal errors which are otherwise hard to detect. If we isolate half the span of the shell as a free body, we can write two equations of equilibrium at the midspan: $\Sigma H = 0$ and $\Sigma M = wL^2/8$, or for the assumed sinusoidal variations $\Sigma M = wL^2/\pi^2$.

1. The sum of the longitudinal forces must be zero. In the shell

$$\int N_x\, ds \approx \left[\frac{10}{57.3}\,(26.6)\right]\frac{1}{3}\,[-2.08 - 4(13.53) - 2(13.02)$$

$$- 4(7.52) - 3.04] + \frac{2.32}{3}\,[2(-2.38) - 3.04] = -184.51\ \text{kips}$$

and in the edge beam where the stress is at the bottom (tension positive)

$$f_b = 227.7 - 530.84 V_b + \frac{37.63}{2} S_b = 173.5 \, \text{kips/ft}$$

and at the top (compression negative)

$$f_t = -2.08 \frac{1.0 \, \text{ft}}{0.25 \, \text{ft}} = -8.32 \, \text{kips/ft}$$

The total force is

$$\left(\frac{173.50 + 8.32}{2} - 8.32 \right) 2.25 = 185.82 \, \text{kips}$$

The sum of the two is close enough to zero to provide a good check.

2. The bending moment of the internal forces about any horizontal axis must equal the moment of the total load at midspan or wL^2/π^2 where w in this case is the total load or

$$w \, (\text{shell}) = \frac{4}{\pi} \, 80 \, \text{psf} \, \frac{\pi r}{4} = 2128 \, \text{lb/lin ft}$$

$$w \, (\text{beam}) = \frac{4}{\pi} \, (338) = 429 \, \text{lb/lin ft}$$

$$w_T = 2557 \, \text{lb/lin ft}$$

and

$$M_T = \frac{2.557 \times (66.5)^2}{\pi^2} = 1141 \, \text{ft-kips}$$

Taking the moment of internal forces about the horizontal line through the shell edge, we obtain the values given in Table 6-5, where the distance to the centroid of each shell segment is given by

$$y = r \, [\cos (\phi_k - \phi) - \cos \phi_k]$$
$$M \, (\text{shell}) = 842 \, \text{ft-kips}$$
$$M \, (\text{beam}) = \frac{173.50}{2} \, (2.25)(1.50) - 8.32 \frac{2.25}{3} \, (1.125)$$
$$= 293.42 - 7.00 = 286.42 \, \text{ft-kips}$$
$$M_T = 842 + 286 = 1128 \, \text{ft-kips}$$

Figures 6-6 and 6-7 show the effects of the torsional and horizontal stiffness of the edge beam on the internal forces and moments.

The curves have been plotted from solutions, using shallow-shell theory, and show that the effect of the edge member is important only for M_ϕ.

Table 6-5 INTERNAL MOMENT IN SHELL AT MIDSPAN*

Point	ϕ	$\phi_k - \phi$	$\cos(\phi_k - \phi)$	y	$N_x(\text{av})$	$M/\Delta S$
1	5	40	0.766	1.57	−7.80	+12.246
2	15	30	0.866	4.23	−13.27	+56.132
3	25	20	0.940	6.20	−10.27	+63.674
4	35	10	0.985	7.39	−5.28	+38.650
5	42.5	2.5	0.999	7.79	−2.71	+10.542

* Refer to Fig. 6-6 (ϕ measured from edge).

Note: $\Delta S = (10/180)\ \pi(26.6) = 4.64$ ft.

$M/\Delta S = -yN_x(\text{av}) = 181.244$

$M = 181.244 \times 4.642 = 842$ ft-kips.

6-6 ANALYSIS OF SIMPLY SUPPORTED MULTIPLE BARRELS

Figure 6-3 shows the system considered here.

Primary System Consists of a series of free-edge shells, each like the example analyzed in Sec. 6-4.

Errors There are no edge beams and the loading and system are taken as symmetrical. Therefore, the edges of two adjoining shells, considered free in the primary system, will deflect vertically the same amount and will not develop any longitudinal shears along their free edges. The shells will, however, displace laterally and rotate relative to each other. Rotation at the edge due to membrane forces only is negligible. These errors can be expressed (from Eq. 5-21 and Table 1B in the Appendix) as

$$\Delta_H = \frac{L^4}{r^3 hE}\left[0.500pr\left(\frac{r}{L}\right)^4 + 64.69V_L - 44.95H_L + 1.372S_L\right]\sin kx$$

$$\theta_s = \frac{r^2}{EI}(0.09256V_L - 0.06746H_L + 0.00164S_L)\sin kx$$

which become (where $\Delta_H = D_{10}$ and $\theta_s = D_{20}$)

$$D_{10} = \frac{L^4}{r^3 hE}(+0.03 + 87.59 - 60.86 + 4.18)\sin kx$$

$$= \frac{L^4}{r^3 hE}(30.94)\sin kx$$

$$D_{20} = \frac{r^2}{EI}(0.125 - 0.091 + 0.005)\sin kx$$

$$= \frac{r^2}{EI}(0.039)\sin kx$$

Corrections Positive values for H and M in Table 1B in the Appendix (see Fig. 5-13c) result in negative displacements. By setting $H = -X_1 = -1$ and $M = -X_2 = -1$, we obtain from Table 1B

$$D_{11} = 44.95 \frac{L^4}{r^3 hE}$$

$$D_{12} = 207.1 \frac{L^4}{r^4 hE}$$

$$D_{21} = 0.06746 \frac{r^2}{EI}$$

$$D_{22} = 0.4348 \frac{r}{EI}$$

$D_{12} = D_{21}$ but is expressed differently in Table 1B.

Compatibility

$$44.95X_1 + 207.1 \frac{X_2}{r} = -30.94$$

$$0.06746X_1 + 0.4348 \frac{X_2}{r} = -0.039$$

from which

$$-X_1 = H = +0.965 \text{ kip/ft}$$
$$-X_2 = M = -0.060 \times 26.6 = -1.596 \text{ ft-k/ft}$$

The forces in a typical interior shell are compiled in Table 6-6. In this case, a comparison with the beam-arch approximation is interesting. Table 6-7 gives the forces by this method obtained from Table 6-1. The correspondence in this case is seen to be good, mainly because the shell edge is restrained from horizontal displacement.

The values in Table 6-7 are for loads distributed uniformly in the longitudinal direction, whereas those in Table 6-6 are for a sine distribution. Thus the N_x values should be smaller in Table 6-7 by the factor $(\frac{1}{2})/(4/\pi^3) = 0.96$ and the S values larger by $(\frac{1}{2})/(4/\pi^2) = 1.23$. The maximum N_x is smaller but by $41.49/48.88 = 0.85$, and the maximum S is larger by $10/8.07 = 1.24$. On the other hand, the N_ϕ values in Table 6-7, being directly influenced by the local loads, should be smaller by the ratio $\pi/4 = 0.785$, and they are about that, being at the crown $2.96/3.65 = 0.81$ and at the edge $0.56/0.69 = 0.81$ also. Thus the beam-arch approximation would be even closer than Table 6-7 shows if the two loads were taken with the same longitudinal distribution.

6-7 SOLUTIONS BY NUMERICAL METHODS

Thin shells analysts frequently use numerical methods already programmed either specifically for one type of geometry or generally for any type. We have described in Chaps. 2–4 such special-purpose programs for shells of rotation. Scor-

Table 6-6 MULTIPLE-BARREL CYLINDRICAL SHELL

Loading	Multipliers	Angles ϕ (measured from edge)					
		45°	40°	30°	20°	10°	0°
		$N_x = T_x$ at $x = L/2$, kips/ft					
Simple shell	From Table 6-3	−0.337	−2.24	−14.32	−22.71	+2.04	+99.03
H_L	Coefficient (C)*	−3.817	−3.361	−0.150	+3.796	+3.104	−10.78
	$0.965 \times 6.25 \times C$	−23.01	−20.26	−0.90	+22.89	+18.71	−64.99
M_L	Coefficient (C)	−12.50	−11.09	−0.92	+12.56	+11.51	−39.59
	$−0.060 \times 6.25 \times C$	+4.68	4.16	+0.34	−4.71	−4.31	+14.84
Total		−18.66	−18.34	−14.88	−4.53	+16.44	+48.88
		$N_{\phi x} = S$ at $x = 0$, kips/ft					
Simple shell	From Table 6-3	0	−0.11	−1.80	−6.16	−9.36	0
H_L	Coefficient (C)	0	−1.005	−2.079	−1.015	+1.260	0
	$0.965 \times 2.5 \times C$	0	−2.435	−5.037	−2.459	+3.053	0
M_L	Coefficient (C)	0	−3.299	−6.970	−3.637	+4.262	0
	$−0.060 \times 2.5 \times C$	0	+0.495	+1.059	+0.554	−0.639	0
Total		0	−2.05	−5.78	−8.07	+6.95	0
		$N_\phi = T_\phi$ at $x = L/2$, kips/ft					
Simple shell	From Table 6-3	−3.57	−3.58	−3.47	−2.75	−1.13	0
H_L	Coefficient (C)	−0.372	−0.230	+0.703	+1.638	+1.497	+0.707
	$0.965 \times C$	−0.360	−0.222	+0.682	+1.587	+1.451	+0.685
M_L	Coefficient (C)	−4.647	−4.168	−0.971	+2.441	+2.363	0
	$−0.060 \times C$	+0.283	+0.253	+0.059	−0.147	−0.144	0
Total		−3.65	−3.55	−2.73	−1.31	+0.17	+0.69
		M_ϕ at $x = L/2$, kips					
Simple shell	From Table 6-3	−2.62	−2.57	−2.10	−1.17	−0.15	0
H_L	Coefficient (C)	+0.0899	+0.0923	+0.1060	+0.1083	+0.0731	0
	$0.965 \times 26.6 \times C$	+2.317	+2.379	+2.733	+2.794	+1.885	0
M_L	Coefficient (C)	+0.2210	+0.2412	+0.3846	+0.5941	+0.7936	+1.000
	$−0.060 \times 26.6 \times C$	−0.357	−0.390	−0.622	−0.960	−1.283	−1.596
Total		−0.66	−0.58	+0.01	+0.66	+0.45	−1.596

*C from Table 1A in Appendix.

Table 6-7　MULTIPLE-BARREL CYLINDRICAL SHELL (By beam-arch analysis)

		N_x $\dfrac{66.5^2}{26.6} \times 0.08$ $= 13.299$	S -66.5×0.08 $= -5.312$	N_ϕ 26.6×0.08 $= 2.128$	M_ϕ $26.6^2 \times 0.08$ $= 56.608$
$\phi_k = 45°$	Coefficient $(C)*$	-1.610 -21.41	0.000 0	-1.393 -2.96	-0.01096 -0.62
$0.75\phi_k = 33.75°$	C	-1.299 -17.28	$+1.183$ -6.29	-1.165 -2.48	-0.00408 -0.22
$0.50\phi_k = 22.5°$	C	-0.381 -5.07	$+1.882$ -10.00	-0.583 -1.23	$+0.00883$ $+0.50$
$0.25\phi_k = 11.25°$	C	$+1.112$ $+14.78$	$+1.630$ -8.66	$+0.052$ $+0.11$	$+0.00854$ $+0.48$
$0 = 0°$	C	$+3.120$ $+41.49$	0.000 0	$+0.266$ $+0.56$	-0.02437 -1.38

*C from Table 6-1.

delis and his students have developed such programs for cylindrical barrels and folded plates.[7] For simply supported barrels his results do not differ significantly from results given in Secs. 6-4 through 6-6. However, these programs allow study of a wide range of geometric parameters, whereas the tables used here are limited. As described in Sec. 5-12, analysts often use general-purpose programs based on the finite-element method. For cylindrical barrels Scordelis has also used special-purpose versions of that method.

To illustrate the types of results gotten as well as some of the problems associated with finite-element programs, we give here in Table 6-8 results from a series of studies on one-quarter of the two-way symmetrical simply supported single-barrel shell analyzed in Sec. 6-4. The table shows the influence of mesh fineness: the coarse mesh consists of ten elements that are evenly spaced ($\Delta x = 3.325$ ft) in the x direction and six elements that are evenly spaced ($\Delta\phi = 7.5°$) in the ϕ direction; the fine-coarse mesh has a finer mesh near the transverse support, the free edge, and the crown; and the fine mesh divides the quarter shell into 306 elements as shown in Fig. 6-8.

Table 6-8 summarizes the results for N_x and M_ϕ and Fig. 6-9 plots N_x to show that even the coarse mesh results follow closely those found in Table 6-3 using the analytic solution. The fine-mesh results do show minor differences in distribution and modest increases in maximum tensions and compressions, but the influence on design would be negligible.

As for the transverse bending moments M_ϕ, the differences between coarse mesh, fine mesh, and the values from Table 6-3 are again of no design importance. The general-purpose program SAP, which is used here, has only flat elements so that the structure is idealized as a folded plate, which emphasizes the unimportance of

Table 6-8

ϕ	N_x Uniform load			N_x Sine load		M_ϕ Uniform load			M_ϕ Sine load	
	Coarse	Fine-coarse	Fine	Fine	Table 6-3	Coarse	Fine-coarse	Fine	Fine	Table 6-3
45	...	...	...	...	−0.34	...	...	...	...	−2.62
43.75	...	2.24	1.61	1.93	...	...	−2.60	−2.51	−2.49	
41.25	−1.27	1.16	0.56	0.88	...	−2.59	−2.58	−2.48	−2.47	
40	...	...	...	...	−2.24	...	...	...	...	−2.57
38.75	...	−0.91	−1.47	−1.16	...	...	−2.53	−2.43	−2.43	
33.75	−8.47	−8.14	−7.83	−7.64	...	−2.30	−2.29	−2.24	−2.27	
30	...	...	...	...	−14.32	...	...	...	...	−2.10
26.25	−17.77	−19.31	−19.17	−19.56	...	−1.77	−1.81	−1.75	−1.82	
20	...	...	...	...	−22.71	...	...	...	...	−1.17
18.75	−19.77	−22.84	−23.35	−24.72	...	−1.07	−1.10	−1.04	−1.13	
16.25	...	...	−20.84	−22.47	...	...	...	−0.78	−0.86	
13.75	...	...	−15.39	−17.13	...	...	...	−0.53	−0.59	
11.25	−2.07	−4.40	−6.32	−7.89	...	−0.39	−0.36	−0.30	−0.34	
10	...	...	...	...	2.04	...	...	...	...	−0.15
8.75	...	...	7.19	6.21	...	...	...	−0.11	−0.13	
6.25	...	24.13	26.13	26.75	...	...	...	0.04	0.02	0.03
3.75	49.8	51.35	51.65	54.00	...	−0.04	0.09	0.08	0.09	
1.25	...	86.00	84.25	89.85	...	...	0.05	0.04	0.05	
0	...	...	...	...	99.03	...	...	...	...	0

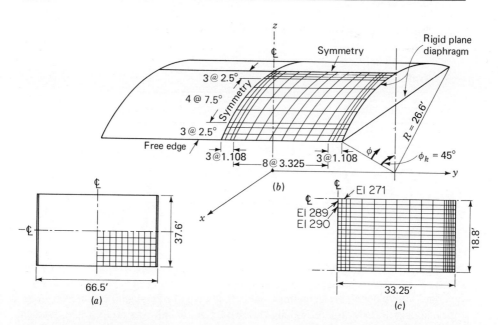

Fig. 6-8   Finite-element mesh for a barrel shell. (*a*) Plan and coarse mesh. (*b*) Fine-coarse mesh. (*c*) Fine mesh.

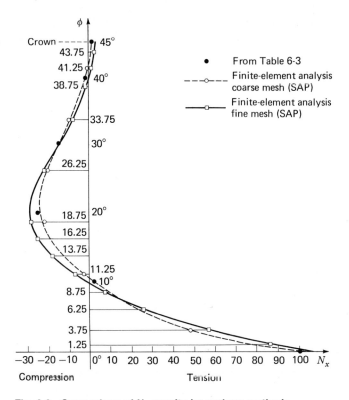

Fig. 6-9 Comparison of N_x results by various methods.

Table 6-9

Point	Location (ϕ)*	N_x†	∂N_x†	$\partial^2 N_x$†
5	11.25	−7.89		
4	8.75	6.21	14.10	
3	6.25	26.75	20.54	6.44
2	3.75	54.00	27.25	6.71
1	ˎ1.25	89.85	35.85	8.60

* In degrees.
† In kips per foot.

minor local deviation in geometry for this type of shell. Also, this type of program only gives internal forces at the center of the elements, while at the nodes and hence the edges it gives only displacements. Thus to get N_x, for example, at the free edge, it is necessary to extrapolate. Taking the results of Table 6-8 as an example, we can extrapolate using finite differences by computing first and second differences in Table 6-9 where the difference in distance $\Delta = r\phi = 26.6\ (2.5/57.3) = 1.16$ ft. Then the value of N_x at the edge $\phi = 0°$ can be estimated as

$$N_{x0} = N_{x1} + \left(\frac{\partial N_x}{\Delta} + \frac{1}{2}\frac{\partial^2 N_x}{\Delta^2}\Delta\right)\frac{1}{2}\Delta$$

$$= N_{x1} + \partial N_x/2 + \partial^2 N_x/4$$

$$= 89.85 + 17.92 + 2.15 = 109.92 \text{ kips/ft}$$

which is about 11 percent higher than the 99.03 kips/ft found in Table 6-3. In the formula for N_{x0} the second term gives the average slope of the N_x curve at point 1, taking the curvature, or changes in slope, into account.

Finally, as one further example of the interpretation of results, we take the center of the shell and show how the finite-element results give an insight into barrel-shell behavior. The third of Eqs. (5-22) gives equilibrium in the radial direction as

$$\frac{\partial Q_x}{\partial x} + \frac{\partial Q_\phi}{r\,\partial\phi} + \frac{N_\phi}{r} = -p_z$$

and the relevant finite-element results (see Fig. 6-8c) for moments in kips are

Element	289	290	271
M_ϕ	−2.490	−2.469	
M_x	−0.5252	....	−0.5290

from which we can compute the changes in Q_x and Q_ϕ (neglecting $M_{x\phi}$ and $M_{\phi x}$) by

$$\left(\frac{\partial Q_x}{\partial x}\right)_{289} = \frac{M_{289} - 2M_{289} + M_{271}}{\Delta_x^2} = \frac{M_{271} - M_{289}}{\Delta^2{}_x}$$

$$= \frac{-0.5290 + 0.5252}{1.23} = -0.0031 \text{ kips/ft}^2$$

$$\left(\frac{\partial Q_\phi}{r\,\partial\phi}\right)_{289} = \frac{M_{289} - 2M_{289} + M_{290}}{\Delta_\phi^2} = \frac{M_{290} - M_{289}}{\Delta^2{}_\phi}$$

$$= \frac{-2.469 + 2.490}{1.35} = +0.0156 \text{ kips/ft}^2$$

In element 289 the value of $N_\phi = -3.27$ kips/ft so that

$$\frac{N_\phi}{r} = -\frac{3.27}{26.6} = -0.123 \text{ kips/ft}^2$$

and the sinusoidal variation in load gives at the center

$$p_z = \frac{4}{\pi}p_d\cos(\phi_k - \phi) = \frac{4}{\pi}(0.08)(1.0) = +0.102 \text{ kips/ft}^2$$

Thus the equilibrium equation becomes

$$-0.0031 + 0.0156 - 0.123 \approx -0.102$$
$$- 0.110 \approx -0.102$$

This error of under 10 percent arises most likely because the M values are not so precise and neither are the finite-difference approximations; also, there are small twisting moments. But we find that even far from the edges, a significant part of the load p_d is carried by bending; i.e., N_ϕ/r is more than 20 percent greater than p_z.

In conclusion, we find that even coarse-mesh approximations give useful results for this specific case, that from a design perspective the analytic and the numerical results are equivalent, and that the finite-element results yield insight into detailed behavior.

6-8   ULTIMATE-LOAD BEHAVIOR OF BARREL SHELLS

This section describes two sets of tests on cylindrical shells loaded through design working loads up to ultimate loads. The first set was carried out in the Netherlands and its goal was a broad study of differing reinforcing patterns for shells whose geometry is similar to those analyzed already in this chapter.

In the first set a series of eleven shells, all of the same cross section, was tested. Seven (A1 to A7) had the same span but varying amounts of reinforcement. One shell had a shorter span (B), two had longer spans (C and E), and one was continuous over three supports (D). A summary of the results is shown in Fig. 6-10. Figure 6-11 shows the crack patterns after failure for two of the A series, and Fig. 6-12 shows the transverse distribution of midspan deflections at various loadings for A2. Figure 6-10 is from Ref. 8, and Figs. 6-11 and 6-12 are from Ref. 9.

Shell A1 was reinforced in the standard way, about as described in Sec. 6-9, and as stated in the report, its "behavior . . . was satisfactory in every respect."[8] This behavior as the load is increased may be described as follows (see Fig. 6-10):

1. At loads of about half design load ($K = 0.5$), the first cracking in the edge beam occurred.
2. Near design load ($K = 1.0$), a crack developed at the crown in the longitudinal direction due to the negative transverse bending moment. Subsequently, similar cracks developed roughly halfway between the crown and edge due to positive transverse bending moment. These cracks, called *linear plastic hinges,* are similar in appearance to the yield lines observed in slabs and are also directly related to bending.
3. At loads of about 3 times design load ($K = 3.0$), the steel in the edge beam began to yield.
4. As the load further increased (to about $K = 3.5$), the cross section lost its coherence because of the large rotations of the linear plastic hinges and failed.

The principal conclusions derived from these tests were:

1. At design load, there existed "reasonably good" agreement between the experimental and calculated values of deflections and transverse curvatures (and thus moments), and "decidedly good" agreement in longitudinal in-plane stress values.

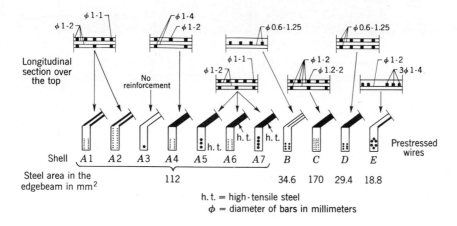

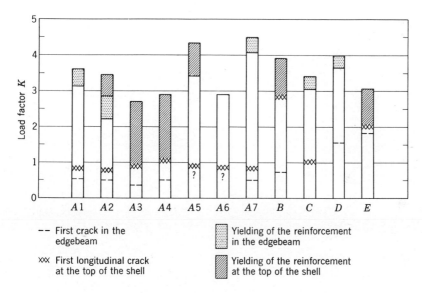

Fig. 6-10   Summary of Dutch tests.

2. Above design load (1500 kg in Fig. 6-12), discrepancies occurred, particularly in the transverse values, due primarily to the yield lines.

3. Reinforcement in the edge beam should be placed as low down as possible. The only difference between $A1$ and $A2$ was the location of the edgebeam steel, and in every respect the behavior of $A2$ was inferior (see Fig. 6-10).

4. Observation 3 leads to the suggestion from the investigators that "It would be worth considering the procedure of determining the amount of reinforcement in the edge beams of these cylindrical shells on the assumption of a cracked tensile zone."

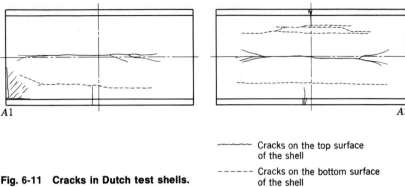

A1   A2

——— Cracks on the top surface of the shell

----- Cracks on the bottom surface of the shell

Fig. 6-11   Cracks in Dutch test shells.

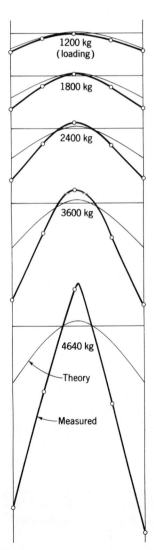

1200 kg (loading)

1800 kg

2400 kg

3600 kg

4640 kg

Theory

Measured

Fig. 6-12   Deflections in Dutch test shells.

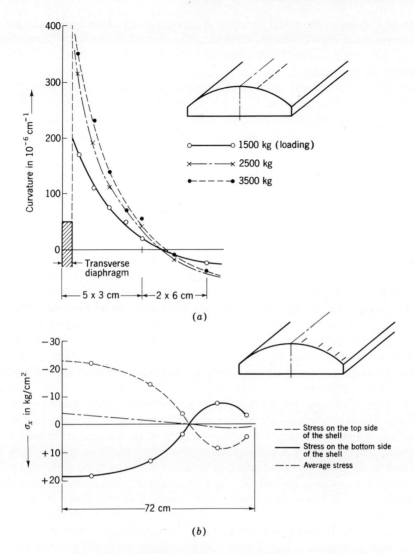

Fig. 6-13 *(a)* **Shell longitudinal curvature for different loadings.** *(b)* **Shell longitudinal stress near the diaphragm for** K = 1.0 (1500 kg).

5. Inadequate reinforcement within the shell had a very adverse effect on shell behavior, even where the edge-beam reinforcement was sufficient.

6. It was possible to determine with reasonable accuracy the magnitude of the longitudinal moments in the zone near the supporting diaphragm by formulas derived from the theory of axisymmetrically loaded cylinders (Sec. 5-10). Figure 6-13 (from Ref. 8) shows measured values both for longitudinal and transverse distribution of this bending effect.

For the continuous shell (D) the longitudinal in-plane stresses were reduced by continuity, about as one would anticipate from an ordinary continuous beam, but the transverse moments were hardly affected by the continuity. The same negative yield line at the crown appeared and extended over the center support nearly the full length of the two spans.

Shell E was prestressed both in the edge beam and in the shell. Cracking first occurred in the edge beam at $K = 1.9$, naturally much above the unprestressed shells. Because of the prestressing, the transverse negative moment was eliminated at working load, and thus no transverse negative moment reinforcement was provided. Hence at about $K = 2.0$, the full transverse working-load moment appeared and a large yield line opened up. Failure occurred in the shell at a load factor about 15 percent below that for $A1$. Thus prestressing, which may lead to improved working-load behavior, does not ensure an adequate factor of safety unless sufficient unprestressed reinforcement is supplied.

For these tests the transverse frames were made to provide very nearly the boundary conditions required by Eq. (5-30) and discussed at the end of Sec. 5-5. The "ribless" shells of Fig. 6-3 behave somewhat differently.

The second set of tests was carried out on a 7-ft-span 4-ft-wide parabolic barrel shell with a rise of 9 in. and a thickness of 0.5 in. thickened over the outer 5 in. to 1.5 in. There were two models, each roughly a 1/10 scale model from the warehouse roof of Fig. 6-3, the first of a single span[10] and the second of two continuous spans.[11] Figure 6-14 shows the under and upper sides of the first model after the ultimate load of about 4.4 times the working design load. The test verified the findings of the Dutch research and showed the good agreement possible even after cracking and yielding if those are accounted for in a numerical analysis.[12] Cracking leads to an equivalent lower average stiffness and yielding can be described by an approximate yield-line analysis. These conclusions are further discussed and generalized in Chap. 9. The major conclusion for design is that the methods of reinforcement outlined in the following section lead to a barrel shell whose safety factor is greater than that required in ordinary reinforced concrete structures. At the same time, there does not seem to be any clear method for basing designs on an ultimate-load analysis, even though some saving in reinforcement might result. Following these tests, studies have arisen that use computer programs to include cracking and yielding throughout the shell surface. These numerical experiments can add insight into behavior and help to establish appropriate safety factors.[13] Reinforcing layout is still determined on the basis of elastic analysis modified where desirable in the light of test results.

The tests on a continuous shell showed that where the interior support is close to a diaphragm in its resistance to displacements in the transverse plane, the two spans act as a continuous beam. For a load uniformly distributed over both spans, the overall moment at the interior support was about 2 percent less than $ql^2/8$, while the overall moment at midspan was about 2 percent higher than half that value. The transverse local moments at midspan, however, were about the same as for the single-span shell.

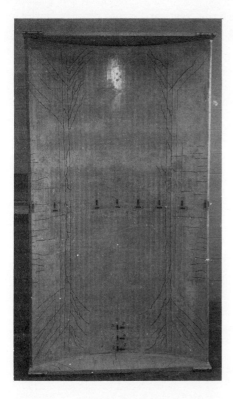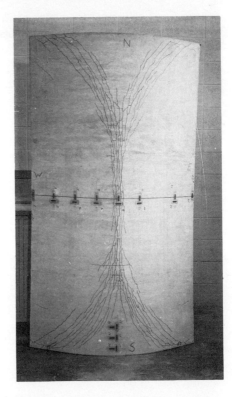

Fig. 6-14 (a) Cracking on the bottom surface of the model. (b) Cracking on the top surface of the model.

When the central support was a tied arch, the two spans behaved much more like two simply supported spans than when that support was a diaphragm. Thus when the tied-arch support shells were carried above design working loads, the behavior of each span was similar to the single-span results, except that the diagonal shear cracks at the interior support progressed more quickly in the continuous shell because of the somewhat higher shear forces there. Again the shells carried 4 times the design working loads without collapsing, although plastic hinges had formed at the crown by then.

6-9 REINFORCEMENT FOR A SIMPLY SUPPORTED SINGLE-BARREL SHELL WITH EDGE BEAMS

The purpose of this section is to present a procedure for reinforcing cylindrical shells, illustrated by the design example presented in Sec. 6-5.

The starting point will be a table of stress resultants and stress couples, such as Table 6-6, and the procedure then consists of the following steps:

1. *Principal stresses* are computed from the values for the in-plane stress resultants, and the *stress trajectories* are drawn based on the standard formulas:

$$N_1 = \frac{N_x + N_\phi}{2} + \sqrt{N_{\phi x}^2 + \left(\frac{N_x - N_\phi}{2}\right)^2}$$

$$N_2 = \frac{N_x + N_\phi}{2} - \sqrt{N_{\phi x}^2 + \left(\frac{N_x - N_\phi}{2}\right)^2}$$

$$\tan 2\theta = \frac{-2N_{\phi x}}{N_x - N_\phi}$$

The results are given in Table 6-10, and the trajectories are plotted in Fig. 6-15. Trajectories are also plotted in Fig. 6-16 for the multiple shell for comparison.

2. *Reinforcement* is provided to take all of the principal tensile stresses. The reinforcement is determined on the basis of an allowable stress, normally 20,000 psi, and may be placed in either of two manners: (1) in the direction of the tensile stress trajectories, or (2) in two directions, usually orthogonal.

At the corners where the shear is a maximum the reinforcement will be placed generally in the direction of the principal tensile stress, at 45°. At $\phi = 0°$, $N_1 = 8730$ lb/ft or 242 psi and with $f_s = 20,000$ psi.

$$A_s = \frac{8730}{20,000} = 0.44 \text{ in.}^2/\text{ft}$$

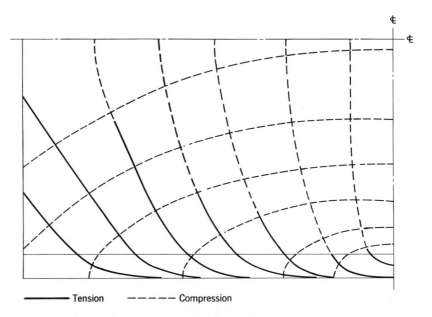

Tension ——— Compression —————

Fig. 6-15 Stress trajectories for shell and edge beams.

Table 6-10 PRINCIPAL STRESS RESULTANTS FOR SHELL WITH EDGE BEAMS, kips/ft

ϕ \ x	0			L/8			L/4			3L/8			L/2		
	N_1	N_2	$\theta°$	N_1	N_2	$\theta°$	N_1	N_2	$\theta°$	N_1	N_2	$\theta°$	N_1	N_2	$\theta°$
Edge beam	0	0		66.4	0	0	122.7	0	0	160.3	0	0	173.5	0	0
0	8.7	−8.7	*	7.7	−8.5	−43.7	5.4	−7.0	−41.9	2.4	−4.5	−37.5	−0.2	−2.1	0
10	6.7	−6.7	*	3.7	−9.5	−34.9	1.0	−11.7	−24.2	−0.9	−13.1	−12.5	−1.6	−13.5	0
20	3.7	−3.7	*	1.0	−6.9	−29.6	−1.0	−10.0	−17.4	−2.1	−12.2	−8.1	−2.5	−13.0	0
30	1.4	−1.4	*	−0.4	−3.5	−27.6	−1.8	−5.6	−15.4	−2.6	−7.0	−6.9	−2.9	−7.5	0
40 †	0.3	−0.3	*	−0.5	−1.8	−44.5	−1.7	−2.6	−43.7	−2.5	−3.0	−41.9	−3.0	−3.0	0
45	0	0		−0.9	−1.1	0	−1.7	−2.1	0	−2.2	−2.7	0	−2.4	−3.0	0

* −45°.

† Near the crown where $\phi = 40°$ the directions of trajectories are sensitive to very small values of shear. The trajectories there denote small stresses insignificant for design; therefore, the plot in Fig. 6-15 is slightly adjusted to give the more realistic trajectories expected in barrel shells with uniformly distributed loadings.

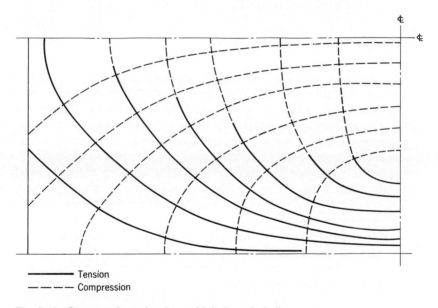

—— Tension
- - - - Compression

Fig. 6-16 Stress trajectories for multiple-barrel shell.

which can be supplied by No. 4 bars spaced 5 in. on centers. At $\phi = 10°$, $A_s = 0.34$ in.²/ft or No. 4 bars at 7 in. on centers (o.c.). At $\phi = 20°$, $A_s = 0.18$ in.²/ft, which would require No. 4 bars at 13 in. o.c.

Note that principal tensile stresses inclined nearly 45° exist at the junction of shell and edge beam. The tension stress goes from 67 psi on 37.5° at $x = 3L/8$ to 242 psi on 45° at $x = 0$. Diagonal bars are, therefore, carried from the edge beams into the shell. At $x = L/4$,

$$A_s = 5.4/20 = 0.27 \text{ in.}^2/\text{ft.}$$

If the bars were placed transversely, more steel would be required,[6]

$$A_s = \frac{N_1}{f_s \cos^2 \theta} = \frac{0.27}{0.745^2} = 0.49 \text{ in.}^2/\text{ft}$$

and at $x = 3L/8$,

$$A_s = \frac{2.4}{20(0.795)^2} = 0.19 \text{ in.}^2/\text{ft}$$

From $x = L/4$ to $3L/8$, we shall use No. 4 at 9 in. at about $40°$, and from $x = 3L/8$ to $L/2$, No. 4 at 11 in. at $0°$ (transverse reinforcement). The diagonal bars from $x = 0$ to $x = L/4$ are carried up into the transverse support. From $x = L/8$ to $L/4$,

$$A_s = 7.7/20 = 0.39 \text{ in.}^2/\text{ft}$$

or No. 4 at 6 in. and from $x = 0$ to $L/8$,

$$A_s = 8.7/20 = 0.44 \text{ in.}^2/\text{ft}$$

or No. 4 at 5 in. Thus the joint between shell and edge beam is more critical than that between shell and transverse support.

3. Transverse bending and in-plane stresses are combined, and *transverse reinforcement* is provided. The longitudinal distribution of this reinforcement would be sinusoidal if we followed the analysis. It has been shown[5] that the distribution actually approaches the sinusoidal only where r/L approaches 0.6. For longer shells, the distribution is more uniform longitudinally and drops off a short distance from the transverse support. From the discussion of the ultimate-load behavior of long and intermediate shells, we saw that the transverse cracking continues nearly to the supports. Thus it appears reasonable to provide the maximum transverse bending reinforcement over the center half of the shell and a reduced amount over the outer quarters of the span.

At the crown $M_\phi = -0.6$ ft-k/ft and $N_\phi = -2.97$ kips. Just as for the case of the dome and wall presented in Sec. 4-10, we shall neglect the compression stress resultant N_ϕ when computing the bending reinforcement. Assuming a cover of ¾ in. and thus a $d \approx 2.0$ in., we obtain

$$A_s = \frac{0.6 \times 12}{20 \times \text{⅞} \times 2.0} = 0.205 \text{ in.}^2/\text{ft}$$

which is taken by No. 3 bars at 6 in. o.c. By $\phi = 20°$ the negative moment is nearly zero, and some positive moment begins as we approach the edge beam. As shown in Sec. 6-5, a relatively large positive moment exists at the edge beam, although it has been neglected in our analysis. Therefore, the No. 4 bars previously provided for principal tensile stress will be placed along the bottom of the shell at its junction with the edge beam (Fig. 6-18) to give positive moment resistance.

4. *Edge-beam reinforcement* is normally provided by computing the magnitude and distribution of the total tensile force in the edge beam and providing reinforcement at an allowable stress, normally 20,000 psi, to take that force. The tests described in the preceding section showed that it is advisable to place the reinforcement as low down as possible in the edge beam.

At the midspan $\theta = 0°$ and all the tension is concentrated in the edge beam. The total tension is

$$T = \left(\frac{173.5 + 8.32}{2} - 8.32\right) 2.25 = 186 \text{ kips}$$

from which

$$A_s = \frac{186}{20} = 9.3 \text{ in.}^2$$

According to the stress distribution, this reinforcement should be placed throughout the edge beam; but on the basis of the discussion of the preceding section it will be located near the bottom. Figures 6-17 and 6-18 show the layout chosen in which eight No. 10 bars are used in three rows. Since tension does exist throughout the edge beam, some steel is included above the three main rows.

At $x = L/4$,

$$T = 0.707(186) = 132 \text{ kips}$$

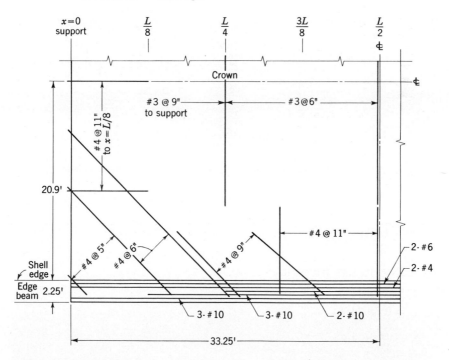

Fig. 6-17 Developed plan of shell reinforcement (welded wire fabric not shown).

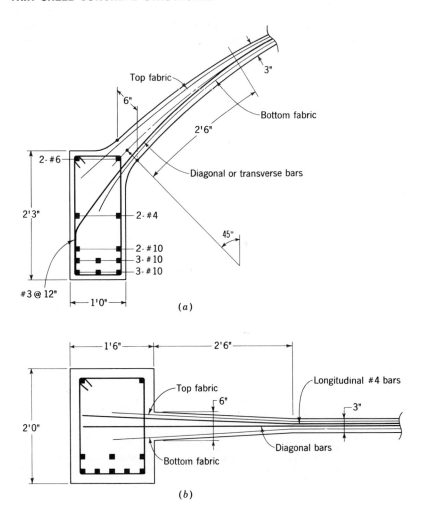

Fig. 6-18 (a) Section at edge beam. (b) Section at transverse support.

and

$$A_s = \frac{132}{20} = 6.6 \text{ in.}^2$$

so that six No. 10 bars are sufficient.

At $x = L/8$,

$$T = 0.383(186) = 72 \text{ kips}$$

$$A_s = \frac{72}{20} = 3.6 \text{ in.}^2$$

or about three No. 10 bars, which are then carried into the support.

5. *Minimum reinforcement* should be placed throughout the shell. In tensile zones a minimum of 0.35 percent in each of two directions has been recommended for crack control, and at other points the ACI Code minimum for slabs of 0.18 percent, spaced no farther apart than 5 times the shell thickness or 18 in., should apply equally to shells. Thus in the 3-in. slab a minimum of 0.065 in.^{2}/ft in each direction would be used, and where tension exists, a minimum of 0.126 in.^{2}/ft.

From $\phi = 0°$ to $\phi = 25°$ a fabric $6 \times 6 - 3/5$ is supplied top and bottom in the shell, and from $\phi = 25°$ to $\phi = 45°$ a fabric $6 \times 6 - 8/8$ is provided top and bottom. The criterion of minimum reinforcement strictly applies only where no other steel is provided. Since other steel is used throughout most of the shell, fabric areas could be reduced from those given above.

6. *Longitudinal bending reinforcement* at the junction of shell and the transverse rib is normally proportioned on the basis of the equations in Sec. 5-10.

If the transverse support were a deep diaphragm, in accordance with the assumption on which the shell analysis was based, then we could determine the longitudinal bending on the basis of Eq. (5-54).

$$M_x = -0.29Eh^2\epsilon_{\phi 0}\psi(\beta x)$$

where

$$\epsilon_{\phi 0} = \frac{N_\phi}{Eh}$$

and

$$\psi(\beta x) = e^{-\beta x}(\cos \beta x - \sin \beta x)$$

at $x = 0$, $\psi(\beta x) = 1$.

The value of N_ϕ at $x = 0$ is zero because of the assumption of a sinusoidal loading, but since the true load is uniform, N_ϕ should be uniform also and at $\phi = \phi_k$ is taken as -2.97 kips/ft, so the shell appears to move in. But the diaphragm permits no movement; hence for compatibility

$$\epsilon_{\phi 0} = \frac{2.97}{Eh}$$

and

$$M_0 = -0.29(0.25)2.97 = -0.215 \text{ ft-k/ft}$$

which requires, with $jd = 1.5$ in.,

$$A_s = \frac{0.215 \times 12}{20 \times 1.5} = 0.086 \text{ in.}^2/\text{ft}$$

We must include the effect of arch deflections. The arch has been analyzed for shell loads as represented by the shearing stress resultants, $N_{x\phi}$ at $x = 0$ as well as for arch dead load. The results are plotted in Fig. 6-19. At the crown, the stress

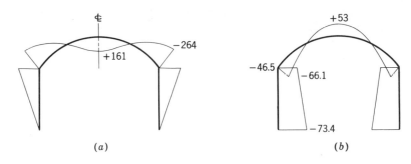

Fig. 6-19 (a) Arch moment in foot-kips. (b) Arch axial forces in kips.

at the junction of shell and arch is tensile,

$$f_\phi = \frac{53,000 \text{ lb}}{24(18) \text{ in.}^2} = +123 \text{ psi}$$

which, from (5-53), gives for $h = 3$ in.

$$M_0 = -0.29 E h^2 \frac{0.123}{E} = -0.32 \text{ ft-k/ft}$$

The two values for M_0 are added since one attempts to account for shell compression and the other for arch tension. If the value for N_ϕ above were equal to, and of the same sign as, $f_\phi h$, there would be no incompatibility and the second of Eqs. (5-53) would give $M_0 = 0$. Total reinforcement required by this highly approximate analysis is

$$A_s = 0.086 \frac{0.535}{0.215} = 0.214 \text{ in.}^2/\text{ft}$$

for which we will provide No. 4 bars at 11 in. o.c. From Fig. 6-13a, we note the rapid damping of M_x in the longitudinal direction, so we provide these bars out to $x = L/8$. Figure 6-13b shows that the moment also diminishes as $\phi_k - \phi$ increases; the bars are thus carried down to $\phi = 25°$, and the fabric will be provided from there to the edge. A cross section is shown in Fig. 6-18.

The arch moments were also computed using the PCA tables, and somewhat higher moments were obtained.[14]

Where interior ribs are not used, part of the shell must act as an arch. As stated in Ref. 15, ". . . The distribution and magnitude of the direct forces (i.e., stress resultants N_x, N_{xy}, and N_ϕ) do not differ greatly from those obtained with shells supported by ribs, with the exception that some intensification of forces occurs near the columns. . . . With respect to the transverse moment, marked changes occur in a band on each side of the column."

The reinforcement calculations presented here are based on working stress design. The 1977 ACI Building Code permits shell reinforcement to be determined by strength methods that could result in less steel for the shell given here. For example, the stress resultants of Table 6-10 are for a load of 80 psf. For strength design we should factor the loads (see page 227) $p = 1.4(50) + 1.7(30) = 121$ psf.

Thus, the stress resultants should be increased by a factor of $121/80 = 1.52$. At the same time the yield stress, say $f_y = 40$ ksi, should be reduced by $\phi = 0.9$ so that the resulting reinforcement

$$\bar{A}_s = \frac{N_1(1.52)}{0.9(2f_s)} = 0.85 A_s$$

The tension reinforcement given above could therefore be reduced by about 15 percent. The goal of this section is not to provide a final design but rather to indicate how results from an elastic analysis can be converted to reinforcement. In any complete design, the designer will decide on reinforcement based either on proven experience or on a thorough study of the geometry, loading, and boundary conditions. A large number of barrel shells have been successfully designed and built following working-load criteria.

REFERENCES

1. Tedesko, A., "Multiple Ribless Shells," *J. Struct. Div.*, ASCE, vol. 87, October 1961.
2. Lundgren, H., *Cylindrical Shells*, Danish Technical Press, Copenhagen, 1960.
3. Chinn, J., "Cylindrical Shell Analysis Simplified by Beam Method," *J. ACI*, vol. 55, May 1959.
4. Zweig, A. [Discussion of ref. 3], *J. ACI*, December 1959, p. 1594.
5. Parme, A. L., and H. W. Conner [Discussion of ref. 3], *J. ACI*, December 1959, p. 1584.
6. "Design of Cylindrical Concrete Shell Roofs," *Man. Eng. Pract.*, ASCE, no. 31, New York, 1952.
7. Scordelis, A. C., "Analysis of Cylindrical Shells and Folded Plates," *Concr. Thin Shells*, ACI, Publ. SP-28, 1971, pp. 207–236.
8. Bouma, A. L., A. C. Van Riel, H. Van Koten, and W. J. Beranek, "Investigations of Models of Eleven Cylindrical Shells Made of Reinforced and Prestressed Concrete," *Proceedings of the Symposium on Shell Research*, Delft, Aug. 30–Sept. 2, 1961, North Holland Publishing Company, Amsterdam, 1961.
9. Bouma, A. L., and J. Van Leeuwen, "Modelonderzoek van tonschalen Vervaardigd van Gewapend Microbeton" (Model Study of Cylindrical Shells Made of Reinforced Microconcrete), *Orgaan van het Instituut T.N.O. voor Boumaterialen en Constructies*, Postbus 49, Delft, Holland.
10. Hedgren, A. W., Jr., and David P. Billington, "Mortar Model Test on a Cylindrical Shell of Varying Curvature and Thickness," *J. ACI*, February 1967, pp. 73–83.
11. Darvall, P. LeP., David P. Billington, and Robert Mark, "Model Analysis of a Continuous Microconcrete Cylindrical Shell," *J. ACI*, November 1971, pp. 832–843.
12. Hedgren, A. W., Jr., and David P. Billington, "Numerical Analysis of Translational Shell Roofs," *J. Struct. Div.*, ASCE, vol. 92, no. ST1, February 1966, pp. 223–244.
13. Fialkow, M. N., "Plastic Collapse of Folded Plate Structures," *J. Struct. Div.*, ASCE, vol. 101, no. ST7, July 1975, pp. 1559–1584.
14. "Design Constants for Circular Arch Bents," Portland Cem. Assoc., Adv. Eng. Bull. no. 7, 1963.
15. "Design Constants for Ribless Concrete Cylindrical Shells," Portland Cem. Assoc., Adv. Eng. Bull. no. 11, 1964.

7

MEMBRANE THEORY
FOR TRANSLATION SHELLS
OF DOUBLE CURVATURE

7-1 DEFINITIONS

Shells generated by translating one plane curve over another plane curve are called *translation shells*. Figure 7-1 illustrates three types of translation shells. Apart from circular cylinders, which are the most common type of translation shells, surfaces formed by translating parabolas are the most frequently encountered translation shells for concrete roof structures.

Figure 7-1a shows a saddle surface formed by translating a convex parabola over a concave parabolic curve. The resulting shell is a hyperbolic paraboloid which has negative gaussian curvature. Figure 7-1b shows a cylindrical surface formed by translating a convex parabola over a straight line. The resulting shell is a parabolic cylinder which has zero gaussian curvature. Figure 7-1c shows a domed surface formed by translating a convex parabola over a convex parabolic curve. This shell is an elliptic paraboloid which has positive gaussian curvature.

The general form for the solution of the stress resultants based on membrane theory will be derived first, followed by separate discussions of hyperbolic and elliptic paraboloids.

For translation shells of positive gaussian curvature in the form of a dome, the membrane theory is often a satisfactory basis for design. The localized bending near the edge ribs can be evaluated approximately[1] by a procedure which uses the equations developed in Sec. 5-10.

A bending theory for translation shells can be formulated from the theory of shallow shells, and for some boundary conditions reasonable solutions have been obtained.[2,3] However, for many practical structures such as hyperbolic paraboloids with straight-line edges (see Fig. 7-11), there are available no solutions of rigor comparable to those for the domes, cylinders, or cylindrical barrels discussed in the previous chapters.

It is an interesting fact that the lack of an adequate bending theory—a great obstacle for the analyst—has not deterred architects, engineers, and builders from designing and constructing numerous imaginative translation shells.

Particularly noteworthy is the work of Felix Candela,[4] best known for his hyperbolic paraboloids, which are analyzed on the basis of membrane theory. The wide variety and often dramatic beauty of Candela's shells attest to the possibilities for these shapes, particularly when created by one who combines design with construction.

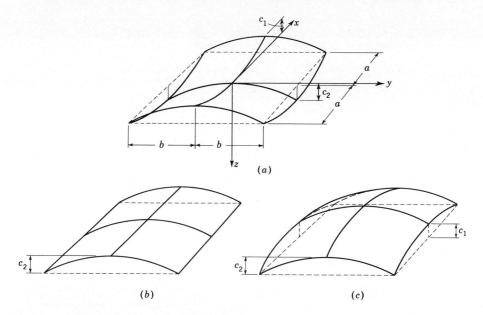

Fig. 7-1 The three types of translation shells.

7-2 GENERAL MEMBRANE THEORY

Equilibrium equations could be derived directly from a curved differential element, but they would be difficult to use; it has been found more convenient to begin by studying the equilibrium of the membrane values projected onto a horizontal xy plane.[5] From these results the true stress resultants can be found.

Figure 7-2 shows a differential element of a translation shell and its projection on the horizontal xy plane. The relationships between the projected sides dx and dy and the curved sides dp and dq are given by

$$dx = dp \cos \phi \qquad\qquad (a)$$
$$dy = dq \cos \theta \qquad\qquad (b)$$

If N_x is projected upon the horizontal plane, the relationship between N_x and the projected stress resultant $\bar{N}_x$ will be

$$\bar{N}_x dy = N'_x \cos \phi \, dq \qquad\qquad (c)$$

or

$$\bar{N}_x = N'_x \frac{\cos \phi}{\cos \theta} \qquad\qquad (7\text{-}1)$$

In a similar manner

$$\bar{N}_y = N'_y \frac{\cos \theta}{\cos \phi} \qquad\qquad (7\text{-}2)$$

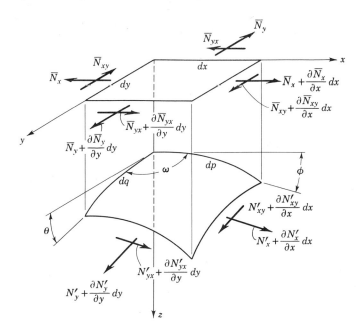

Fig. 7-2 Membrane stresses in translation shells.

and

$$\bar{N}_{xy}\, dy = N'_{xy} \cos \theta\, dq \tag{d}$$

which gives

$$\bar{N}_{xy} = N'_{xy} \tag{7-3}$$

In order to solve for the membrane stress distribution, it is now possible to express equilibrium on the xy plane

$$\Sigma X = \frac{\partial \bar{N}_x}{\partial x} + \frac{\partial \bar{N}_{yx}}{\partial y} + \bar{p}_x = 0 \tag{7-4}$$

and

$$\Sigma Y = \frac{\partial \bar{N}_y}{\partial y} + \frac{\partial \bar{N}_{xy}}{\partial x} + \bar{p}_y = 0 \tag{7-5}$$

where $\bar{p}_x$ and $\bar{p}_y$ are the components of the loads p_x and p_y which act on the projection of the element in the xy plane. The area of the element is

$$dA = \frac{dx\, dy}{\cos \phi \cos \theta} \sin \omega = dx\, dy\, \frac{\sqrt{1 - \sin^2 \phi \sin^2 \theta}}{\cos \phi \cos \theta}$$

Therefore,

$$\bar{p}_x\, dx\, dy = p_x\, dx\, dy\, \frac{\sqrt{1 - \sin^2 \phi \sin^2 \theta}}{\cos \phi \cos \theta}$$

and

$$\bar{p}_x = p_x \frac{\sqrt{1 - \sin^2 \phi \sin^2 \theta}}{\cos \phi \cos \theta}$$

The relationships between $\bar{p}_y$ and p_y and between $\bar{p}_z$ and p_z are the same.

For $\Sigma Z = 0$, it is necessary to obtain the components of the stress resultants in the z direction.

$$N'_x \sin \phi \, dq = \bar{N}_x \frac{\cos \theta}{\cos \phi} \sin \phi \, dq = \bar{N}_x \tan \phi \, dy = \bar{N}_x \frac{\partial z}{\partial x} dy \qquad (e)$$

$$N'_y \sin \theta \, dp = \bar{N}_y \frac{\partial z}{\partial y} dx \qquad (f)$$

$$N'_{xy} \sin \theta \, dq = \bar{N}_{xy} \tan \theta \, dy = \bar{N}_{xy} \frac{\partial z}{\partial y} dy \qquad (g)$$

$$N'_{yx} \sin \phi \, dp = \bar{N}_{yx} \frac{\partial z}{\partial x} dx \qquad (h)$$

Equilibrium in the z direction is achieved by taking terms like

$$-\bar{N}_x \frac{\partial z}{\partial x} dy + \left[\bar{N}_x \frac{\partial z}{\partial x} dy + \frac{\partial}{\partial x} \left(\bar{N}_x \frac{\partial z}{\partial x} \right) dy \, dx \right]$$

and canceling out equal terms of opposite sign. Combining (e) to (h), we obtain

$$\frac{\partial}{\partial x} \left(\bar{N}_x \frac{\partial z}{\partial x} \right) + \frac{\partial}{\partial y} \left(\bar{N}_y \frac{\partial z}{\partial y} \right) + \frac{\partial}{\partial x} \left(\bar{N}_{xy} \frac{\partial z}{\partial y} \right) + \frac{\partial}{\partial y} \left(\bar{N}_{yx} \frac{\partial z}{\partial x} \right)$$
$$+ \bar{p}_x = 0 \quad (i)$$

Carrying out the differentiation and recognizing that $\bar{N}_{yx} = \bar{N}_{xy}$, we next obtain

$$\bar{N}_x \frac{\partial^2 z}{\partial x^2} + \bar{N}_y \frac{\partial^2 z}{\partial y^2} + 2\bar{N}_{xy} \frac{\partial^2 z}{\partial x \, \partial y} + \frac{\partial z}{\partial x} \left(\frac{\partial \bar{N}_x}{\partial x} + \frac{\partial \bar{N}_{xy}}{\partial y} \right)$$
$$+ \frac{\partial z}{\partial y} \left(\frac{\partial \bar{N}_y}{\partial y} + \frac{\partial \bar{N}_{xy}}{\partial x} \right) + \bar{p}_z = 0 \quad (j)$$

From (7-4) and (7-5), the two bracketed terms in (j) are equal to $-\bar{p}_x$ and $-\bar{p}_y$, respectively; thus

$$\bar{N}_x \frac{\partial^2 z}{\partial x^2} + \bar{N}_y \frac{\partial^2 z}{\partial y^2} + 2\bar{N}_{xy} \frac{\partial^2 z}{\partial x \, \partial y} = -\bar{p}_z + \bar{p}_x \frac{\partial z}{\partial x} + \bar{p}_y \frac{\partial z}{\partial y} \qquad (7\text{-}6)$$

The solution of the stress resultants can be conveniently handled by introducing a stress function. The normal displacement is w. Such a function is commonly used in plate theory where all of the bending moments are expressed by various derivatives of w as used in Chap. 5 to develop equations for shallow shells. Also following that development in Sec. 5-4, we now introduce the stress function F,[6] serving the same mathematical purpose as w—namely permitting a solution in terms of one unknown and then allowing the evaluation of the needed quantities as previously defined in Eqs. (5-25). Then the stress function F must be solved by differentiation. As previously defined in Eqs. (5-25), F satisfies all three of the general expressions of static equilibrium—(7-4), (7-5), and (7-6). It can easily be shown that the following functions solve Eqs. (7-4) and (7-5):

$$\bar{N}_x = \frac{\partial^2 F}{\partial y^2} - \int \bar{p}_x \, dx \qquad \bar{N}_y = \frac{\partial^2 F}{\partial x^2} - \int \bar{p}_y \, dy$$

$$\bar{N}_{xy} = -\frac{\partial^2 F}{\partial x \, \partial y} \tag{7-7}$$

If these functions are then introduced into Eq. (7-6), the following differential equation governing the stress function F is obtained:

$$\frac{\partial^2 F}{\partial x^2} \frac{\partial^2 z}{\partial y^2} - 2 \frac{\partial^2 F}{\partial x \, \partial y} \frac{\partial^2 z}{\partial x \, \partial y} + \frac{\partial^2 F}{\partial y^2} \frac{\partial^2 z}{\partial x^2} = q \tag{7-8}$$

in which

$$q = -\bar{p}_z + \bar{p}_x \frac{\partial z}{\partial x} + \bar{p}_y \frac{\partial z}{\partial y} + \frac{\partial^2 z}{\partial x^2} \int \bar{p}_x \, dx + \frac{\partial^2 z}{\partial y^2} \int \bar{p}_y \, dy \tag{k}$$

Where only uniform vertical loads are acting, $\bar{p}_x = \bar{p}_y = 0$ and $q = -\bar{p}_z$.

Equation (7-8) can be thought of as a general equation for the membrane theory of shallow shells when

$$\frac{\partial^2 z}{\partial x^2} \approx \frac{1}{r_x}$$

$$\frac{\partial^2 z}{\partial y^2} \approx \frac{1}{r_y} \tag{l}$$

$$\frac{\partial^2 z}{\partial x \, \partial y} \approx \frac{1}{r_{xy}}$$

Gaussian curvature (Chap. 1) is defined in general as

$$K = \frac{1}{r_x} \frac{1}{r_y} - \frac{1}{r_{xy}^2} \tag{m}$$

so that

$$\frac{\partial^2 z}{\partial x^2} \frac{\partial^2 z}{\partial y^2} - \left(\frac{\partial^2 z}{\partial x \, \partial y}\right)^2 = K \tag{n}$$

and when K is positive, zero, or negative, the surface is elliptic, parabolic, or hyperbolic and Eq. (7-8) is of the elliptic, parabolic, or hyperbolic type as defined in the theory of partial differential equations.

The cylindrical and conical shells of Chaps. 2, 3, 5, and 6 have parabolic surfaces with $K = 0$ because $1/r_x = 1/r_{xy} = 0$. The dome surfaces of Chap. 3 are elliptical. In this chapter we shall consider hyperbolic and elliptical surfaces which are generated by translating a parabolic curve along another parabolic curve to get a doubly curved surface.

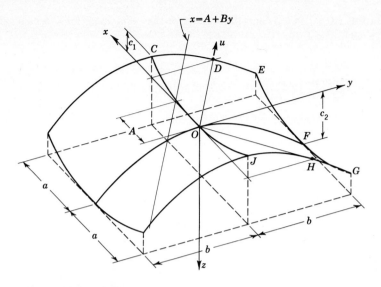

Fig. 7-3 Saddle shell.

7-3 HYPERBOLIC PARABOLOIDS BOUNDED BY PARABOLAS

There are many ways in which hyperbolic paraboloid shells have been used. In this section, several types will be discussed briefly. It is beyond the scope of this text to investigate more than a few of the many interesting possibilities with these surfaces.

Figure 7-3 shows a hyperbolic paraboloid "saddle" surface. The boundary lines formed by passing vertical planes parallel to the xz and yz planes at $x = \pm a$ and $y = \pm b$ are parabolas. The equation for this surface is

$$z = \frac{y^2}{h_2} - \frac{x^2}{h_1} \tag{7-9}$$

where

$$h_1 = \frac{a^2}{c_1} \tag{a}$$

$$h_2 = \frac{b^2}{c_2}$$

When (7-9) is introduced into Eq. (7-8), we get

$$\frac{\partial^2 F}{\partial x^2}\frac{2}{h_2} - \frac{\partial^2 F}{\partial y^2}\frac{2}{h_1} = q \tag{7-10}$$

Consider the simple case of loading uniform over the horizontal projection of the surface. In this case $q = -\bar{p}_z$ and there is more than one value for F which can be used (page 184 of Ref. 5):

$$
\begin{aligned}
F &= -\tfrac{1}{4}\bar{p}_z h_2 x^2 \\
F &= +\tfrac{1}{4}\bar{p}_z h_1 y^2 \\
F &= -\tfrac{1}{8}\bar{p}_z (h_2 x^2 - h_1 y^2)
\end{aligned}
\tag{7-11}
$$

Substituting the first solution into (7-7), we obtain

$$
\begin{aligned}
\bar{N}_x &= 0 \\
\bar{N}_y &= -\frac{\bar{p}_z h_2}{2} \\
\bar{N}_{xy} &= 0
\end{aligned}
\tag{7-12}
$$

Thus the entire load is being carried in the shell along parabolas parallel to the yz plane and fully supported at the base where $y = \pm b$. In the same manner, if we choose the second of (7-11):

$$
\begin{aligned}
\bar{N}_x &= +\frac{\bar{p}_z h_1}{2} \\
\bar{N}_y &= 0 \\
\bar{N}_{xy} &= 0
\end{aligned}
\tag{7-13}
$$

wherein the entire load is carried by parabolas parallel to the xz plane and hanging from the parabolic arch supports at $x = \pm a$. Finally, the third of (7-11) gives

$$
\begin{aligned}
\bar{N}_x &= +\frac{\bar{p}_z h_1}{4} \\
\bar{N}_y &= -\frac{\bar{p}_z h_2}{4} \\
\bar{N}_{xy} &= 0
\end{aligned}
\tag{7-14}
$$

and shows the forces equally split between the two systems of parabolas.

The apparently arbitrary nature of this choice of stress distribution depends upon our choice of edge supports. Where there is a full support along $y = \pm b$ and no appreciable stiffening along the edge parabolas at $x = \pm a$, there will be no appreciable support for the hanging parabolas and the system is a barrel arch. Where arches are provided at $x = \pm a$ and no support along $y = \pm b$, the hanging parabolas will carry the load to the end arches. Note that the end arches must resist heavy loading normal to their planes; hence some bracing is necessary. Where both pairs of edges are equally stiff, the third solution can be considered applicable. Obviously the support conditions must be realized or none of the solutions shown are valid. If, for example, it were proposed to support the structure solely by vertical arches at $x = \pm a$, it is obvious that we have no solution which is satisfactory without introducing new systems of edge forces or bending in the shell. Also where the arch rises are large, the simplified assumption of loading uniform over the horizontal projection can no longer be considered a satisfactory approximation of dead loading.

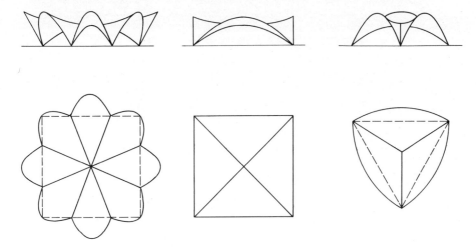

Fig. 7-4 Example of assembly of groined vault segments. (From Scordelis.)

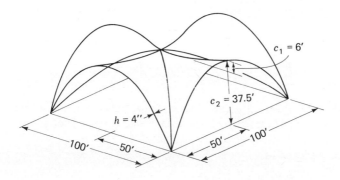

Fig. 7-5 Rectangular-plan hyperbolic groined vault.

Another type of hyperbolic paraboloid bounded by parabolas is the groined vault where two or more saddles intersect. Figure 7-4 shows a series of possible forms and Fig. 7-5 defines the geometry of a rectangular-plan vault for which Eq. (7-9) applies. For such deep vaults $q \equiv -\bar{p}_z = $ a constant is no longer a reasonable approximation for dead load since the projected shell area near the support is very much greater than at the crown.

For a live load, taken to be uniformly distributed over the horizontal projection, the solutions (7-11) still apply. Here, where the quarter vault (Fig. 7-5) is bounded by a parabola at $x = a$, the $\bar{N}_x = 0$ at that edge, and the solution (7-12) applies. The vault thus carries live loads by arching forces in the y direction.

For dead loads these simple solutions no longer apply because $\bar{p}_z$ can no longer be taken as constant. Rather a more complex stress function is required. Rather than develop this solution here, we will use a simpler structure to illustrate the essential

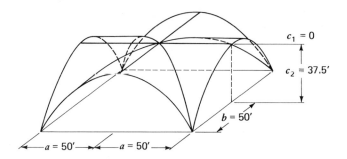

Fig. 7-6 Rectangular-plan cylindrical groined vault.

vault behavior. Where analyses are required for large-scale groined vaults, use will normally be made of computer programs based on numerical methods.[7]

We take the simpler case where $h_1 = \infty$ and the vault is made up of intersecting parabolic cylinders (Fig. 7-6). To simplify the solution further, we take the loading to be

$$q = -\bar{p}_{zo}\left(k_1 + k_2 \cos\frac{\pi y}{2b}\right) \tag{7-15}$$

where

$$k_1 = \sqrt{1 + (2c_2/b)^2} = (1/\cos\theta) \text{ when } y = b$$
$$k_2 = 1 - k_1$$

In this way where $y = 0$, $q = -\bar{p}_{zo}$, the surface load at the crown, and where $y = b$, $q = -k_1\bar{p}_{zo}$, the horizontal projection of the surface load at the corner support. This loading assumes the shell thickness to be constant.[8]

Equation (7-10) reduces to

$$\frac{\partial^2 F}{\partial x^2}\frac{2}{h_2} = -\bar{p}_{zo}\left(k_1 + k_2 \cos\frac{\pi y}{2b}\right) \tag{7-16}$$

This equation is satisfied by the stress function

$$F = -\tfrac{1}{4}\bar{p}_{zo}h_2\left[x^2 k_1 + (a - x)^2 k_2 \cos\frac{\pi y}{2b}\right] \tag{7-17}$$

This solution represents an extension of the first of (7-11) and implies that the load is mainly carried by arching forces in the y direction. Substitution of (7-17) in (7-7) gives

$$\bar{N}'_x = \frac{\pi^2 \bar{p}_{zo}h_2}{16}\left(\frac{a}{b}\right)^2\left(1 - \frac{x}{a}\right)^2 k_2 \cos\frac{\pi y}{2b}$$

$$\bar{N}'_y = -\frac{\bar{p}_{zo}h_2}{2}k_1\left(1 + \frac{k_2}{k_1}\cos\frac{\pi y}{2b}\right) \tag{7-18}$$

$$\bar{N}'_{xy} = \frac{\pi\bar{p}_{zo}h_2}{4}\frac{a}{b}\left(1 - \frac{x}{a}\right)k_2 \sin\frac{\pi y}{2b}$$

The second of (7-18) is a modification of the second of (7-12) but now there will be $\bar{N}'_x$ and $\bar{N}'_{xy}$ values. This is so because the parabolic arch form now deviates from the funicular polygon of the loading (7-15). The only way that this nonuniform loading can be carried by membrane forces alone is if shears arise and hence also membrane forces in the direction perpendicular to the parabolic arches.

The actual stress resultants in the shell come from substituting (7-18) into (7-1) to (7-3) where $\phi = 0°$ and where

$$\cos \theta = \frac{1}{\sqrt{1 + (2y/h_2)^2}}$$
$$N'_x = \bar{N}'_x \cos \theta$$
$$N'_y = \bar{N}'_y/\cos \theta$$
$$N'_{xy} = \bar{N}'_{xy}$$

(7-19)

As an example, we take the vault dimensions shown in Fig. 7-6 where $a = b = 50$ ft, $c_1 = 0$, $c_2 = 37.5$ ft, and $\bar{p}_{z0} = 60$ psf. Then

$$k_1 = \sqrt{1 + (75/50)^2} = 1.8$$
$$k_2 = 1 - k_1 = -0.8$$
$$a/b = 1.0$$
$$h_2 = b^2/c_2 = 2500/37.5 = 66.7 \text{ ft}$$
$$k_2/k_1 = -0.444$$

from which the values of $\cos \theta$, N'_x, N'_y, and N'_{xy} are computed and summarized in Table 7-1 for various locations on one-half of a typical quarter vault. Table 7-2 gives the comparable values for the hyperbolic paraboloid groined vault of Fig. 7-5 where $a = b = 50$ ft, $c_1 = 6$ ft, $c_2 = 37.5$ ft, and $\bar{p}_{z0} = 60$ psf. The load was assumed to be the same as in Eq. (7-15) and the values given as $K = \cos \theta/\cos \phi$. This vault is bounded by the same parabolic arches as those of the cylindrical vault, but now the height at $x = y = 0$ is $c_2 - c_1 = 31.5$ ft.

Comparison of results of Table 7-1 with those of Table 7-2 shows that the shell stresses are nearly the same and thus the two systems behave nearly the same. Furthermore, Scordelis has shown that the results of Table 7-2 are almost the same as results obtained from a more exact analysis. The greatest differences are in the shearing stresses, which are anyway very small, i.e., a maximum of 33 psi for the dead load of 60 psf and a live load of 30 psf (Ref. 7, page 565).

With the shell stress resultants computed, the forces acting on the groin arches can be found. These forces come to the groins where they meet similar forces from the adjacent x-direction arches (Fig. 7-5). Components perpendicular to the groin cancel, and the resultant forces give vertical and horizontal loads on the groin arch in the plane of that groin. The groin arch consists of a valley thickening plus some v-shaped flange area from the shell. Scordelis has proposed that this flange area be taken as 8 times the shell thickness on either side.

To study the case of a groined hyperbolic paraboloid, we can make use of Table 7-3 obtained for the load of Eq. (7-15) and based upon a solution to (7-10) more complicated than (7-17). Table 7-2 was prepared from the values in Table 7-3. The mem-

Table 7-1 RESULTS FOR DEAD LOAD ON CYLINDRICAL GROINED VAULT

y/b	x/a					
	0	0.2	0.4	0.6	0.8	1.0
			$\cos \theta$			
0	1.0	1.0	1.0	1.0	1.0	1.0
0.2		0.958	0.958	0.958	0.958	0.958
0.4			0.857	0.857	0.857	0.857
0.6				0.743	0.743	0.743
0.8					0.640	0.640
1.0						0.555
			$N_x'{}^*$			
0	−1974	−1260	−711	−316	−79	0
0.2		−1150	−648	−288	−72	0
0.4			−493	−219	−54.8	0
0.6				−138	−34.5	0
0.8					−15.6	0
1.0						0
			$N_y'{}^*$			
0	−2002	−2002	−2002	−2002	−2002	−2002
0.2		−2172	−2172	−2172	−2172	−2172
0.4			−2696	−2696	−2696	−2696
0.6				−3580	−3580	−3580
0.8					−4860	−4860
1.0						−6490
			$N_{xy}'{}^*$			
0	0	0	0	0	0	0
0.2		−622	−466	−311	−195	0
0.4			−877	−592	−296	0
0.6				−814	−407	0
0.8					−478	0
1.0						0

* In pounds per foot.

brane theory upon which these tables are based gives reasonable results because the loads are carried by arches directly to groins which, being arches themselves, provide relatively rigid supports. It is essential, however, that the four corner supports be rigidly restrained against horizontal movement as well as against differential vertical movements. The supports for the St. Louis Air Terminal vaults are tied together by the stiff floor in which heavy steel ties were embedded around the four sides between adjacent hinged supports (Fig. 7-7).[9]

Table 7-2 RESULTS FOR DEAD LOAD ON HYPERBOLIC
PARABOLOID GROINED VAULT

			x/a			
	0	0.2	0.4	0.6	0.8	1.0
y/b			$[1 - (x/a)]\,\sqrt{c_1/c_2}$			
	0.40	0.32	0.24	0.16	0.08	0
			K			
0	1.000	1.001	1.005	1.010	1.018	1.028
0.2		0.959	0.962	0.968	0.975	0.985
0.4			0.861	0.866	0.873	0.882
0.6				0.751	0.757	0.764
0.8					0.652	0.658
1.0						0.570
			$N_x'^*$			
0	−1910	−1238	−706	−317	−80	0
0.2		−1128	−643	−289	−73	0
0.4			−489	−220	−56	0
0.6				−139	−35	0
0.8					−16	0
1.0						0
			$N_y'^*$			
0	−2305	−2195	−2101	−2030	−1977	−1945
0.2		−2363	−2272	−2196	−2143	−2110
0.4			−2783	−2709	−2652	−2614
0.6				−3581	−3523	−3481
0.8					−4769	−4720
1.0						−6316
			$N_{xy}'^*$			
0	0	0	0	0	0	0
0.2		−596	−455	−307	−155	0
0.4			−866	−585	−295	0
0.6				−805	−406	0
0.8					−477	0
1.0						0

* In pounds per foot.

Note: Coefficient K is dimensionless and $c_1/c_2 = 0.16$.

Table 7-3 INTERNAL FORCES IN GROINED VAULTS FOR DEAD LOADS

$\bar{p}_{z0}$ = intensity of distributed load at the crown (pounds per square foot)

$$K = \sqrt{\frac{1 + [(2c_1/a)(x/a)]^2}{1 + [(2c_2/b)(y/b)]^2}}$$

$$k_1 = \sqrt{1 + (2c_2/b)^2}$$

$$k_2 = 1 - k_1$$

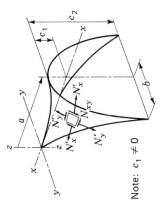

Note: $c_1 \neq 0$

$$N'_x = \frac{k_2 a^2}{2c_1} \, \bar{p}_{z0} K \text{ (coefficient)}$$

										$(1 - x/a)\sqrt{c_1/c_2}$								
y/b	0	0.04	0.08	0.12	0.16	0.20	0.24	0.28	0.32	0.36	0.40	0.45	0.50	0.55	0.60	0.65	0.70	0.75
0	.0000	.0020	.0079	.0177	.0314	.0489	.0702	.0952	.1237	.1557	.1910	.2396	.2929	.3506	.4122	.4775	.5460	.6173
0.10	.0000	.0019	.0078	.0175	.0310	.0483	.0694	.0940	.1222	.1538	.1886	.2366	.2893	.3462	.4071	.4716	.5393	.6097
0.20	.0000	.0019	.0075	.0168	.0299	.0465	.0668	.0905	.1176	.1481	.1816	.2279	.2786	.3334	.3920	.4541	.5193	.5871
0.30	.0000	.0018	.0070	.0158	.0280	.0436	.0626	.0848	.1102	.1387	.1702	.2135	.2610	.3123	.3673	.4255	.4865	.5500
0.40	.0000	.0016	.0064	.0143	.0254	.0396	.0568	.0770	.1001	.1259	.1545	.1938	.2370	.2836	.3335	.3863	.4417	.4994
0.50	.0000	.0014	.0056	.0125	.0222	.0346	.0497	.0673	.0875	.1101	.1350	.1694	.2071	.2479	.2915	.3376	.3861	.4365
0.60	.0000	.0012	.0046	.0104	.0185	.0288	.0413	.0559	.0727	.0915	.1123	.1408	.1722	.2060	.2423	.2807	.3209	.3628
0.70	.0000	.0009	.0036	.0080	.0143	.0222	.0319	.0432	.0562	.0707	.0867	.1088	.1330	.1591	.1871	.2168	.2479	.2803
0.80	.0000	.0006	.0024	.0055	.0097	.0151	.0217	.0294	.0382	.0481	.0590	.0740	.0905	.1083	.1274	.1476	.1687	.1908
0.90	.0000	.0003	.0012	.0028	.0049	.0077	.0110	.0149	.0193	.0244	.0299	.0375	.0458	.0548	.0645	.0747	.0854	.0966
1.00	.0000	.0000	.0000	.0000	.0000	.0000	.0000	.0000	.0000	.0000	.0000	.0000	.0000	.0000	.0000	.0000	.0000	.0000

$$N'_v = -\frac{k_1 b^2}{2c_2}\frac{\bar{p}_{z0}}{K}\left[1 + \frac{k_2}{k_1}\,(\text{coefficient})\right]$$

0	1.0000	.9980	.9921	.9823	.9686	.9511	.9298	.9048	.8763	.8443	.8090	.7604	.7071	.6494	.5878	.5225	.4540	.3827
0.10	.9877	.9857	.9799	.9702	.9567	.9393	.9183	.8937	.8655	.8339	.7991	.7510	.6984	.6415	.5805	.5161	.4484	.3780
0.20	.9511	.9492	.9436	.9342	.9212	.9045	.8843	.8605	.8334	.8030	.7694	.7232	.6725	.6177	.5590	.4969	.4318	.3640
0.30	.8910	.8892	.8840	.8752	.8630	.8474	.8284	.8062	.7808	.7523	.7208	.6775	.6300	.5787	.5237	.4655	.4045	.3410
0.40	.8090	.8074	.8026	.7947	.7836	.7694	.7522	.7320	.7089	.6831	.6545	.6152	.5721	.5254	.4755	.4227	.3673	.3096
0.50	.7071	.7057	.7015	.6946	.6849	.6725	.6574	.6398	.6196	.5970	.5721	.5377	.5000	.4592	.4156	.3695	.3210	.2706
0.60	.5878	.5866	.5831	.5774	.5693	.5590	.5465	.5318	.5151	.4963	.4755	.4470	.4156	.3817	.3455	.3071	.2668	.2249
0.70	.4540	.4531	.4504	.4459	.4397	.4318	.4221	.4108	.3978	.3833	.3673	.3452	.3210	.2948	.2668	.2372	.2061	.1737
0.80	.3090	.3084	.3066	.3035	.2993	.2939	.2873	.2796	.2708	.2609	.2500	.2350	.2185	.2007	.1816	.1615	.1403	.1183
0.90	.1564	.1561	.1552	.1537	.1515	.1488	.1454	.1415	.1371	.1321	.1266	.1190	.1106	.1016	.0919	.0817	.0710	.0599
1.00	.0000	.0000	.0000	.0000	.0000	.0000	.0000	.0000	.0000	.0000	.0000	.0000	.0000	.0000	.0000	.0000	.0000	.0000

$$N'_{xy} = \frac{k_2\,ab}{2\sqrt{c_1 c_2}}\,\bar{p}_{z0}\,(\text{coefficient})$$

0	.0000	.0000	.0000	.0000	.0000	.0000	.0000	.0000	.0000	.0000	.0000	.0000	.0000	.0000	.0000	.0000	.0000	.0000
0.10	.0000	.0098	.0196	.0293	.0389	.0483	.0576	.0666	.0754	.0838	.0920	.1016	.1106	.1190	.1266	.1334	.1394	.1445
0.20	.0000	.0194	.0387	.0579	.0768	.0955	.1138	.1316	.1489	.1656	.1816	.2007	.2185	.2350	.2500	.2635	.2753	.2855
0.30	.0000	.0285	.0569	.0851	.1129	.1403	.1671	.1933	.2187	.2433	.2669	.2943	.3210	.3452	.3673	.3871	.4045	.4194
0.40	.0000	.0369	.0737	.1101	.1462	.1816	.2164	.2503	.2832	.3150	.3455	.3817	.4156	.4470	.4755	.5012	.5237	.5430
0.50	.0000	.0444	.0886	.1325	.1759	.2185	.2603	.3011	.3407	.3789	.4156	.4592	.5000	.5377	.5721	.6029	.6300	.6533
0.60	.0000	.0508	.1014	.1516	.2012	.2500	.2978	.3445	.3897	.4335	.4755	.5254	.5721	.6152	.6545	.6898	.7208	.7474
0.70	.0000	.0559	.1117	.1670	.2216	.2753	.3280	.3794	.4292	.4774	.5237	.5787	.6300	.6775	.7208	.7597	.7939	.8232
0.80	.0000	.0597	.1192	.1782	.2365	.2939	.3501	.4049	.4582	.5096	.5590	.6177	.6725	.7232	.7694	.8109	.8474	.8787
0.90	.0000	.0620	.1238	.1851	.2456	.3052	.3636	.4205	.4758	.5292	.5806	.6415	.6984	.7510	.7991	.8421	.8800	.9125
1.00	.0000	.0628	.1253	.1874	.2487	.3090	.3681	.4258	.4818	.5358	.5878	.6494	.7071	.7604	.8090	.8526	.8910	.9239

Fig. 7-7 Air terminal groined vaults, St. Louis. (Courtesy Anton Tedesko.)

7-4 HYPERBOLIC PARABOLOIDS BOUNDED BY STRAIGHT LINES

Consider that through the saddle surface of Fig. 7-3 we pass a vertical plane which intersects the xy plane with the line

$$x = A + By \tag{7-20}$$

The resulting line formed by the intersection of the vertical plane with the saddle surface is obtained by substituting (7-20) into (7-9):

$$z = \frac{y^2}{h_2} - \frac{A^2 + 2ABy + B^2y^2}{h_1}$$

or

$$zh_1 = y^2 \frac{h_1}{h_2} - A^2 - 2ABy - B^2y^2 \tag{a}$$

which will be a straight line when

$$B^2 = \frac{h_1}{h_2} \tag{b}$$

The term B signifies the slope of the intersection line on the xy plane, and from (b)

$$\tan \alpha = B = \sqrt{\frac{h_1}{h_2}} \tag{7-21}$$

Thus the hyperbolic paraboloid can also be generated by translating straight lines at a constant slope when projected onto the xy plane but at a changing angle to the xy plane. When (b) is substituted into (a),

$$z = -\frac{A^2}{h_1} - \frac{2AB}{h_1} y \qquad (c)$$

where $2AB/h_1$ is the slope, $\partial z/\partial y$, in which the term $2B/h_1$ is constant, and A, the point at which the line intersects the xz plane, constantly changes.

Figure 7-8 shows a section cut from the saddle surface by the two straight-line generators passing through the origin, where $A = 0$ and hence $z = 0$ and $\partial z/\partial y = 0$. Considering new axes u and v as shown in Fig. 7-9, lines OD and OH in Fig. 7-3, we can describe the surface by

$$z = \frac{d}{b_0} v$$

where

$$d = \frac{c_0}{a_0} u$$

from which

$$z = \frac{c_0}{a_0 b_0} uv = kuv \qquad (7\text{-}22)$$

The constant $k = c_0/a_0 b_0$, the twist of the surface, $\partial^2 z/\partial u\, \partial v$, and is often called the warping. From Fig. 7-9 we observe that

$$x = u \sin \alpha - v \sin \alpha$$
$$y = u \cos \alpha + v \cos \alpha$$

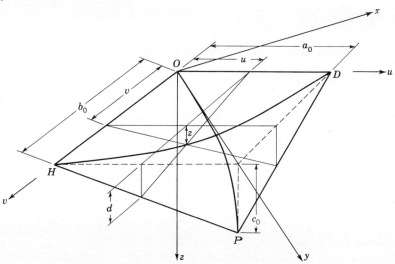

Fig. 7-8 Hyperbolic paraboloid bounded by straight lines.

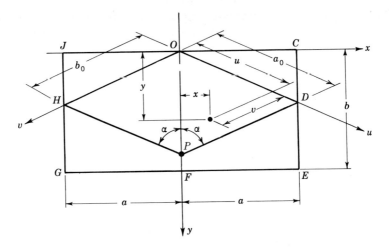

Fig. 7-9 Transformation of coordinates.

from which

$$v = \tfrac{1}{2}\left(\frac{y}{\cos \alpha} - \frac{x}{\sin \alpha}\right) \qquad\qquad (d)$$

and

$$u = \tfrac{1}{2}\left(\frac{y}{\cos \alpha} + \frac{x}{\sin \alpha}\right) \qquad\qquad (e)$$

Substituting (d) and (e) into (7-22), we obtain

$$z = \frac{k}{4}\left(\frac{y^2}{\cos^2 \alpha} - \frac{x^2}{\sin^2 \alpha}\right) \qquad\qquad (7\text{-}23)$$

where $c_0 = 4a^2/h_1$ and $a_0 = b_0 = a/\sin \alpha$ (Fig. 7-9), then

$$k = \frac{4}{h_1 + h_2} = \frac{4\cos^2 \alpha}{h_2} \qquad\qquad (f)$$

which, when substituted into (7-23), gives

$$z = \frac{1}{h_2}\left(y^2 - \frac{x^2}{\tan^2 \alpha}\right) \qquad\qquad (g)$$

From (7-21) $\tan^2 \alpha = h_1/h_2$ and thus (g) becomes the familiar

$$z = \frac{y^2}{h_2} - \frac{x^2}{h_1} \qquad\qquad (7\text{-}9)$$

which was used originally to define the surface in Sec. 7-3.

Where $\alpha = 45°$, the u and v axes are at right angles to each other and can be considered as new x and y axes. Then

$$z = kxy$$

which, when substituted into (7-8), yields

$$-2 \frac{\partial^2 F}{\partial x\, \partial y} k = q \tag{7-24}$$

From (7-7)

$$\bar{N}_{xy} = -\frac{\partial^2 F}{\partial x\, \partial y} = \frac{q}{2k} \tag{h}$$

Differentiating (h) with respect to x, we obtain

$$\frac{\partial^3 F}{\partial x^2\, \partial y} = -\frac{1}{2k} \frac{\partial q}{\partial x}$$

and integrating with respect to y, we get the expression

$$\frac{\partial^2 F}{\partial x^2} = -\frac{1}{2k} \int \frac{\partial q}{\partial x}\, dy + f_2(x) = \bar{N}_y + \int \bar{p}_y\, dy \tag{i}$$

and by a similar derivation,

$$\frac{\partial^2 F}{\partial y^2} = -\frac{1}{2k} \int \frac{\partial q}{\partial y}\, dx + f_1(y) = \bar{N}_x + \int \bar{p}_x\, dx \tag{j}$$

If we take the simple case where $q = -\bar{p}_z = $ constant, then (h), (i), and (j) reduce to

$$\bar{N}_{xy} = -\frac{\bar{p}_z}{2k}$$

$$\bar{N}_y = f_2(x)$$
$$\bar{N}_x = f_1(y)$$

$f_1(y)$ and $f_2(x)$ are constants of integration which depend upon the boundary conditions. If we assume the boundaries incapable of sustaining normal forces, then $f_1(y) = f_2(x) = 0$ and hence $\bar{N}_y = \bar{N}_x = 0$. The membrane theory thus gives results which imply that the entire vertical load is taken by shearing stress resultants.

The principal tension and compression stress resultants appear at sections rotated $45°$ from the straight boundaries, and their values are

$$\bar{N}_c = -\frac{\bar{p}_z}{2k}$$

$$\bar{N}_t = \frac{\bar{p}_z}{2k} \tag{k}$$

From (f) with $\alpha = 45°$, so that $a = b$, $h_1 = h_2$ and $c_1 = c_2 = c$, then

$$k = \frac{2}{h_2}$$

and hence

$$\bar{N}_c = -\frac{\bar{p}_z h_2}{4} = -\frac{\bar{p}_z a^2}{4c}$$

In the same manner

$$\bar{N}_t = \frac{\bar{p}_z a^2}{4c}$$

Thus the membrane theory leads to the conclusion that the vertical load is carried by pure arch action, just as the saddle shell that was described in Sec. 7-3. Figure 7-10 shows that these arch stress resultants combine at the straight edges to give pure shear for which an edge beam has often been supplied. This description of behavior does not fit with results from more accurate analyses including bending, and the reason is that the straight-line boundaries cannot absorb the arching forces without significant displacements. For this reason the membrane theory cannot be used as the sole basis for design in all hyperbolic paraboloids. Whereas it gives reasonable results for the curved boundary cases discussed in Sec. 7-3, it does not so do for the gabled shells of this section. The following analysis, used as part of the basis for designing the largest gabled hyperbolic paraboloid built in the United States up to 1960,[10] indicates the low shell stresses derived from membrane theory. These low values increase in bending theory results but not enough to affect the shell reinforcement appreciably. A discussion of more accurate results will be given after the following description of the membrane theory values.

The analysis based on loading uniform over the horizontal projection of the surface is a reasonable approximation of uniform shell dead load for flat hyperbolic paraboloids, i.e., small values of k. Where k increases, the actual membrane values under dead load should be investigated. For a shell of uniform thickness,

$$q = -\bar{p}_z = -\frac{p_z}{\cos \phi \cos \theta} \sin \omega$$

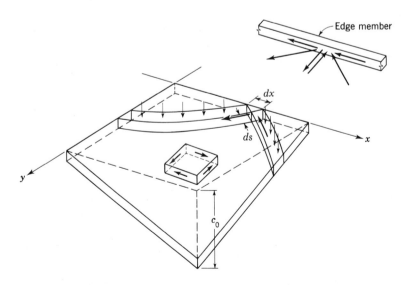

Fig. 7-10   Simplified behavior of hyperbolic paraboloids.

Fig. 7-11 Hyperbolic paraboloid thin shell at Denver. (Courtesy of Roberts and Schaefer Company, Inc.)

or

$$\bar{p}_z = p_z \sin\omega \ \sqrt{[1 + (kx)^2][1 + (ky)^2]} \tag{l}$$

Substituting (l) into Eq. (7-24) gives

$$-2 \frac{\partial^2 F}{\partial x \, \partial y} k = 2\bar{N}_{xy}k = -p_z \sin\omega \ \sqrt{[1 + (kx)^2][1 + (ky)^2]}$$

from which, after neglecting the term $k^4 x^2 y^2$, which gives $\sin\omega = 1$, we find

$$\bar{N}_{xy} = -\frac{p_z}{2k} \sqrt{1 + k^2 x^2 + k^2 y^2} = N'_{xy} \tag{7-25}$$

$\bar{N}_x$ and $\bar{N}_y$ can be determined from (7-4) and (7-5) by differentiating $\bar{N}_{xy}$ and then integrating the result.

$$\bar{N}_x = \frac{p_z y}{2} \log\left(\frac{kx + \sqrt{1 + k^2 x^2 + k^2 y^2}}{\sqrt{1 + k^2 y^2}}\right)^2 \tag{7-26}$$

$$\bar{N}_y = \frac{p_z x}{2} \log\left(\frac{ky + \sqrt{1 + k^2 x^2 + k^2 y^2}}{\sqrt{1 + k^2 x^2}}\right)^2 \tag{7-27}$$

The actual values of N'_x and N'_y are obtained from (7-1) and (7-2).

It is of interest to compare the results of an accurate membrane analysis for dead loads with the approximation of uniform load on the horizontal projection of the roof.

Consider the roof of Fig. 7-11 in which $k = 0.0076$ ft^{-1}, where $c_0 = 28$ ft, $a_0 = 56$ ft, and $b_0 = 66$ ft; at $x = a_0$ and $y = b_0$, $\tan \phi = c_0/a_0$ and $\tan \theta = c_0/b_0$, so that $\phi = 26.5°$ and $\theta = 23°$.[10]

In the case of the approximation, the principal tension stress in the 3-in.-thick shell under dead load would be, from (k), with $p_z = 37.5$ psf,

$$\bar{N}_t = \frac{\bar{p}_z}{2k} = 66\bar{p}_z = 2470 \text{ lb/ft}$$

$$f = \frac{N_t}{36} = 69 \text{ psi} \qquad \text{(tension)}$$

For the more precise analysis and for the maximum value of shear at $x = a_0$ and $y = b_0$,

$$\sqrt{1 + k^2 x^2 + k^2 y^2} = 1.2$$

$$\sin \omega = \sqrt{1 - \sin^2 \phi \, \sin^2 \theta} = 0.985 \approx 1.0$$

$$N'_{xy} = -1.2 \frac{p_z}{2k} = 79 p_z$$

$$N'_x = p_z \frac{66}{2} \log \left(\frac{0.0076 \times 56 + 1.2}{\sqrt{1 + 0.25}} \right) \frac{\cos \theta}{\cos \phi} = 12.7 p_z$$

$$N'_y = p_z \frac{56}{2} \left(\frac{0.89}{0.92} \right) \log \left(\frac{0.0076 \times 66 + 1.2}{\sqrt{1 + 0.18}} \right) = 12.2 p_z$$

The principal stress resultants will be given by the usual formula:

$$N'_a = \tfrac{1}{2}(N'_x + N'_y) \pm \tfrac{1}{2} \sqrt{(N'_x - N'_y)^2 + 4N'^2_{xy}}$$

$$= (+12.5 \pm 79) p_z = \begin{cases} 92 p_z \text{ lb/ft} \\ -66 p_z \text{ lb/ft} \end{cases}$$

where $h = 3$ in., $p_z = 37.5$ psf, and

$$f = \frac{N'_a}{12 \times 3} = \frac{92 \times 37.5}{36} = 96 \text{ psi} \qquad \text{(tension)}$$

The angle which this principal stress makes with respect to one system of generators is

$$\tan 2\alpha_0 = \frac{2N'_{xy}}{N'_x - N'_y} = \frac{158}{0.5} = 316$$

$$\alpha_0 \approx 45°$$

Clearly the approximation can be used with no appreciable error for this example when only uniform loads are considered. It must be remembered, however, that there will now exist forces normal to the edge because $N'_x \neq 0$, and $N'_y \neq 0$. Mathematically this fact can be deduced from Eqs. (i) and (j), which show one constant of integration available for each force even though there are two free edges for each. The approximation no longer gives a complete statical analysis, and the normal forces must be carried either by an edge support capable of taking normal forces or by bending in the shell. In a shell with greater warping this bending may become critical.

As shown in Fig. 7-10, the membrane theory assumes that the shell is supported at the straight-line boundaries by an edge member which provides an edge shearing reaction which, when $\bar{p}_z$ is constant, $\alpha = 45°$,

$$S \, dx = (-\bar{N}_c \sin \alpha + \bar{N}_t \sin \alpha) \, ds$$

$$S \, dx = \frac{\bar{p}_z}{k} \sin \alpha \, ds$$

$$S = \frac{\bar{p}_z}{k} \sin \alpha \frac{ds}{dx} = \frac{\bar{p}_z}{k} \sin \alpha \cos \alpha$$

$$S = \frac{\bar{p}_z}{2k}$$

In these edge members the axial compression consists of the summation of the edge shears. In the horizontal member along the x axis

$$C = (a - x)S$$

which, for the shell in Fig. 7-11, would give a maximum value at $x = 0$ of

$$C = 56 \times 0.0375/0.0152 = 138 \text{ kips}$$

These membrane theory results assume that the straight-line boundaries provide the proper support, but, as shown by more complete solutions using numerical methods, the straight edge beams do not prevent horizontal displacements. The result is that the inverted arch action cannot develop the tension $\bar{N}_t$ forces and the compression arch action thus carries most of the load toward the corner supports. The $\bar{N}_c$ values rise to about double the membrane theory predictions and the edge shears drop. Therefore, the compression in the edge members is radically different from that found by the membrane theory.

The gabled shell in Fig. 7-12 demonstrates the differences between membrane theory and bending theory for the behavior of the shell, of the horizontal ridge beams, and of the gabled edge beams.[11, 12]

For the shell under uniform load on the horizontal projection the membrane theory would give shear stresses everywhere of $N'_{xy} = \bar{p}_z/2k$ where $\bar{p}_z = 37.5$ psf and $k = 8/40^2 = 0.005$, so that $N'_{xy} = 37.5/0.01 = 3.75$ kips/ft or $\tau_{xy} = 3.75/(12 \times 3) = 104$ psi. This stress is numerically equal to the principal tensile and compressive stresses through (k), and hence, according to the membrane theory in each quadrant, the shell carries the load—half by parabolic arching slabs parallel to the diagonal between the central crown and the support and half by parabolic hanging slabs parallel to the diagonal between the gable crowns at the edges. Shaaban and Ketchum showed that the hanging slabs do not carry any significant load; rather it is nearly all taken by the arching slabs which have thereby twice the stress that is gotten by membrane theory.[12] Figure 7-13 shows this behavior and further emphasizes how the diagonal arching slab part that is directly between the crown and support takes very much higher stresses near that support. Thus the gabled hyperbolic paraboloid shell acts

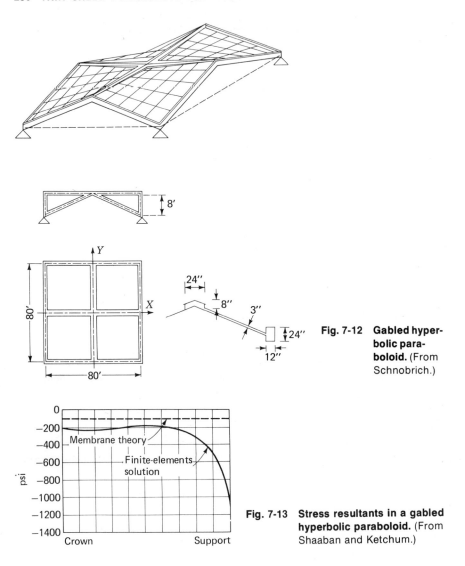

Fig. 7-12 Gabled hyperbolic paraboloid. (From Schnobrich.)

Fig. 7-13 Stress resultants in a gabled hyperbolic paraboloid. (From Shaaban and Ketchum.)

primarily as two diagonal arches crossing at the crown and delivering nearly all the roof load directly to the corner supports. It does not act as a set of arching and hanging slabs bringing load to the straight-line edges as shown in Fig. 7-10 and as derived from the membrane theory.

This behavior arises because those edges cannot carry the tensions assumed by the membrane theory; rather they move inward as shown in Fig. 7-14 and thereby force the diagonal arches to carry the load. We can see the same behavior reflected in the forces carried by the edge beams. Figure 7-15 shows the forces in the edge beams derived from both the membrane theory and the bending theory. The former gives results closely approximated by $C = (a - x)S = 40 \times 3.75 = 150$ kips at the support where $x = 0$ and $a = 40$ ft. The figure shows a slightly higher value because it is based upon dead load rather than a uniform load on the horizontal projection of the surface. The bending theory gives smaller values which are nearly constant up until about halfway down the gable. Then its force increases rapidly until at the support it is about equal to the membrane theory result. Thus Figs. 7-13 and 7-15 show that the shell rather than the edge beam carries much of the load except near the corners at the supports

These results, while useful as insight, do not have a major bearing on the layout of shell reinforcement, since the stresses are so low. However, they have a major influence on design of the edge beams for two reasons. First, the much-reduced edge-beam loads suggest that those beams need not be large. Second, the reduced compression makes such beams less able to carry bending, because they are less precompressed than the membrane theory assumes.

Indeed, Schnobrich has shown that the edge beams and the ridge beams actually load the shell rather than the other way around and, therefore, the design needs to reduce their size.[11] This is especially true for the ridge beams since they are horizontal and hence can be subjected to substantial deflections and bending in regions where the shell is so flat that it provides little support. Figure 7-16 shows this effect clearly where the 8 × 24 in. ridge beams deflect by a constant amount, whereas the lighter beams (6 × 16) even rise slightly at midspan and the heavier beams (12 × 48) deflect over twice as much and bend substantially, inducing moments of over 40 in.-kips at

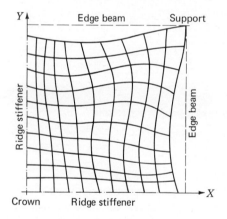

Fig. 7-14 Displacements in a gabled hyperbolic paraboloid. (From Shaaban and Ketchum.)

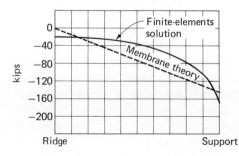

Fig. 7-15 Forces in edge beam of a gabled hyperbolic paraboloid. (From Shaaban and Ketchum.)

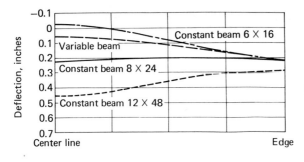

Fig. 7-16 Deflections in ridge beam of a gabled hyperbolic paraboloid. (From Schnobrich.)

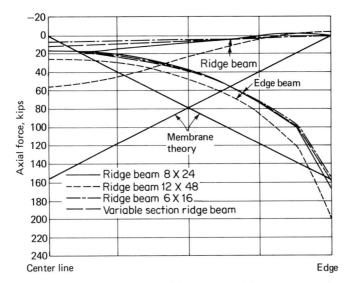

Fig. 7-17 Force in ridge and edge beams of a gabled hyperbolic paraboloid. (From Schobrich.)

the midspan, whereas the 8 × 24 in. beam has almost no dead-weight moment (Ref. 11, Fig. 9). Even more important is the distribution of compressive thrust in that ridge beam. Figure 7-17 shows that for the 8 × 24 in. beam the midspan thrust is about 20 kips, whereas the membrane theory gives over 150 kips. The heavier beam carries more thrust but still far less than assumed in the membrane theory.

These ridge beams are essential to stiffen the shell in its very flat central section, but for that they need not be overly heavy. Rather they need substantial reinforcement, some depth, and a location above the shell. This last requirement is essential since the compression force enters at the level of the shell and thus, by having the shell below the beam center line, that compressive force causes a negative moment which counteracts the positive moment resulting from beam dead weight. In the example shown in Fig. 7-12, all but the heavy beam are so balanced that almost no bending results under dead load of shell and beam together.

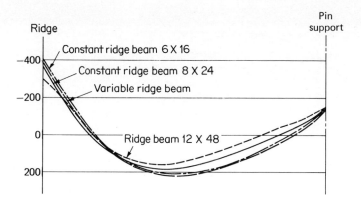

Fig. 7-18   Bending moments in edge beam of a gabled hyperbolic paraboloid. (From Schnobrich.)

The ridge beam, being in compression, does need enough reinforcing steel to control the tendency for flat shells to creep and hence be in danger of showing large displacements. The Denver shell had 2 percent of steel in its ridge beams.

The gabled edge beams also bend under their own weight; they act somewhat like continuous beams on three supports where the middle support is the gable crown. Schnobrich plotted these moments (Fig. 7-18) which clearly show this behavior. If the spans are taken as $l = 40$ ft—the load is the edge-beam weight $16 \times 12 \times 150/144 = 200$ lb/ft—then the moments at crown $M_c = ql^2/8 = -480$ in.-kips and near midspan $M_c = +ql^2/16 = 240$ in.-kips are very close to the values found from a full analysis of shell and edge beams. Thus the gabled edge beam is mostly carrying its own dead weight by bending and even to some extent loading the shell (by the difference between its own bending moment and that found from the full analyses). Twelve years before the publication of Fig. 7-18, Tedesko noted the same behavior in the gabled edge beams of the shell at Denver (Fig. 7-11) where "Fine hair cracks developed at the top of the ribs at the peaks, such as might occur at the center support of a two-span beam" (Ref. 10, pages 411–412).

7-5   ELLIPTICAL PARABOLOIDS

The equation for an elliptical paraboloid is the same as Eq. (7-9) for hyperbolic paraboloids except that the signs of the x and y parabolas are the same.

$$z = \frac{x^2}{h_1} + \frac{y^2}{h_2} \tag{7-28}$$

The constants h_1 and h_2 have the same meaning here as in (7-9), namely

$$h_1 = \frac{a^2}{c_1} \quad \text{and} \quad h_2 = \frac{b^2}{c_2}$$

which clearly define the two sets of parabolas (Fig. 7-19). If a horizontal plane is

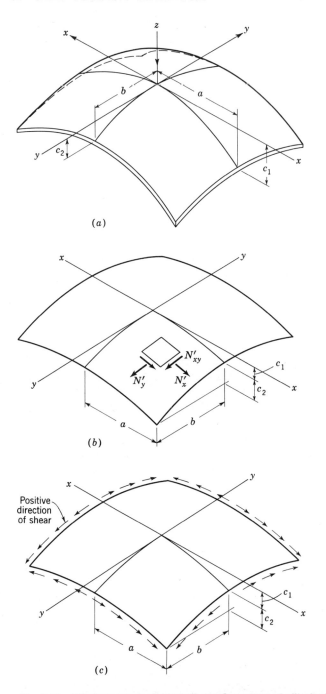

Fig. 7-19 Elliptical paraboloid shell. (*a*) Geometry of elliptical paraboloid. (*b*) Shell stress resultants. (*c*) Tangential shear along boundary. (After Ref. 13.)

passed through the surface at z_c = constant, then

$$z_c = \frac{x^2}{h_1} + \frac{y^2}{h_2} = \text{constant}$$

which defines an ellipse with semiaxes of

$$a_1 = \sqrt{z_c h_1} \quad \text{and} \quad b_1 = \sqrt{z_c h_2}$$

hence the name elliptical paraboloid. In the special case where $h_1 = h_2$, then

$$z = \frac{x^2 + y^2}{h_1} \tag{7-29}$$

which is a paraboloid of revolution since at z_c = constant,

$$z_c h_1 = x^2 + y^2$$

Thus planes normal to the z axis intersect the surface in a circle of radius $r = \sqrt{z_c h_1}$.

Introducing Eq. (7-28) into the differential equation of equilibrium (7-8), we obtain

$$\frac{2}{h_2} \frac{\partial^2 F}{\partial x^2} + \frac{2}{h_1} \frac{\partial^2 F}{\partial y^2} = q \tag{7-30}$$

Again we shall make the simplifying assumption of loading uniformly distributed over the horizontal projection of the surface, $q = -\bar{p}_z$. The problem now is to find a solution to Eq. (7-30) which will at the same time satisfy reasonable boundary conditions. For example, if the boundaries at $y = \pm b$ are to be supported by vertical arches which are not stiff laterally, a solution can be made by setting

$$F_0 = \frac{\bar{p}_z h_1}{4} (b^2 - y^2) \tag{7-31}$$

Using (7-31) in (7-7), we obtain

$$\bar{N}_x = - \frac{\bar{p}_z h_1}{2} = - \frac{\bar{p}_z a^2}{2c_1}$$

$$\bar{N}_y = 0 \tag{7-32}$$

$$\bar{N}_{xy} = 0$$

We observe that this assumption leads to an internal stress pattern in which all the load is carried by the "arches" parallel to the x axis. There is really no need for stiff arches along the edges $y = \pm b$, but stiff supports must be provided at $x = \pm a$ to carry the $\bar{N}_x$ forces, or "arch" thrusts, which can be written in the familiar form by recognizing that $\bar{N}_x = H$ and $a = L/2$, so that the first of (7-32) can be written:

$$H = - \frac{pL^2}{8c_1}$$

Such a system is a form of barrel arch in which the arches do not spring from a straight line in plan as a cylindrical barrel arch but from an elliptical curve. Obviously by assuming

$$F_0 = -\frac{ph_2}{4}(a^2 - x^2) \qquad (7\text{-}33)$$

we can direct the system to carry loads by the "arches" parallel to the y axis. Neither system shows the great structural potential of this form of shell, which is its ability to carry loads without any appreciable lateral restraint except at the four low corners. In order to determine the internal stress distribution for such a case we must choose F such that $\bar{N}_x = 0$ at $x = \pm a$ and $\bar{N}_y = 0$ at $y = \pm b$. Since Eqs. (7-31) or (7-33) satisfy only one or the other of these requirements, an additional force function is needed, which will be taken in the form of the infinite series:

$$F_1 = \sum_{n=1,3,5\ldots}^{\infty} A_n \cosh \beta x \cos \lambda y$$

where

$$\beta = \frac{n\pi}{2a}\sqrt{\frac{c_1}{c_2}} \quad \text{and} \quad \lambda = \frac{n\pi}{2b}$$

$$\frac{\partial^2 F_1}{\partial x^2} = \sum_{n=1,3,5\ldots}^{\infty} A_n \beta^2 \cosh \beta x \cos \lambda y = \bar{N}_y$$

$$\frac{\partial^2 F_1}{\partial y^2} = \sum_{n=1,3,5\ldots}^{\infty} -A_n \lambda^2 \cosh \beta x \cos \lambda y = \bar{N}_x \qquad (a)$$

If F_1 is combined with F_0, the result must give $\bar{N}_y = 0$ at $y = \pm b$, which F_0 does alone and which F_1 gives as well $\left(\cos\dfrac{n\pi y}{2b} = 0 \text{ for } y = b\right)$, and it must also give $\bar{N}_x = 0$ at $x = a$. From Eq. (7-32),

$$\bar{N}_{x0} = -\frac{\bar{p}_z a^2}{2c_1}$$

at $x = a$, and F_1 gives at $x = a$

$$\bar{N}_{x_1} = \sum_{n=1,3,5\ldots}^{\infty} -A_n \lambda^2 \cosh \beta a \cos \lambda y$$

$\bar{N}_{x0} + \bar{N}_{x_1}$ must be zero at $x = a$. First we express $\bar{N}_{x0}$ in terms of a Fourier series where

$$\bar{N}_{x0} = -\left(\frac{\bar{p}_z a^2}{2c_1}\right)\frac{4}{\pi}\sum_{n=1,3,5\ldots}^{\infty}\frac{(-1)^{(n-1)/2}\cos \lambda y}{n}$$

In order to satisfy the boundary value of $\bar{N}_x = 0$ at $x = a$

$$\sum_{n=1,3,5\ldots}^{\infty} -A_n \lambda^2 \cosh \beta a \cos \lambda y = \left(\frac{\bar{p}_z a^2}{2c_1}\right)\frac{4}{\pi}\sum_{n=1,3,5\ldots}^{\infty}\frac{(-1)^{(n-1)/2}\cos \lambda y}{n}$$

The function $\Sigma \cos \lambda y$ cancels out and

$$A_n = - \frac{2\bar{p}_z a^2}{c_1 \pi n} \frac{1}{\lambda^2} \frac{(-1)^{(n-1)/2}}{\cosh \beta a}$$

Substituting this value for A_n into Eqs. (a) and also solving for $\bar{N}_{xy}$, we obtain, by combining (7-32) with (a),

$$\bar{N}_x = \frac{\bar{p}_z a^2}{c_1} \left(\frac{2}{\pi} \sum_{n=1,3,5\ldots}^{\infty} \frac{(-1)^{(n-1)/2} \cosh \beta x}{n \cosh \beta a} \cos \lambda y - \frac{1}{2} \right)$$

$$\bar{N}_y = - \frac{\bar{p}_z b^2}{c_2} \left(\frac{2}{\pi} \sum_{n=1,3,5\ldots}^{\infty} \frac{(-1)^{(n-1)/2} \cosh \beta x}{n \cosh \beta a} \cos \lambda y + 0 \right) \qquad (7\text{-}34)$$

$$\bar{N}_{xy} = - \frac{ab\bar{p}_z}{\sqrt{c_1 c_2}} \left(\frac{2}{\pi} \sum_{n=1,3,5\ldots}^{\infty} \frac{(-1)^{(n-1)/2} \sinh \beta x}{n \cosh \beta a} \sin \lambda y + 0 \right)$$

Since the values within the parentheses contain only the parameter c_1/c_2, they have been tabulated to simplify the computations. Equations (7-34) can be rewritten in terms of the actual shell forces as

$$N_x' = - \frac{\bar{p}_z a^2 K}{c_1} \qquad \text{(coefficient)}$$

$$N_y' = - \frac{\bar{p}_z b^2}{K c_2} \qquad \text{(coefficient)}$$

$$N_{xy}' = - \frac{\bar{p}_z ab}{\sqrt{c_1 c_2}} \qquad \text{(coefficient)}$$

The coefficients are given in Table 7-4 from Ref. 13 where

$$K = \frac{\cos \theta}{\cos \phi} = \sqrt{\frac{1 + [(2c_1/a)(x/a)]^2}{1 + [(2c_2/b)(y/b)]^2}}$$

This internal force pattern satisfies our criterion that no forces normal to the edges be allowed. As shown in Fig. 7-19, all the load is, therefore, carried along the free edges by tangential shear down to the corner supports, in the same manner as the hyperbolic paraboloid with straight-line generators as boundaries. Table 7-5 from Ref. 13 gives the tangential shear at the boundaries. Both tables show one apparent impossibility, namely, that at the corners, $x = \pm a$ and $y = \pm b$, the shear goes to infinity. The third of Eqs. (7-34) will show this to be the case.

Physically this may be interpreted as follows.[13] In a translational surface any element bounded by generators will have no twist in its surface. The tangential shearing forces on either side of the element are exactly parallel; hence the shears do not contribute to vertical equilibrium. This fact can be seen from Eq. (7-30) or from Eq. (7-10) for hyperbolic paraboloids, in which the term $\partial^2 F/\partial x \, \partial y$ does not appear because $\partial^2 z/\partial x \, \partial y = 0$.

At the corners where $N_x = N_y = 0$, nothing is available to resist the vertical loads except the shearing force. Thus transverse shearing forces and bending moments will occur to carry these loads, and the value of N_{xy} will be finite.

Table 7-4 COEFFICIENTS FOR COMPUTING STRESS RESULTANTS IN ELLIPTICAL PARABOLOIDS*

VALUE OF y/b

x/a	Stress result-ants	$c_1/c_2 = 1.0$					$c_1/c_2 = 0.8$				
		0	0.25	0.50	0.75	1.0	0	0.25	0.50	0.75	1.0
0.00	N'_y	0.250	0.233	0.182	0.101	0	0.289	0.270	0.213	0.119	0
	N'_x	0.250	0.267	0.318	0.399	0.500	0.211	0.230	0.287	0.381	0.500
	N'_{xy}	0	0	0	0	0	0	0	0	0	0
0.25	N'_y	0.267	0.250	0.199	0.111	0	0.304	0.285	0.228	0.130	0
	N'_x	0.233	0.250	0.301	0.389	0.500	0.196	0.215	0.272	0.370	0.500
	N'_{xy}	0	0.029	0.068	0.096	0.108	0	0.034	0.069	0.100	0.114
0.50	N'_y	0.318	0.301	0.250	0.150	0	0.347	0.331	0.277	0.169	0
	N'_x	0.182	0.199	0.250	0.350	0.500	0.153	0.169	0.223	0.331	0.500
	N'_{xy}	0	0.068	0.140	0.210	0.244	0	0.065	0.139	0.215	0.255
0.75	N'_y	0.399	0.389	0.350	0.250	0	0.416	0.406	0.369	0.270	0
	N'_x	0.101	0.111	0.150	0.250	0.500	0.084	0.094	0.131	0.230	0.500
	N'_{xy}	0	0.096	0.210	0.356	0.465	0	0.091	0.201	0.353	0.480
1.0	N'_y	0.500	0.500	0.500	0.500	0	0.500	0.500	0.500	0.500	0
	N'_x	0	0	0	0	0	0	0	0	0	0
	N'_{xy}	0	0.108	0.243	0.465	∞	0	0.101	0.229	0.443	∞

x/a	Stress result-ants	$c_1/c_2 = 0.6$					$c_1/c_2 = 0.4$				
		0	0.25	0.50	0.75	1.0	0	0.25	0.50	0.75	1.0
0.00	N'_y	0.336	0.316	0.252	0.143	0	0.395	0.374	0.307	0.180	0
	N'_x	0.164	0.184	0.248	0.357	0.500	0.105	0.126	0.193	0.320	0.500
	N'_{xy}	0	0	0	0	0	0	0	0	0	0
0.25	N'_y	0.348	0.329	0.267	0.155	0	0.403	0.383	0.319	0.192	0
	N'_x	0.152	0.171	0.233	0.345	0.500	0.097	0.117	0.181	0.308	0.500
	N'_{xy}	0	0.031	0.067	0.103	0.120	0	0.026	0.060	0.101	0.125
0.50	N'_y	0.383	0.367	0.312	0.197	0	0.425	0.410	0.357	0.235	0
	N'_x	0.117	0.133	0.188	0.304	0.500	0.075	0.090	0.143	0.265	0.500
	N'_{xy}	0	0.060	0.132	0.216	0.265	0	0.049	0.115	0.208	0.274
0.75	N'_y	0.436	0.426	0.392	0.296	0	0.459	0.451	0.419	0.331	0
	N'_x	0.064	0.074	0.108	0.204	0.500	0.041	0.049	0.081	0.169	0.500
	N'_{xy}	0	0.081	0.185	0.342	0.494	0	0.065	0.156	0.316	0.506
1.00	N'_y	0.500	0.500	0.500	0.500	0	0.500	0.500	0.500	0.500	0
	N'_x	0	0	0	0	0	0	0	0	0	0
	N'_{xy}	0	0.089	0.208	0.413	∞	0	0.070	0.173	0.363	∞

x/a	Stress result-ants	$c_1/c_2 = 0.2$				
		0	0.25	0.50	0.75	1.0
0.00	N'_y	0.462	0.446	0.388	0.248	0
	N'_x	0.038	0.054	0.112	0.252	0.500
	N'_{xy}	0	0	0	0	0
0.25	N'_y	0.465	0.451	0.396	0.261	0
	N'_x	0.035	0.049	0.104	0.239	0.500
	N'_{xy}	0	0.014	0.040	0.088	0.128
0.50	N'_y	0.473	0.462	0.414	0.303	0
	N'_x	0.027	0.038	0.086	0.197	0.500
	N'_{xy}	0	0.027	0.074	0.174	0.280
0.75	N'_y	0.485	0.480	0.456	0.383	0
	N'_x	0.015	0.020	0.044	0.117	0.500
	N'_{xy}	0	0.034	0.098	0.246	0.510
1.00	N'_y	0.500	0.500	0.500	0.500	0
	N'_x	0	0	0	0	0
	N'_{xy}	0	0.038	0.108	0.262	∞

* From Ref. 13.

Table 7-5 SHEAR ALONG EDGES OF ELLIPTICAL PARABOLOIDS*

y/b	c_1/c_2				
	1.0	0.8	0.6	0.4	0.2
	At $x = \pm a$				
0.0	0.0000	0.0000	0.0000	0.0000	0.0000
0.1	0.0419	0.0389	0.0342	0.0307	0.0137
0.2	0.0854	0.0793	0.0701	0.0550	0.0286
0.3	0.1319	0.1231	0.1096	0.0872	0.0481
0.4	0.1836	0.1721	0.1546	0.1254	0.0731
0.5	0.2432	0.2294	0.2081	0.1728	0.1075
0.6	0.3204	0.3066	0.2859	0.2493	0.1818
0.7	0.4071	0.3897	0.3627	0.3173	0.2296
0.8	0.5363	0.5178	0.4887	0.4400	0.3443
0.85	0.6279	0.6090	0.5791	0.5292	0.4306
0.9	0.7570	0.7378	0.7074	0.6667	0.5659
0.95	0.9777	0.9582	0.9276	0.8763	0.7741
1.0	∞	∞	∞	∞	∞
x/a	At $y = \pm b$				
0.0	0.0000	0.0000	0.0000	0.0000	0.0000
0.1	0.0419	0.0444	0.0468	0.0488	0.0500
0.2	0.0854	0.0903	0.0950	0.0990	0.1014
0.3	0.1319	0.1391	0.1460	0.1519	0.1553
0.4	0.1836	0.1930	0.2019	0.2095	0.2140
0.5	0.2432	0.2545	0.2652	0.2743	0.2798
0.6	0.3204	0.3317	0.3425	0.3516	0.3571
0.7	0.4071	0.4213	0.4348	0.4463	0.4532
0.8	0.5363	0.5515	0.5659	0.5782	0.5855
0.85	0.6279	0.6434	0.6582	0.6707	0.6782
0.9	0.7570	0.7728	0.7878	0.8005	0.8081
0.95	0.9777	0.9935	1.0087	1.0215	1.0290
1.0	∞	∞	∞	∞	∞

* From Ref. 13.

REFERENCES

1. Salvadori, M. G., *Trans. ASCE,* vol. 123, 1958, p. 1018.
2. Bouma, A. L., "Some Applications of the Bending Theory Regarding Doubly Curved Shells," *Proceedings of the Symposium on the Theory of Thin Elastic Shells,* Delft, 1959, North Holland Publishing Company, Amsterdam, 1959, pp. 202–230 and 387.
3. Apeland, K., "Stress Analysis of Translational Shells," *J. ASCE,* Eng. Mech. Div., vol. 87, February 1961, pp. 111–139.
4. Faber, C., *Candela: the Shell Builder,* Reinhold Publishing Corporation, New York, 1963.
5. Flügge, W., *Stresses in Shells,* Springer-Verlag OHG, Berlin, 1960.
6. Pucher, A., "Über den Spannungszustand in Gekrümmten Flächen," *Beton u. Eisen,* vol. 33, 1934.

7. Scordelis, Alexander C., "Analysis and Design of HP Groined Vaults," *Proceedings,* IASS World Congr. Space Struct., Montreal, 1976, pp. 561–568.
8. "Elementary Analysis of Hyperbolic Paraboloid Shells," Portland Cem. Assoc., 1960, p. 10.
9. Becker, William C. E., "Intersecting Ribs Carry Concrete Roof Shell," *Civ. Eng.,* July 1955, p. 58–61. See also Tedesko, Anton, "The St. Louis Air Terminal Shells", *Proceedings,* World Congr. Shell Struct., San Francisco, October 1962, pp. 469–474.
10. Tedesko, A., "Shell at Denver—Hyperbolic Paraboloid Structure of Wide Span," *J. ACI,* October 1960, p. 403.
11. Schnobrich, William C., "Analysis of Hipped Roof Hyperbolic Paraboloid Structures," *J. Struct. Div.,* ASCE, vol. 98, no. ST-7, July 1972, pp. 1575–1583.
12. Shaaban, Ahmed, and Milo Ketchum, "Design of Hipped Hypar Shells," *J. Struct. Div.,* ASCE, vol. 102, no. ST-11, November 1976, pp. 2151–2161.
13. Parme, A. L., "Shells of Double Curvature," *Trans. ASCE,* vol. 123, 1958, p. 989.

8
BEHAVIOR AND DESIGN
OF FOLDED PLATES

8-1 INTRODUCTION

Folded plates are normally considered together with thin shells because their behavior is similar to that of thin shells of the same overall dimensions. Comparing a segment of a cylinder with a series of thin plates, we shall find that the two systems have about the same stresses, and as the thin plates become narrower, we approach the cylindrical shell as a limit.

A full bibliography on folded plates appeared in the report of the ASCE Task Committee on Folded Plates.[1] This report recommended a simplified analysis, which was given in the first edition of this text[2] and can be found in the chapter on thin shells of the *Structural Engineering Handbook.*[3]

Since that 1963 report, computer-based analyses have become common. In fact the finite-element results given in Chap. 6 for the barrel shell are actually based upon flat-plate elements. Thus, the barrel was treated as a folded plate.

The simplified analysis presented earlier was used to emphasize the behavior of the plate systems. Not all engineers will want to accept large-scale computer system results until they have comparable results gained by tracing through a simplified analysis, each part of which relates directly to elementary structural forms. In that earlier analysis, those forms were one-way slabs and thin inclined rectangular beams. A large number of folded-plate structures have been designed successfully on the basis of that simplified analysis. At the same time, for research and for some designs it is more practical to use numerical solutions such as the solution described in Sec. 6-7. In this chapter, after describing behavior, we shall discuss the results of such a numerical solution for the more complex problems of folded plates continuous over three supports.

8-2 DEFINITIONS AND ASSUMPTIONS

A *folded-plate structure* is defined as a system of thin plates spanning longitudinally and monolithically joined to each other along longitudinal joints at some angle other than 180°. Figure 8-1 defines the typical elements of a folded-plate structure and Fig. 8-2 shows a series of typical folded-plate cross sections.

Figure 8-2a suggests a single-barrel cylindrical shell. The differences lie in the elements of the system, not in the system itself. When we replace the plates shown

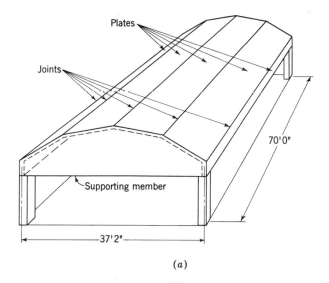

(a)

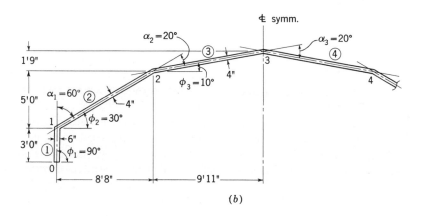

(b)

Fig. 8-1 Typical folded-plate roof.

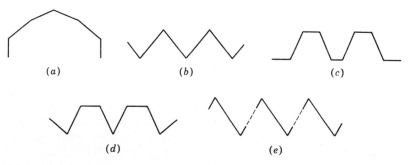

(a)   (b)   (c)

(d)   (e)

Fig. 8-2 Various folded-plate cross sections.

in Fig. 8-2a by a great many more plates, the section approaches closely a cylindrical cross section. In fact Professor Yitzhaki has developed a method of analyzing cylindrical shells based on a folded-plate approximation.[4]

The dimensions of folded-plate systems are often similar to those of long-barrel cylindrical shells. We can expect, therefore, that the beam-arch method will apply here as it did in the curved shell. Longitudinal stresses will approximate those in a beam of curved cross section. The transverse forces will be different here because of the lack of curvature and the sharp discontinuities at the joints. The arch analysis really becomes a continuous-slab analysis where the slabs span between joints and hence are assumed to carry loads transversely solely by flexure. Transverse bending moments will be significantly larger in folded plates than in comparable cylindrical barrels because of the discontinuities at the joints.

Referring again to Fig. 8-2, we see the section in a approximates a single-barrel shell with vertical edge beams. Figures 8-2b, c, and d are similar to multiple-barrel shells. In b the section is a series of inclined thin beams, whereas in c the section can be thought of as I beams with inclined webs. In Fig. 8-2d the section is somewhat similar to T beams. Section b is limited both by compression stresses which must be taken by the restricted joint and by available space for tension reinforcement. Section c has been frequently used for longer spans. Each of b, c, and d is advantageous for repetitive forming. Section e is a Northlight folded plate which has been used in industrial applications.

There are two important reasons for the frequent use of folded-plate structures: first, they can span great distances with little material and with relatively simple formwork, and second, they have a striking appearance. Theoretically they require more materials than a smoothly curved shell, but often they are more economical to build because of the lack of curvature.

Folded plates were first used as storage bins where the plates were deep with respect to their span and where they were often restrained by plates on three or four sides (see Fig. 8-3). Rigorous analysis[5] for such plates involves two general studies: first, the analysis for the loading in the plane of the plates, which are deep beams (Sec. 4-13), and second, the analysis of the loading perpendicular to the plates.

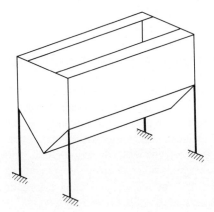

Fig. 8-3 Folded-plate storage bin.

Table 8-1   COMPARISON BETWEEN SIMPLIFIED AND
MORE RIGOROUS ANALYSES

	L = 105 ft			L = 70 ft			L = 35 ft		
	A*	B	C	A	B	C	A	B	C
f_3	−637†	−685	−685	−147	−186	−189	25	1	−1
f_2	−535	−548	−547	−281	−282	−279	−112	−100	−99
f_1	396	500	496	34	93	87	18	0	1
f_0	2517	2310	2320	1456	1330	1340	382	403	400
M_3	−3861‡	−3326	−3426	−2067	−1954	−1994	−958	−922	−941
M_2	−1950	−1568	−1655	−948	−857	−844	−591	−607	−595
M_1	0	343	289	0	238	214	0	−96	−63

* A, simplified analysis (Ref. 2, Chapter 8); B, more rigorous analysis with a Poisson's ratio of zero (Ref. 5); C, more rigorous analysis with a Poisson's ratio of 0.25 (Ref. 5).
† Stresses in pounds per square inch.
‡ Moments in foot-pounds per foot.

Such a general treatment of folded plates is normally not needed for the long-span roof and floor structures commonly used. Unlike the original folded plates, these roof or floor systems are made up of plates with longitudinal spans which are much greater than their total height. In the simplified analysis, two basic approximations are made:

1. The deep-beam analysis is replaced by a one-dimensional analysis based on the simple flexural theory.
2. Since the plates are long and relatively narrow, they can be considered essentially as restrained only along their longitudinal edges. Just as in cylindrical barrels, the effect of restraint at the supporting ribs or diaphragms is considered separately from the basic analysis.

Generally, where the span is at least 3 times the plate depth, these two simplifications are justified.

Table 8-1 gives results for the structure shown in Fig. 8-1 loaded by its own weight plus an additional load of 30 psf over the roof surface. This table shows a comparison between those values gotten from the simplified analyses and similar values obtained from a two-dimensional analysis based on the general formulations presented in the paper by Goldberg and Leve.[5] In each case shown in Table 8-1, the simplified analysis gives reasonable values upon which to base a design. Only the moment at joint 1 is very different but it is small compared to the moments at the other points. The more rigorous analysis also includes values for twisting moments, longitudinal moments, and in-plane stresses in the transverse direction. Also, the longitudinal distribution of all values is easily obtained. For example, the transverse moment at joint 3, where $L = 70$ ft and $\nu = 0.25$, reduces from −1994 ft-lb/ft at midspan only to −1642 ft-lb/ft at the quarterspan and to −1101 ft-lb/ft at a point 7 ft from the support. Analyses such as those given in Ref. 5 will normally not be used so long as finite-element programs are readily available because those numerical programs allow study of different transverse boundaries and variations in geometry that are difficult to include in the analytic solution. Because the simplified analysis gives reliable results, we shall base our discussion of behavior on studies using it. It is thus useful to review its assumptions.

The simplified analysis is made by considering the plates as acting in two distinct ways: transversely as slabs and longitudinally as beams. Because of the assumption of restraint only along the longitudinal edges, the plates can be considered as one-way slabs spanning transversely between joints. Because of the assumption of the validity of the simple flexural theory, the plates are considered as thin inclined beams. This analysis is therefore referred to here as the *slab-beam analysis.*

We have avoided the complicated differential equations of Sec. 5-4 and made a physical simplification—replacing the system by a different one which nevertheless behaves in nearly the same manner. It is important therefore to summarize the assumptions of this simplification in order to emphasize the limitations of this approach.

1. The material is linearly elastic, isotropic, and homogeneous.
2. The longitudinal distribution of all loads on all plates is the same.
3. The plates carry loads transversely only by bending normal to their planes (transverse one-way slab action).
4. The plates carry loads longitudinally only by bending within their planes (longitudinal beam action).
5. Longitudinal strains and hence stresses vary linearly over the depth of each plate.
6. The supporting members (diaphragm, frames, beams, etc.) are infinitely stiff in their own planes and completely flexible normal to their own planes.
7. Plates have no torsional stiffness normal to their own planes.
8. Displacements due to forces other than bending moments are neglected.
9. The cross section is constant throughout its span length; i.e., the folded plates are prismatic.

As a further limitation it has been recommended[1] that the angle between adjoining plates be more than 15° and less than 165° to avoid excessive lateral deflections which might invalidate the assumption of superposition.

On the basis of the above assumptions, a number of writers[6-9] have presented procedures for carrying out the slab-beam analysis. The one used to get the results presented here is due mainly to Yitzhaki[10] and consists of two separate analyses:

1   *Elementary transverse slab analysis,* in which the loads are assumed to be carried transversely to the joints by one-way slab bending only. The interior joints are considered unyielding along their entire lengths and the analysis gives transverse moments at the joints or ridges and vertical reactions with the same longitudinal distribution as the loads. No longitudinal beam stresses are developed.

2 *Slab-beam analysis for joint loads,* in which the reactions from the elementary slab analysis are applied as loads to the combined slab-beam system. As a result of the joint loads, longitudinal beam stresses and transverse bending moments will occur, and it is their computation which constitutes the difficulty in folded-plate analysis. The resulting bending moments must be combined with those obtained in the elementary slab analysis.

8-3 BEHAVIOR AND DESIGN

Insight into structural behavior can come from discussing a specific design problem. We shall assume that a longitudinal span of 70 ft is required, with columns spaced transversely at roughly half the span. The live load is 20 psf, and the roofing, insulation, and mechanical loading are taken as 10 psf. Both of these loads will be considered as uniformly distributed over the surface of the roof.

The choice of the type of cross section depends, to a large extent, upon functional and often aesthetic considerations. From an engineering standpoint we may say that type *a* (Fig. 8-2*a*) permits the greatest overall depth with the least concrete. However, the transverse moments are high because the section is flexible transversely. Type *b* (Fig. 8-2*b*) requires more surface area and hence probably more concrete than type *a*, but the transverse moments are substantially less. Lastly, type *c* (Fig. 8-2*c*) requires even more concrete and surface area than type *b*, but the tension and compression "flanges" permit easier steel placement and substantially reduced stresses.

The example used here is type *a* (Fig. 8-1), which has been taken from the ASCE report.[2] This structure is close enough to the cylindrical segment used in Chap. 6 so that we can compare the behavior of the two systems.

Having chosen the type of system, we must decide upon the dimensions for the various elements. The principal design parameters studied here are:

1 *Ratio of span length to the total width of the cross section, L/B.* Table 8-2 summarizes the maximum stresses and moments for the cross section of Fig. 8-1 with span and plate thickness as variables. From Fig. 8-4, showing the results plotted over the midspan cross section, we see that as L/B increases, the longitudinal stresses tend to straighten out and approach those values which would be obtained from a beam analysis of the entire cross section. Note how rapidly the transverse moments increase with L/B. We can observe that this type of cross section is not efficiently used as $L/B \to 1$, and that for long spans the transverse moments will be critical. All longitudinal stresses are low except those at the bottom of the edge member. In these examples B is held constant at 37 ft 2 in.

2 *Ratio of span length to the total depth of the cross section, L/D.* Table 8-3 shows how the critical values vary with changes in the total section depth D as well as in the depth d_1 of the edge plates and the slope ϕ_2 of the first interior plates. The flatter system results in higher maximum longitudinal stresses and higher transverse moments at joint 3, while the steeper system produces much higher moments at joint 2. Evidently a slope between 30° and 45° for plate 2 will result in a better

Table 8-2 CRITICAL VALUES FOR VARIOUS SPAN LENGTHS AND PLATE THICKNESSES

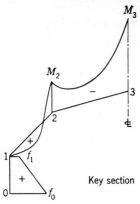

Key section

Span, ft	Plate thickness, in.		Longitudinal stresses, psi		Transverse moments, ft-lb	
	h_1	$h_2 = h_3$	f_0	f_1	M_2	M_3
35	4	2	542	72	−680	−740
	4	4	571	25	−631	−961
	4	6	563	19	−742	−1220
	6	2	356	61	−675	−740
	6	4	382	18	−591	−958
	6	6	383	10	−640	−1173
	8	2	263	53	−673	−740
	8	4	288	15	−572	−956
	8	6	291	6	−588	−1150
70	4	2	2378	−16	−546	−1274
	4	4	2060	100	−1196	−2396
	4	6	1605	200	−2213	−4123
	6	2	1618	−46	−488	−1236
	6	4	1456	34	−948	−2067
	6	6	1196	122	−1757	−3425
	8	2	1230	−56	−457	−1217
	8	4	1131	2	−811	−1889
	8	6	958	78	−1488	−3017
105	4	2	4985	55	−901	−2056
	4	4	3339	640	−2412	−4608
	4	6	2281	760	−3527	−6545
	6	2	3493	−94	−727	−1824
	6	4	2517	396	−1950	−3861
	6	6	1827	557	−2971	−5613
	8	2	2705	−153	−630	−1700
	8	4	2032	260	−1670	−3415
	8	6	1528	428	−2607	−5004

Table 8-3 CRITICAL VALUES FOR VARIOUS CHANGES IN
PLATE SLOPES AND DEPTHS

d_1, ft	ψ_2	D, ft	Longitudinal stresses		Transverse moments	
			f_0, psi	f_1, psi	M_2, ft-lb	M_3, ft-lb
4	45°	14.39	1042	−104	−1610	−635
3	45°	13.39	1479	−118	−2087	−984
2	45°	12.39	2126	−31	−3103	−1839
4	30°	10.73	984	47	−546	−1649
3	30°	9.73	1456	34	−948	−2067
2	30°	8.73	2230	119	−1862	−3124
4	25°	9.96	1011	91	−39	−2116
3	25°	8.96	1526	72	−429	−2580
2	25°	7.96	2411	149	−1352	−3786

balance of the moments at joints 2 and 3. On the other hand, the steeper slopes are more difficult to cast. The compromise chosen here is 30°. For this type of cross section we could combine 1 and 2 into a ratio of L/r where r would be an equivalent radius of curvature. This equivalence is discussed in Sec. 8-5.

3 *The plate thicknesses.* The thicker edge plate reduces both moments and longitudinal stresses substantially (Table 8-2). For a minimum interior plate thickness of 2 in. with an 8-in.-thick edge plate, the maximum working-load moment will be 1217 ft-lb/ft. On the basis of working-load criteria, the maximum moment capacity of a 2-in. slab ($d = 1$ in.) of 3000 psi concrete will be

$$M = 236 \times 1^2 \times 1 = 236 \text{ ft-lb/ft}$$

The 2-in. slab is clearly unsatisfactory. From a practical standpoint such a thin slab is difficult and thereby costly to cast. The 4-in. slab is thus chosen as a practical thickness which provides sufficient moment capacity. The edge plate could be made 8 in. wide, but the additional concrete may not be justified by the relatively small reductions in stresses.

This discussion is not intended to indicate a unique solution to the dimensioning, but rather to show how certain variables affect the results of analysis. The final dimensions, as usual, will be chosen on the basis both of analysis and of practical and localized conditions. The choices in this example are reasonable but not necessarily optimum.

8-4 BEHAVIOR OF A FOLDED-PLATE ROOF

For the design example described in the previous section, the simplified analysis gives at midspan the longitudinal stresses and transverse moments shown in Fig. 8-4 for $L = 70$ ft. Table 8-1 shows that these simplified analysis results,

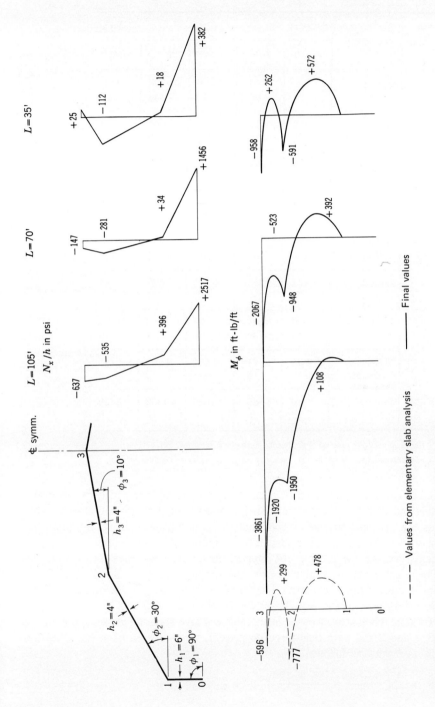

Fig. 8-4 Folded-plate behavior for various spans.

Table 8-4 PRINCIPAL STRESSES*

Plate	Location	Support				Midspan
		$x = 0$	$L/8$	$L/4$	$3L/8$	$L/2$
3	Joint 3	0 0	0 −64†	0 −110	0 −138	0 −147
	Mid-height of plate	+51 −51	+13 −107	+4 −164	+1 −199	0 −214
	Joint 2	+123 −123	+45 −167	+17 −227	+4 −268	0 −281
2	Joint 2	+123 −123	+45 −167	+17 −227	+4 −268	0 −281
	Mid-height of plate	+179 −179	+110 −164	+56 −146	+16 −130	0 −122
	Joint 1	+193 −193	+153 −137	+111 −83	+68 −34	+34 0
1	Joint 1	+129 −129	+105 −89	+79 −51	+53 −19	+34 0
	Mid-height of plate	+95 −95	+341 −15	+564 −4	+700 0	+745 0
	Joint 0	0 0	+635 0	+1090 0	+1362 0	+1456 0

* Based on parabolic variation of all shearing forces (see Table 9-4, Ref. 2).

† All stresses in pounds per inch²; + indicates tension, − indicates compression.

compared to the somewhat smaller values from a more rigorous analysis, provide a reasonable and slightly conservative basis for proportioning the reinforcement. Table 8-4 gives the principal stresses throughout the folded plate and Fig. 8-5 plots the principal stress trajectories. At midspan the longitudinal beam flexural stresses are the principal stresses, whereas at the support the shearing stresses are the principal ones. These latter values come also from the simplified analysis (see Ref. 2, Chap. 9) and must balance the total vertical load of 80 psf over the two sloping plates, 35 × 20 × .08 = 56 kips, plus the edge-beam weight, 35 × 3 × 0.5 × 0.15 = 8 kips, for a total of 64 kips. The shear stresses from Table 8-4 give shearing forces in the planes of plates 2 and 3 and a vertical force on plate 1, all estimated from the tabulated values by averaging as follows:

$$S_1 = \frac{1}{2}\left(\frac{0 + 95}{2} + \frac{95 + 129}{2}\right)(36 \times 6) = 17 \text{ kips}$$

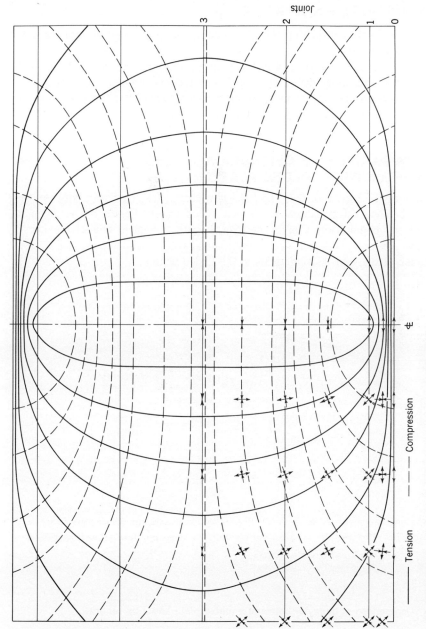

Fig. 8-5 Stress trajectories in the developed folded plate.

———— Tension   – – – – Compression

$$S_2 = \frac{1}{2}\left(\frac{193 + 179}{2} + \frac{179 + 123}{2}\right)(120 \times 4) = 81 \text{ kips}$$

$$S_3 = \frac{1}{2}\left(\frac{123 + 51}{2} + \frac{51 + 0}{2}\right)(120 \times 4) = 27 \text{ kips}$$

with the total vertical component being

$$S_1 + S_2 \sin 30° + S_3 \sin 10° = 63 \text{ kips}$$

a satisfactory check, considering that the shearing stresses used are approximate (see Table 9-4 of Ref. 2).

As Fig. 8-4 shows, the overall plate system acts somewhat like a beam with compression in the upper plates and tension in the lower ones. The transverse moments, however, deviate significantly from those found for a one-way slab because of the transverse flexibility of the overall cross section.

8-5 COMPARISON OF CYLINDRICAL SHELLS WITH FOLDED PLATES

A study of Figs. 6-6, 6-7, and 8-4 will reveal a close similarity between the structural behavior of cylindrical shells and folded plates. Figure 8-6 is given to emphasize this comparison. Clearly the overall systems behave similarly insofar as longitudinal stress N_x/h is concerned, but the transverse moments are quite different because of the joint discontinuities in the folded plates.

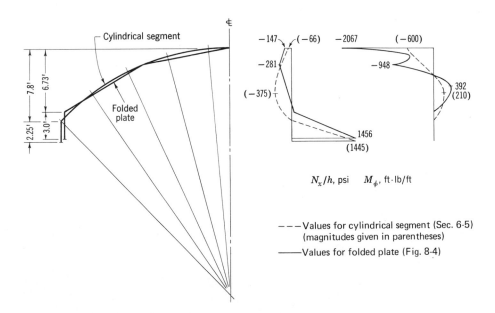

Fig. 8-6 Comparison of behavior between a folded plate and a barrel shell.

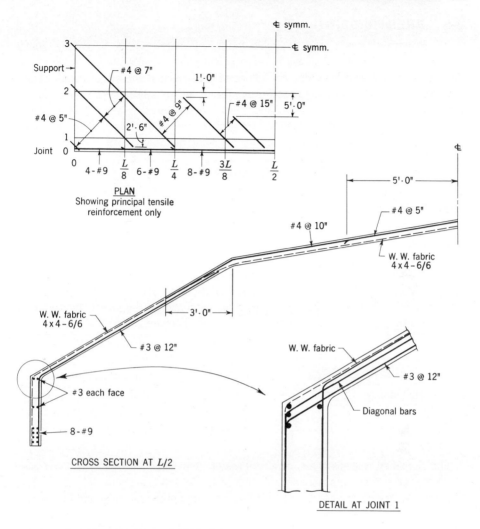

Fig. 8-7 Reinforcement for a folded plate.

In this comparison a 45° arc is replaced by only two plates; hence we should not expect the moments to correlate well. However, with more plates, the moment lines would approach each other.

The finite-element analysis discussed briefly in Sec. 5-12 and illustrated in Sec. 6-8 for barrel shells was in reality a folded-plate analysis because the elements used in the SAP program are plates and not shells. Thus, when such programs are used, the analyses for the two geometries do not differ. It is for this reason that Scordelis discusses both geometries together.[11] Because of the similarity in behavior between cylindrical barrels and folded plates, the reinforcing layouts will also be similar. Figure 8-7 shows such a layout for the folded-plate roof of Fig. 8-1.

8-6 PRESTRESSING

Prestressing is clearly desirable for folded plates where large longitudinal tensions occur. Figure 8-8 shows several typical prestressed systems. To illustrate the behavior, we shall consider the effect of prestressing the edge beams of the example of Fig. 8-1. Since the prestressing forces all are applied in the planes of the plates, there are no joint reactions and no elementary slab analyses. Just as in the case of cylindrical shell edge-beam prestressing, formulated in Sec. 5-8, we consider separately the two effects of prestressing: bending and shortening.

Considering the vertical edge beam alone, the prestressing produces an upward load due to the curvature of the tendon profile

$$P_1 = \frac{8Fe}{L^2} = \frac{8(1)F}{70^2} = 0.00163F = 0.163 \, \text{kip/ft}$$

where the tendon profile is assumed to have a parabolic variation from an eccentricity $e = 0$ at the supports to $e = 1$ ft at midspan and the force F taken as 100 kips. Figure 8-9 shows the edge stresses due to the bending moment Fe and the axial force F when the edge beam is considered alone as in Sec. 5-8. The longitudinal distributions of these two effects are different: parabolic for the stress due to bending, and constant for the stress due to shortening. For bending,

$$f'_{01} = -f'_{10} = \frac{0.00163F(70^2)12}{8(1296)} = -0.00925F = -925 \, \text{psi}$$

and for shortening,

$$f''_{01} = f''_{10} = -\frac{F}{A} = -\frac{F}{216} = -0.00462F = -462 \, \text{psi}$$

The simplified analysis gives, as a final result, the values for longitudinal stresses and transverse moments shown in Tables 8-5 and 8-6, respectively, where the prestressing results are combined with those already given for dead and live loads.

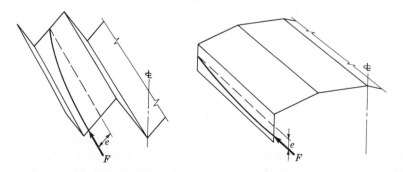

Fig. 8-8 Prestressed folded plate.

Table 8-5 LONGITUDINAL STRESSES AT MIDSPAN*

Joints	0	1	2	3
Bending:				
Stresses in edge beam alone	−925	925	0	0
Final stresses	−421	127	43	−82
Final stresses (F = 146 kips)	−617	185	63	−119
Shortening:				
Stresses in edge beam alone	−462	−462	0	0
Final stresses	−572	−147	15	25
Final stresses (F = 146 kips)	−839	−216	22	36
Total stresses (F = 146 kips)	−1456	−31	85	−83
Stresses from loads (Table 8-1)	1456	34	−281	−147
Combined stresses	0	3	−196	−230

* All stresses are in pounds per square inch; + indicates tension, − indicates compression.

Table 8-6 TRANSVERSE BENDING MOMENTS AT MIDSPAN*

	Joint 1	Plate 2	Joint 2	Plate 3	Joint 3
Final values (Fig. 8-4)	0	+392	−948	−523	−2067
Total prestressing	0	+411	+822	+888	+954
Final values	0	+803	−126	+365	−1113

* All moments in foot-pounds per foot.

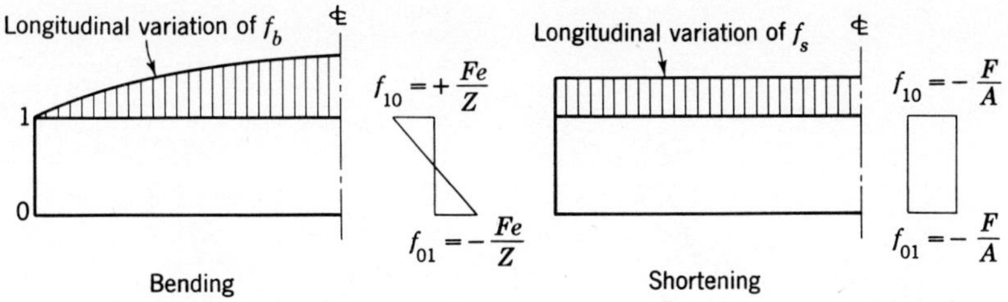

Fig. 8-9 Stresses from prestressing.

The solution based on $F = 100$ kip gives a maximum longitudinal tension of −993 psi. The final value for F of 146 kips is then set by requiring that the prestressing eliminate that maximum tension stress of + 1456 psi due to dead and live loads.

From Tables 8-5 and 8-6 we see that the longitudinal tension is practically eliminated and transverse joint moments are greatly reduced. Note, however, that the moment at the center of plate 2 is more than doubled. Also, the prestressing values given are final ones so that an overload of 50 percent (40 psf) gives an almost 100 percent increase in M_3. Thus for design, load factors should be used.

8-7 ULTIMATE-LOAD BEHAVIOR OF FOLDED PLATES

Of the few reinforced-concrete folded-plate model tests, we describe here briefly those carried out on two models 70 in. in span, 30 in. wide, and 0.51 in. thick (Fig. 8-10). The longitudinal design stresses in pounds per square inch are shown for one model in Fig. 8-10b and c based upon beam theory (Mc/I) and upon folded-plate theory (the more rigorous analysis of Ref. 5), respectively. The design loads giving those stresses are 57.5 psf (37.5 psf for a prototype 3-in.-thick shell plus 20 psf for live load) and in the models they were applied as equivalent line loads along the ridges.[12]

One model was reinforced on the basis of beam theory and one on the basis of folded-plate theory. Both shells performed well up to design loads and both failed at an ultimate load of 4½ times the design load. Thus the researchers concluded that either theory would be a satisfactory basis for design for the type of folded plate

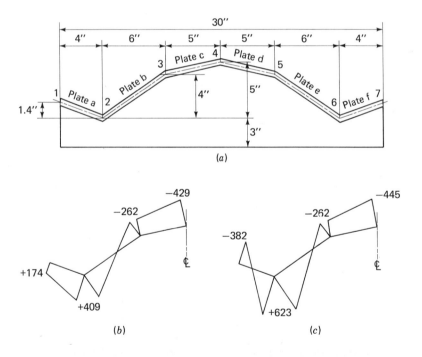

Fig. 8-10 Dimensions and behavior of a folded-plate model. (*a*) Basic dimensions of models A and B. (*b*) Midspan longitudinal stresses (per square inch) under full design load as obtained by beam theory. (*c*) Midspan longitudinal stresses (per square inch) under full design load as obtained by folded-plate theory. (From Scordelis and Gerasimenko.)

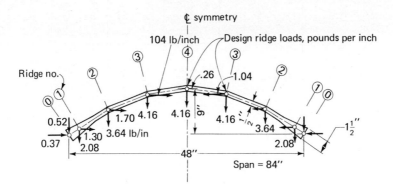

Fig. 8-11   Folded-plate structure of Ref. 13.

tested. They caution against generalizing that conclusion when conditions and geometry differ from those in the tests.

These tests and the one by Hedgren described in Sec. 6-8 were restudied by Fialkow who used a limit analysis based upon ideas developed in plasticity theory.[13] Figure 8-11 shows Fialkow's folded-plate idealization of Hedgren's model. His plastic collapse analysis indicated a load factor of 2.74, whereas Hedgren found that the mechanism of a longitudinal yield line at the crown began at a factor of 2.8. Fialkow also showed that following such yielding, further load could be carried up to at least the factor of 4.4 times design load which Hedgren determined as failure. The overall conclusion for folded plates of the geometries tested appears to be the same as for barrel shells, i.e., that proportioning of reinforcement based on design loads gives a structure that performs well at those loads and has a safety factor higher than that normally used in framed structures.

8-8   FOLDED PLATES CONTINUOUS OVER THREE SUPPORTS

Folded plates, continuous in the longitudinal direction, have also been analyzed in the same manner as similar cylindrical shells, where the simply supported system is a base and ratios are derived from a continuous-beam analysis.[10] Thus for two uniformly loaded 70-ft spans with the cross section of Fig. 8-4, the stress f_0 at the midspan would be ½(1456) psi. At the support, f_0 would be -1456 psi because in a two-span continuous beam the support moment is of equal magnitude but of opposite sign to the midspan moment in a simply supported beam.

Pultar developed for continuous folded plates a method of analysis based upon thin shell theory, and his results are presented in Table 8-7 for various types of interior support.[14] The exterior supports are taken to be rigid diaphragms as shown in Fig. 8-1. For the case of an interior rigid diaphragm (A5) the results show that $f_0 = 53.41$ kips/ft, which for the 6-in. plate gives $f_0 = 53.41/72 = 742$ psi at the midspan and $-84.99/72 = -1180$ psi at the interior support, whereas simple ratios from the single-span solution would give $f_0 = 728$ psi and -1456 psi. The transverse moments are -1.424 kips at midspan, whereas for the single span they are -2.067 kips.

Table 8-7

Joint	Y*	Midspan (x = 35.0)					Interior support (x = 70.0)				
		A1	A2	A3	A4	A5	A1	A2	A3	A4	A5
		(a) Longitudinal stress resultant†									
3	0.0	-38.28	-12.29	-5.51	-4.56	-3.63	-66.98	-34.02	-5.68	-1.18	4.49
	5.0	-29.73	-13.46	-7.11	-6.20	-5.17	-44.00	-13.66	0.84	2.95	5.53
2	10.0	-21.25	-15.12	-9.20	-8.38	-7.43	-22.55	7.98	12.17	12.34	12.09
	15.0	4.41	-1.10	-2.64	-2.86	-2.88	27.61	15.64	7.67	6.49	4.90
1	20.0	30.07	12.24	4.26	3.02	1.25	82.40	29.45	6.81	4.10	1.79
0	23.0	125.57	80.40	60.16	57.44	53.41	-22.58	-68.54	-80.00	-82.12	-84.99
		(b) Transverse stress couple M_Y‡									
3	0.0	0.654	-0.379	-1.182	-1.293	-1.424	1.483	2.036	0.797	0.497	0.040
	5.0	1.580	0.609	0.093	-0.085	-0.052	2.943	2.226	0.314	0.128	-0.028

Joint	Y										
2	10.0	0.625	-0.379	-0.542	-0.577	-0.644	2.673	0.961	0.243	0.197	0.003
	15.0	1.197	0.549	0.583	0.580	0.571	1.709	-3.360	-0.421	-0.262	-0.061
1	20.0	0.117	-0.071	0.081	0.095	0.110	-0.875	-11.779	-2.645	-1.539	0.029
0	23.0	0	0	0	0	0	0	0	0	0	0

(c) Vertical displacement§

Joint	Y										
3	0.0	0.599	0.220	0.044	0.019	-0.010	0.908	0.405	0.098	0.053	0
	5.0	0.570	0.217	0.060	0.037	0.010	0.855	0.346	0.080	0.043	0
2	10.0	0.469	0.192	0.079	0.062	0.043	0.659	0.298	0.043	0.023	0
	15.0	0.335	0.166	0.101	0.092	0.081	0.367	0.124	0.011	0.006	0
1	20.0	0.155	0.122	0.105	0.103	0.101	0.001	0	0	0	0
0	23.0	0.154	0.122	0.105	0.103	0.101	0	0	0	0	0

* Y is measured along the plate middle surface beginning at joint 3 and ending at joint 0.
† Units in kips per foot.
‡ Units in kip-feet per foot.
§ Units in feet.

Therefore, just as with continuous barrel shells, the transverse moments are not as much reduced by continuity as are the longitudinal stresses.

The results of Table 8-7 for the rigid diaphragm were independently confirmed by Scordelis and De Ngo using their programs developed for box girder bridges.[15] In this table the interior support conditions located at the center of the 140-ft total length are as follows:

A1 Flexible columns under the edge beams at joint 0 (Fig. 8-1) provide only vertical restraint. There is neither gable nor diaphragm.
A2 As in A1 except columns fixed at joint 0.
A3 Flexible gable with a constant 1-ft depth.
A4 Flexible gable with a constant 2-ft depth.
A5 Rigid diaphragm.

The results in Table 8-7 show that as the interior support gets stiffer (from A1 to A5), the maximum midspan and quarterspan tensile stress resultants decrease but the maximum midspan compression stresses increase. The continuous folded plate on flexible columns behaves somewhat as two simply supported structures with maximum tensile stresses of $125.57/72 = 1740$ psi. On the other hand, the differences between the gabled supports A3 and A4 and the diaphragm A5 are not of great significance for design in this example.

REFERENCES

1. Phase I Report of the Task Committee on Folded Plate Construction, *J. Struct. Div.*, ASCE, vol. 90, December 1963.
2. Billington, D. P., *Thin Shell Concrete Structures,* chap. 8, McGraw-Hill Book Company, New York, 1965.
3. Billington, D. P., "Thin-Shell Concrete Structures," in E. H. Gaylord, Jr., and C. N. Gaylord (eds.), *Struct. Eng. Handb.*, 2nd ed., sec. 20, McGraw-Hill Book Company, 1979.
4. Yitzhaki, D., *Prismatic and Cylindrical Shell Roofs,* Haifa Science Publishers, Haifa, 1958.
5. Goldberg, J. E., and H. L. Leve, "Theory of Prismatic Folded Plate Structures," *International Association of Bridge and Structural Engineering,* vol. 17, 1957.
6. Gaafar, I., "Hipped Plate Analysis Considering Joint Displacement, *Trans. ASCE,* vol. 119, 1954.
7. Girkmann, K., *Flächentragwerke,* 5th ed., Springer-Verlag OHG, Vienna, 1959.
8. Scordelis, A. C., "Matrix Formulation of the Folded Plate Equations," *J. Struct. Div.*, ASCE, vol. 86, October 1960.
9. Simpson, H., "Design of Folded Plate Roofs," *J. Struct. Div.*, ASCE, vol. 84, January 1958.
10. Yitzhaki, D., and M. Reiss, "Analysis of Folded Plates," *J. Struct. Div.*, ASCE, vol. 88, October 1962.
11. Scordelis, A. C., "Analysis of Cylindrical Shells and Folded Plates," *Concr. Thin Shells,* ACI, Publ. SP-28, 1971, pp. 207–236.
12. Scordelis, A. C., and P. V. Gerasimenko, "Strength of Reinforced Concrete Folded Plate Models," *J. Struct. Div.*, ASCE, vol. 92, no. ST-1, February 1966, pp. 351–363.
13. Fialkow, M. N., "Plastic Collapse of Folded Plate Structures," *J. Struct. Div.*, ASCE, vol. 101, No. ST-7, July 1975, pp. 1559–1584.
14. Pultar, M., D. P. Billington, and J. D. Riera, "Folded Plates Continuous Over Flexible Supports," *J. Struct. Div.*, ASCE, vol. 93, no. ST-5, October 1967, pp. 253–277.
15. Scordelis, A. C., and De Ngo [Discussion of Ref. 14], *J. Struct. Div.*, ASCE, vol. 94, no. ST-6, June 1968, pp. 1639–1642.

9
STABILITY AND SAFETY

9-1 GENERAL IDEAS

The two main types of thin shells, walls and roofs, behave differently enough to merit separate general discussion when considering ultimate-load capacity. The preceding chapters centered on the calculation of stresses and displacements derived from classical thin shell theory supplemented by results from numerical analyses.

This present chapter considers stability and safety characterized by less well defined behavior, which still is essential to consider in design. Such behavior cannot be as easily or as accurately predicted as that derived from classical thin shell theory for working loads. It is essential for the designer to recognize the incompleteness of the classical theory when applied to thin shell concrete structures and to take less well defined factors into consideration. These factors can sometimes be studied by more rigorous analyses, but mainly they are best incorporated into design by simplified methods used by designers with substantial experience in full-scale concrete thin shells. This chapter does not seek to provide rules for design but rather to make readers aware of the types of problems encountered with thin shells in the past.

The classical theory focuses on perfectly formed, thin, linearly elastic shells with small deflections. Among other things, that theory implies the following assumptions:

1. A perfectly formed thin shell is one whose thickness is small with respect to its principal radii of curvature, one whose geometry is accurately defined, and hence one whose behavior is predominately in-plane.
2. By linear-elastic theory the material is defined by two constants E and ν neither of which vary with stress level.
3. Small deflections mean that the deformations after loading are so small that the resulting changes in geometry do not influence the static equilibrium.

When any of these three basic ideas are violated then the stresses computed in preceding chapters may not provide a reliable measure for safety. At the same time analyses which consider behavior deviating from these general ideas are less well defined than those already presented. This is so for three reasons basic to the design of thin shell concrete structures.

First, construction tolerances cannot be overly refined without making field costs prohibitive. Thus some imperfections are always present in full-scale shells and the problem mainly is to set limits and recognize how such imperfections influence behavior and thereby control the resulting response by adequate reinforcement.

Second, reinforced concrete is not a refined homogeneous material but rather is subject to local cracking and variations in properties throughout. Hardy Cross showed years ago that such variations in arches had little overall effect if the quality of workmanship was reasonable.[1] Nevertheless cracking and even some yielding can be predicted in shells and needs to be controlled.

Third, thin flat shells under permanent load normal to their surfaces may undergo deflection large enough to change equilibrium and make instability a possibility. This danger is accentuated in concrete shells because of creep especially when shells are decentered too soon, are made of lightweight concrete, or are resting on yielding supports.

Therefore, three general questions need be considered in design beyond the classical stress analysis: first, shell geometry and imperfections; second, concrete cracking and steel yielding; and third, creep and large deflections. For shell geometry, buckling formulas based on tests provide guidance aimed at preventing instability, i.e., ensuring that the shell safety is controlled by strength not stiffness. Therefore, we shall begin with an introduction into shell buckling divided into walls and roofs.

When the primary roof loads are vertical, the shells tend to be relatively flat horizontal forms, whereas for the walls when the primary loads for buckling are horizontal, the shells tend to be essentially vertical forms. The major difference between the two types is that the primary roof loading is permanent (gravity), whereas the primary wall loading is transient (wind or seismic). Thus for roof shells, creep under compression is of major importance and the influence of deflections on buckling can be crucial, while for walls, the distribution of the load is of significance and the influence of cracking on buckling can be great. Following the discussion on buckling three examples of shell systems will be presented, each one chosen to illustrate one of these three basic problems: imperfections, cracking and yielding, and creep. A detailed discussion of these problems appears in Ref. 2.

9-2 BUCKLING ANALYSIS FOR THIN SHELL WALLS

To emphasize the buckling behavior of shell walls, we shall give a brief introduction into the mathematical analysis of shell buckling, beginning with simple classical ideas of Euler columns and ending with the buckling of wind-loaded cylinders and hyperboloids.

Columns with Lateral Springs

We begin with a slightly bent vertical column restrained laterally by springs and derive expressions for the energy where the internal strain energy of bending[3]

$$\Delta U_1 = \int_0^L \frac{M^2 \, dx}{2EI} \tag{9-1}$$

or since $M/EI = d^2y/dx^2$,

$$\Delta U_1 = \int_0^L \frac{EI}{2}\left(\frac{d^2y}{dx^2}\right)^2 dx \tag{9-2}$$

This bending goes together with a decrease in the potential energy ΔT of the vertical load P owing to a vertical displacement $\delta_x = \int_0^L (ds - dx)$ where $ds = \sqrt{dx^2 + dy^2} = dx\sqrt{1 + (dy/dx)^2} \approx dx\,[1 + \frac{1}{2}(dy/dx)^2]$ and thus

$$ds - dx = dx + \frac{1}{2}\left(\frac{dy}{dx}\right)^2 dx - dx = \frac{1}{2}\left(\frac{dy}{dx}\right)^2 dx$$

so that $\Delta T = P\delta_x$ gives

$$\Delta T = P\int_0^L \frac{1}{2}\left(\frac{dy}{dx}\right)^2 dx \tag{9-3}$$

Finally the strain energy associated with a uniform lateral restraint arises from the spring constant β in kips per square foot which for a lateral displacement y gives a distributed force $\beta y\,dx$ and an energy of $\frac{1}{2}(\beta y\,dx)y$ so that

$$\Delta U_2 = \int_0^L \frac{\beta y^2}{2} dx \tag{9-4}$$

In order to determine the bifurcation buckling load P_{cr} it is necessary to assume a deflected shape, to evaluate the energy integrals, and to solve for P_{cr} from the energy equation

$$\Delta U_1 + \Delta U_2 = \Delta T \tag{9-5}$$

Knowing the Euler buckling shape to be sinusoidal, we take

$$y = a_m \sin\frac{m\pi x}{L}$$

which implies simple supports at top and bottom, and find for $EI = $ constant,

$$\Delta U_1 = \frac{EI}{2}\int_0^L \left(-a_m\frac{m^2\pi^2}{L^2}\sin\frac{m\pi x}{L}\right)^2 dx = \frac{EI}{4}a_m^2\frac{m^4\pi^4}{L^3}$$

$$\Delta U_2 = \frac{\beta}{2}\int_0^L \left(a_m\sin\frac{m\pi x}{L}\right)^2 dx = \frac{\beta L}{4}a_m^2$$

$$\Delta T = P\,a_m^2\,\pi^2\,m^2/4L$$

so that Eq. (9-5) gives

$$P_{cr} = \frac{EI}{L^2}m^2\pi^2 + \frac{\beta L^2}{m^2\pi^2} \tag{9-6}$$

The amplitude a_m cancels out and the critical load depends upon two stiffnesses: EI for the column and β for the elastic restraint. Unlike the case of the column alone [the first term in Eq. (9-6)], the lowest m does not necessarily give the lowest P_{cr}.

Cylindrical Shells under Axial Load

The preceding example can be used directly for the case of an axially loaded cylindrical shell once we recognize that a vertical strip of shell with stiffness of $D = Eh^3/12(1 - \nu^2)$ corresponds to EI and that the lateral restraint on that strip is provided by the ring stiffness of the shell with radius a, $k_r = Eh/a^2$ which corresponds to β so that Eq. (9-6) can be modified for shells that are simply supported at top and bottom to give (Ref. 3, page 458)

$$N_{cr} = \frac{D}{L^2} m^2\pi^2 + \frac{EhL^2}{a^2 m^2\pi^2} \tag{9-7}$$

where N_{cr} is the axial stress resultant in kips per foot since both D and Eh/a^2 are stiffnesses per foot. Thus cylindrical shell buckling under axial load involves a combination of vertical flexural stiffness $D\pi^2 m^2/L^2$ and circumferential axial (or ring) stiffness $(Eh/a^2) L^2/m^2\pi^2$. These results are significant because they demonstrate the two types of resistance that a cylindrical shell presents against buckling, but they are dangerous as a design guide because axially loaded cylinders actually buckle under much less load than N_{cr} because they are imperfection-sensitive (Ref. 3, page 469).

Cylinders under Uniform Lateral Pressure

For cylinders that are simply supported on top and bottom and loaded only by a uniform radial pressure q, the critical circumferential stress resultant $N_{\theta cr}$ or critical load q_{cr} will be in the same form as Eq. (9-7) (Ref. 3, page 478),

$$N_{\theta cr} = aq_{cr} = \frac{D}{a^2}\left(n^2 - 1 + \frac{2n^2 - 1 - \nu}{1 + A}\right)$$
$$+ Eh\left(\frac{1}{(n^2 - 1)(1 + A)^2}\right) \tag{9-8}$$

where $A = (nL/\pi a)^2$ and wherein we observe the same structure as in Eq. (9-7) with the flexural term containing n^2 in the numerator and the axial term having it in the denominator. The differences between Eqs. (9-7) and (9-8) are crucial, however, because the term in D now refers essentially to horizontal ring flexural stiffness, whereas the term in Eh refers to the axial stiffness of the vertical cylinder strip. As the ratio of L/a increases, A increases rapidly and hence all terms with A in the denominator become progressively insignificant until at about $L/a = 50$,

$$q_{cr} \approx \frac{D}{a^3}(n^2 - 1) \tag{9-9}$$

with a minimum value

$$q_{cr} = 3\frac{D}{a^3} \tag{9-10}$$

which coincides with the buckling pressure for a ring of unit width when $\nu = 0$

(Ref. 3, page 291). Equation (9-8) can be written in compact form as

$$q_{cr} = \lambda \frac{D}{a^3} = CE \left(\frac{h}{a}\right)^3$$

with $C = \lambda/(12)(1 - \nu^2)$ and where λ depends upon n, L/a, ν, and a/h. Values for λ given in Tables 1 and 2 of Ref. 4 (pages 1010–1013), show that for cooling tower–type cylinders with $L/a = 4$, $a/h = 100$, and $\nu = 0$, $\lambda = 27.3$ and $n = 4$. Taking a cylinder that roughly corresponds to the Trojan Tower dimensions (i.e., $L/a = 452/116 = 3.9$, $a/h = 116/0.83 = 140$), we find, with $E = 5.2 \times 10^5$ ksf,

$$q_{cr} = \lambda \frac{D}{a^3} = \lambda \frac{E}{12} \left(\frac{h}{a}\right)^3 = \frac{27.3 \times 5.2 \times 10^5}{12} \times \frac{10^{-6}}{1.4^3}$$

$$= 432 \text{ psf}$$

This value should be somewhat higher since λ was calculated in Table 2 for $a/h = 100$ and λ increases with a/h (in Table 2, $\lambda = 81.6$ when $a/h = 1000$ and $L/a = 4$, for example). Values obtained from eq. (9-8) have been found in good agreement with experiments and therefore do give some guidance for design.

Next, the boundary conditions will be taken as clamped at the base and free at the top, and the results for cylinders appear in Tables 3 and 4 of Ref. 4 (page 1015). Taking our same example as before, we find $\lambda = 16.95$ and $n = 3$ for $a/h = 100$, $L/a = 4$, and $\nu = 0$, so that

$$q_{cr} = \frac{16.95}{12} \times 5.2 \times 10^5 \left(\frac{1}{100}\right)^3 = 735 \text{ psf}$$

which compares to the uniformly loaded, simply supported shell given above but for $a/h = 100$,

$$q_{cr} = \frac{27.3}{12} \times 5.2 \times 10^5 \left(\frac{1}{100}\right)^3 = 1180 \text{ psf}$$

Thus the free top lowers the buckling load under uniform lateral pressure.

Cylinders under Wind Loading

Cylinders, free at the top and fixed at the base, respond to wind loads in a manner similar to uniform loads. Figure 9-1 gives curves taken from Ref. 4 (Table 6). Taking again $a/h = 100$, $L/a = 4$, $\nu = 0$ (Fig. 9-1 is based on $\nu = 0$), and $c = 0.5$, we get $\lambda = 24.17$ or from Fig. 9-1, $C = {\sim}2.0$ $({\approx}\lambda/12)$, so that from (9-8a)

$$q_{cr} = 2 \times 5.2 \times 10^5 \times \left(\frac{1}{100}\right)^3 = 1040 \text{ psf}$$

where $c = 0$ (no internal suction), $\lambda = 46.06$ or from Fig. 9-1, $C = {\sim}3.9$ $({\approx}\lambda/12)$ so that

$$q_{cr} = \frac{3.9}{2.0} \, 1040 = 2030 \text{ psf}$$

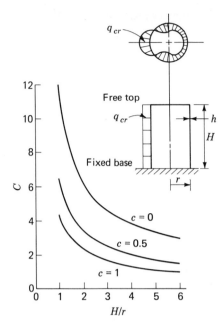

Fig. 9-1 Buckling of a cylinder under wind load. Note: c = internal suction coefficient as a multiple of q_{cr}.

This shows that the buckling under wind load is less severe than that under uniform lateral pressure load. It is apparently for this reason that uniform load has sometimes been used as a conservative estimate for wind-buckling estimates.

For wind-loaded cylinders Johns and his colleagues have proposed the formula[5]

$$q_{cr} = CE \frac{a}{l} \left(\frac{h}{a}\right)^{2.5} \tag{9-11}$$

where Johns has suggested $C = 0.66$ for wind-loaded cylinders with internal suction coefficient of $c = 0.6$.

$$q_{cr} = 0.66 \times 5.2 \times 10^5 \times \tfrac{1}{4} \times 10^{-5} = 855 \text{ psf}$$

compared to 950 psf found from Ref. 4, where $\lambda = 21.87$ for $c = 0.6$.

Hyperboloids under Wind

Approximate numerical studies of hyperboloids under wind loading have given results that are between 1.5 and 2.2 times the values found from wind tunnels. On the other hand, the same models analyzed numerically under uniform lateral pressure give results which are between 0.9 and 1.3 times the wind tunnel values.[6] When the hyperboloids are taken as cracked owing to thermal gradients, the reduction in buckling capacity is in the same range as found by tests.[7] The cracking was estimated as follows (Ref. 7):

There are two constants that must be computed to represent orthotropic material properties: (1) The ratio of circumferential to meridional Young's modulus, $k =$

E_θ/E_s; and (2) the shear modulus correction factor, $S = G_{s\theta}(1 - kv^2)/E_s$. There are also two stiffness equivalences that should be satisfied for orthotrophy. First, the membrane stiffness in the circumferential direction of the equivalent orthotropic shell should be equivalent to the membrane circumferential stiffness of the meridionally cracked concrete shell

$$\frac{E_\theta h}{1 - k_m v^2} = \frac{E_c h_{\theta \text{ eff}}}{1 - v^2} \tag{9-12}$$

in which k_m = the modulus ratio for membrane action; E_c = the modulus of concrete; and $h_{\theta \text{ eff}}$ = the effective thickness of the cracked shell. Second, a similar equivalence for the bending stiffness is

$$\frac{E_\theta h^3}{12(1 - k_b v^2)} = \frac{E_c I_{\theta \text{ eff}}}{1 - v^2} \tag{9-13}$$

in which k_b = the modulus ratio for bending action; and $I_{\theta \text{ eff}}$ = the effective moment of inertia per unit width of the cracked shell.

Figure 9-2a shows the strain distribution in the shell just prior to cracking. As the section cracks, the curvature, ϕ, must remain the same. Thus, the tensile force is transferred to the steel. Since the steel alone has a lower bending capacity than the uncracked section, the compression zone must be reduced. Therefore, the neutral axis is shifted away from the steel so that it is at a distance $\bar{k}h$ from the compression face (Fig. 9-2b).

Design specifications for cooling tower shells state that the minimum cross-sectional area of steel, A_{st}, in each face of the shell shall be 0.175 percent in each direction. If one can consider the Trojan Tower as typical, one can make some rough calculations to obtain an approximate value of k. Clear cover on each face is 1-½ in. and bar radius is about ½ in. Thus, for a minimum thickness of 10.0 in., the distance from one face is 8 in., or, in Fig. 9-2b, $z = 0.8$. If p' = the ratio of cross-sectional area of the circumferential tension steel to the area of the shell and n = the ratio of E_{st}/E_c (taken herein as 8.35), $p'n$ is 0.0146. Eq. 50 in Ref. 8,

$$\bar{k} = -p'n + [p'n(p'n + 2z)]^{1/2} \tag{a}$$

can now be solved to obtain $\bar{k} = 0.14$.

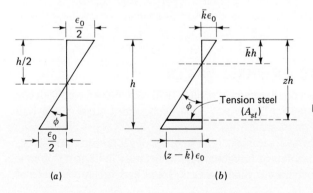

(a)
(b)

Fig. 9-2 Strains in a shell wall from thermal gradients (a) Uncracked section. (b) Cracked section. (From Cole, et al.)

The membrane stiffness equivalence (9-12) then gives

$$\frac{E_\theta h}{1 - k_m \nu^2} = \frac{E_c(\bar{k}h + nA_{st})}{1 - \nu^2}$$ (9-14)

in which the compression steel has been neglected; and $nA_{st} = p'nh$ = an equivalent concrete thickness for the tension steel. The denominator terms in Eq. (9-14) are negligibly different from unity for ordinary ranges of ν for concrete (1/6 to 1/9). Therefore, substituting $E_\theta = k_m E_c$ into (9-14) gives the approximate relationship

$$k_m \approx \bar{k} + p'n$$ (b)

which yields a value of $k_m = 0.15$.

Similarly, the bending stiffness equivalence (9-13) when compression steel is neglected gives

$$\frac{E_\theta h^3}{12(1 - k_b \nu^2)} = \frac{E_c}{1 - \nu^2} \left\{ np'h[(z - \bar{k})h]^2 + \bar{k}h\frac{(\bar{k}h)^2}{4} + \frac{(\bar{k}h)^3}{12} \right\}$$ (c)

with simplifications analogous to those applied to Eq. (9-14) one obtains

$$k_b \approx 12 \left[p'n(z - \bar{k})^2 + \frac{\bar{k}^3}{3} \right]$$ (d)

which gives $k_b = 0.09$. A weighted average of these two values may be calculated using Hayman and Chilver's analysis[9] of the relative importance of membrane and bending actions:

$$k = (k_m^{0.35})(k_b^{0.65})$$ (e)

This gives $k = 0.11$.

However, there is no analytical method to calculate an average S, the shear modulus correction factor. For isotropic materials, $S = (1 - \nu)/2$, or for $\nu = 1/9$, $S = 4/9$. Similarly, for $\nu = 1/6$, $S = 5/12$. For a cracked shell, it would appear reasonable to reduce this value somewhat. In the absence of experimental evidence to provide a firmer basis for the selection of S, the Trojan Tower was analyzed for two different values of S to examine the sensitivity of the response to variations of this parameter. For example, when S was changed from 0.4 to 0.3, the critical wind pressure for buckling dropped by about 11 percent (Ref. 7, page 1214).

Hayman's later research lends further support to this analysis for cracking and to its use in buckling studies.[10]

9-3 BUCKLING TESTS ON SHELL WALLS

A series of tests carried out between 1965 and 1976 have helped establish the validity of formulas like Eq. (9-11). Three sets in particular merit study by anyone seriously engaged in the design of shell walls.

The first of these, not in chronological order but in order of geometric complexity, is the extensive work of D. L. Johns at Loughborough University in England. Stimulated by the collapse in wind of a metal cylindrical tank, Johns had directed a series of tests including 75 plastic (polyvinyl chloride) models with $E \approx 0.45 \times 10^6$ psi, $\nu = 0.3$, l/a from 100 to 500, and with height up to 24 in. (see Ref. 5).

Fixed at their bases and free at their open tops, the models were tested in an open-jet-type wind tunnel from which the wind loading was found (in other experiments) to vary circumferentially roughly as shown in Table 2-1 and with coefficients,[11] (which include a suction of $0.607 \, p_H$) of $A_0 = 0.220$, $A_1 = 0.338$, $A_2 = 0.533$, $A_3 = 0.471$, $A_4 = -0.166$, $A_5 = -0.066$, and $A_6 = 0.055$. Johns measured only the wind velocity at buckling and then compared those results to an analytic bifurcation solution.[12] The experiments found that the shells buckled at values between 15 and 30 percent lower than the bifurcation predictions, which Johns and his colleagues attributed to the influence of dynamic response due to vortex shedding and other separated flow effects. As further evidence supporting this interpretation, they cited the fact that the longer shells, i.e., those most sensitive to dynamic effects, buckled at values about 30 percent below bifurcation whereas the shorter ones at only about 15 percent below (Ref. 5).

The second series, carried out by Ihsan Mungan in Bochum, West Germany, first studied cylinders[13] and then torroids,[14] which later series we shall describe here. The models were of a cold-mix epoxy resin with hardener and accelerator and had an $E = 500,000$ psi with $\nu = 0.38$. The five torroidal models were about 4 ft high with throat radii of about 17 in. and with throat thicknesses of from 1.9 to 1.1 mm. Mungan ran one set of tests on models symmetrical about the throat and then cut off the tops to make a second set of tests on cooling tower – type shapes of about 32 in. in height.

The models were fixed at their bases and partially fixed at their tops. That top boundary plays a significant role in the buckling and consisted of a fixity against all motion except axial displacement under axisymmetrical loading. Thus for prebuckling analysis that motion is allowed, but for buckling no axial motion can occur.

The shell models received axisymmetrical axial tensions or compressions and axisymmetrical pressures or suctions. In this way a wide combination of axisymmetrical loadings was achieved. The procedure was to set one ratio of axial to lateral load and then increment the combined loading, keeping the ratio constant. At each increment, there was applied a small dynamic disturbance in the lateral pressure with a frequency of 30 cycles per minute. At low loadings nothing happened, but at some critical load the shell began to vibrate and after 30 to 60 seconds it buckled.

All tests were carried out under water to slow down the snap-through, to lessen the risks of shell failure, and to keep a constant shell temperature during the tests.

The axial load was measured by a load cell, the lateral pressure by a mercury manometer, and the shell response by electrical resistance strain gages in both meridional and circumferential directions at five points around the throat. The strain gage readings increased almost linearly with load until the critical load above which load the readings changed more rapidly.

The experimental results are from 25 to 38 percent lower than numerical bifurcation predictions, which is a greater difference than found with other such axisymmetrically loaded shells.[15] Possibly the lower experimental values are due to the dynamic influence as in Johns's case.

Mungan plotted his results in an interaction diagram giving critical combinations of

σ_ϕ and σ_θ stresses. His results are for shells in which the state of stress is almost fully constant throughout. He stated that he "sees a possibility for application of the present results on cooling tower shells under combined dead and wind load, at least for purposes of preliminary design." Such a method is certainly convenient for preliminary design, but its validity as a true measure of safety is not fully established. The tests, being well designed, do serve as a valuable basis for determining the reliability of numerical analyses for axisymmetrically loaded shell buckling studies.

The third set of tests directed by T. J. Der and R. Fidler in the National Physical Laboratories at Teddington, England, represent the most comprehensive yet published.[16] Stimulated by the Ferrybridge failures, these tests were on models of electroformed copper and of polyvinyl chloride. A restudy of their interpretation was carried out in conjunction with R. Fidler in 1977 (see Ref. 6, page 52).

The models were about 15.6 in. high with radii of about 6.2 in. at the base and 4.26 in. at the top. For the plastic models $E \approx 4.5 \times 10^5$ psi and $\nu = 0.4$, while for the copper $E = 1.4 \times 10^6$ psi and $\nu = 0.3$. The thicknesses were nearly constant for the copper but varied for the plastic of from 0.028 in. and 0.044 in. at the throat to about 0.023 in. and 0.035 in. at the edges, respectively. Fixed at their bases and free at their tops, the wind coefficients estimated were taken similar to those of Johns's but based upon earlier wind tunnel measurements.[17] The models were tested in the compressed air wind tunnel, and Der and Fidler interpreted their results by the following formula:

$$q_{cr} = CE \left(\frac{h}{a}\right)^\alpha \tag{9-15}$$

for which they estimated $\alpha = 2.3$. A later theoretical study established $\alpha = 7/3$, and estimates for C have varied between 0.052^{18} to $0.086.^{19,20}$ On the basis of the 1977 restudy of these tests[6] it appears as if the lowest value should be about 0.052 even taking the more reasonable value of $E \approx 4.5 \times 10^5$ psi, instead of 5.5×10^5 psi as found in the 1968 Ref. 16.

9-4 THE BUCKLING OF DOMES

The classical analysis for the bifurcation buckling of spherical domes under axisymmetrical radial pressure q_{cr} was long ago found to be (Ref. 3, pages 512 – 519)

$$q_{cr} = CE \left(\frac{h}{a}\right)^2 \tag{9-16}$$

with

$$C_c = 2/\sqrt{3(1 - \nu)^2}$$

Numerous tests have shown that the classical value C_c is too high. Improved theoretical studies, based on theories that relax the restriction of small deflections, show values of C of between 0.31 and 0.365.[21] Finally a comprehensive set of experiments carried out on concrete models under the direction of D. Vandepitte at Ghent University in Belgium had established by 1971 that C was about half C_c and exceeded considerably 0.365, that in all their tests buckling occurred in the same way, and that minor thickness variations did not significantly affect the buckling load (Ref. 21).

The models were 7 mm thick with a radius of 243.1 cm, a plan diameter of 190 cm, and a rise of 19.33 cm. All seven models reported on in 1971 were of micro concrete with a strength of about 450 kg/cm². The models were loaded upside down under uniform radial pressure and were supported by a steel ring beam which under later tests was prestressed.

A report on the complete work from 1967 to 1976 appeared in 1977 and included a description of testing of 83 more microconcrete domes of the same dimension as the earlier seven. The conclusions, consistent with those in 1971, were extended to include the following: first, the value of $C = 0.542$ with a standard deviation of 9.5 percent, which for prestressed ring beams rose to 0.618 with a standard deviation of 14.6 percent; second, that creep caused a drastic reduction in the buckling pressure. Some domes failed at pressures as low as ⅓ of q_{cr}.[22] Eighteen of the domes were tested for creep buckling and the observation of a reduction to ⅓ of q_{cr} is consistent with the recommendations of the Committee on Concrete Shell Design and Construction of the American Concrete Institute.[23]

Thus buckling capacity for uncracked spherical domes without imperfections could be reasonably estimated by taking $C = (0.905 \times 0.542)/3 = 0.16$ in Eq. (9-16).[24] Scordelis summarized methods for including imperfections and cracking;[25] he gives graphs taken from the 1979 *Recommendations* of the International Association for Shell and Spatial Structures.[26] As Scordelis notes, the *Recommendations* give buckling capacities much lower than does the *Report* of ACI Committee 344.[27] In any case, designers need to consider imperfections and cracking in estimating dome buckling capacity so that the design will not be controlled by buckling.

9-5 IMPERFECTIONS AND CRACKING

The significance of imperfections including cracking came out clearly in the experience with cooling towers in Britain. In his "Résumé of the Report on the Failures," I. W. Hannah summarized the findings of the committee of inquiry on the 1965 collapse of three cooling towers at Ferrybridge, England, by noting three design defects: (1) an underestimation of the wind loads, (2) a lack of overload factors for the wind, and (3) a reliance on the balancing of wind and dead loadings with no provision for overloads.[28] From these defects the committee then concluded that "the prime mode of failure in each case was a vertical tensile failure within the lower part of the structure." They also noted that "shear, buckling, and foundation failure were eliminated as prime causes, but the possibility of forced vibrations having aggravated the weakness could not be established or ruled out." Finally, they listed the best estimates of the wind speeds that could have caused the meridional steel to yield and to fail; these speeds were close to those estimated to have occurred at the time of failure.

Clearly the meridional reinforcement was insufficient, as the committee emphasized, but so was the horizontal hoop reinforcement about which they made no recommendation.[29]

Hannah had noted at the beginning of his paper (Ref. 28, page 1) that "Vertical hairline cracks, extending through the shell and cornice were located on all the standing towers." In a companion paper, A. H. Chilver emphasized the significance of these cracks and of others by noting that (Ref. 28, page 13) "When a number of vertical cracks are present, the bending stiffening of the shell in the circumferential direction is reduced considerably. Consequently, the stability of the tower under wind loading is reduced. . . . "

It was, therefore, well recognized in 1967 that vertical cracking was a common occurrence in high cooling towers, that such cracking endangered the safety of the towers, and that the remedy was to provide a substantial circumferential reinforcement to the shell. Unfortunately so much emphasis had been put on the lack of meridional reinforcement that the concerns expressed by Chilver and Hannah about the circumferential problem somehow got lost. One can see clearly that this was so from a brief history of the second major tower failure in Great Britain, that at Ardeer in 1973.

This tower, having exactly the same height as those at Ferrybridge (375 ft), was originally designed as a 5-in.-thick shell with a single central layer of reinforcement (Ref. 30, page 2). By the time the Ferrybridge towers collapsed (November 1, 1965), the Ardeer design had been completed but construction had not begun, so the designers were directed to revise their design in the light of those failures. Because of the emphasis on increasing wind loads and on insufficient meridional reinforcement, the revision consisted mainly of raising the wind loading, of thickening the shell to 6 in., and of adding steel over the bottom 20 ft of shell.

Construction proceeded on the shell in the spring of 1966, but during the fall, large imperfections began to be reported in the shell geometry, reaching as high as 21 in. of misadjustments in the shell surface (Ref. 30, Appendix C, Figure 1). In December the owner decided that "the strength or stability of the tower had to be corroborated by an independent authority." Following a study, the report concluded that the design was adequate (Ref. 30, page 3).

Because of the large imperfections and of the concern arising from Ferrybridge, the Ardeer tower was very well documented in such a way that not only a full record of its imperfect geometry but also a careful recording of observed cracking was included. This cracking was vertical and rather extensive around the lower area of the shell surface. The tower began operation in November of 1968.

Then on September 27, 1973, with wind gusting between 55 and 68 mph, the tower collapsed in spite of the fact that earlier in its life it had sustained higher gusts including velocities up to 104 mph in January of 1968, before operation began. A committee of inquiry was set up and a detailed report prepared. The major conclusion was as follows (Ref. 30, page 16):

> That the primary causes were a combination of the serious imperfections in the shell's shape, meridional cracks extending in the area of maximum imperfection, and the relatively small amount of horizontal reinforcement (0.15% of the cross sectional area).

It was also noted that for the first 5 months of the cooling tower's life only the northern half of the tower was in commission and that during this time cracking occurred only in this half. The Committee have therefore concluded that meridional cracks, which are fairly common in this type of cooling tower, are probably caused by the thermal stressing brought about by the changes in operating conditions.

In short, then, the importance of thermal stresses, the need for substantial circumferential reinforcement, and the significance of meridional cracking were identified, along with the large construction imperfections built into the shell originally.

Kemp and Croll studied this problem of imperfections and showed how an axisymmetrical approximation of the main deviations in geometry could explain the failure.[31] Figure 9-3 shows this idealization and Fig. 9-4 demonstrates the strong influence of such a large imperfection on the N_θ values, even under gravity loads where the perfect geometry gives about 2500 lb/ft or for the 6-in. shell $\sigma_\theta = -2500/6 \times 12 = -35$ psi. For the imperfection of 1 ft, the compressive N_θ increases by a factor of four while now a tension of about 70 psi arises. Under a wind pressure of 100 mph, a tensile stress resultant of $490 \times 25 = 12,200$ lb/ft for a stress of 170 psi. (Fig. 6a from Ref. 27). This stress could be carried by uncracked concrete, but since the concrete can be cracked (by thermal gradients), it must then be carried by the reinforcement which circumferentially was only $0.0015 \times 6 \times 12 = 0.108$ in.^{2}/ft for a yield capacity ($f_y = 50,000$ psi) of 5400 lb/ft. When the dead load tension of about 5000 lb/ft is added on, the yield of the steel is far exceeded as is the ultimate stress of 70,000 psi. Thus even low wind loads will fail the circumferential steel in the imperfect tower when the ratio is as low as 0.15 percent. In its report, the joint ACI-ASCE committee on concrete shell construction recommended a minimum of 0.35 percent steel in both directions, thus providing in the case of a 6-in. shell, $0.0035 \times 72 = 0.252$ in.^{2}/ft for a yield capacity at $fy = 50,000$ psi of 12,600 lb/ft and an ultimate-load capacity of $(7/5)12,600 = 17,600$ lb/ft. Possibly therefore the shell would not have failed with that steel, but just as important such imperfection should

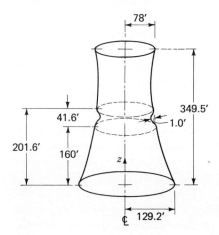

Fig. 9-3 Cooling tower imperfections. (From Kemp and Croll.)

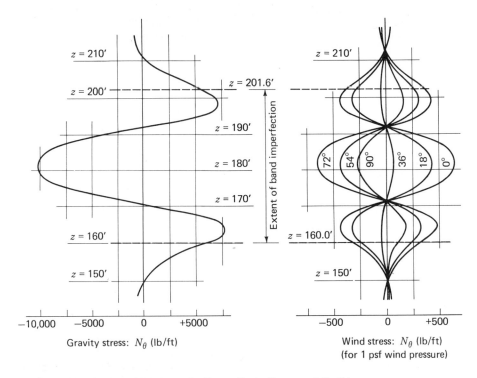

Fig. 9-4 N_θ stresses due to imperfections. (From Kemp and Croll.)

not have been permitted. The focus in 1967 was on meridional behavior, and Chilver's warning about circumferential behavior went unheeded.

9-6   CRACKING AND YIELDING

If one problem with shell walls lay in local overstressing and circumferential integrity, with barrel shells one question has been the ultimate-load capacity. The classical design, begun by Finsterwalder in the 1920s, gives elastic stresses and hence little direct information on overload capacity. For that reason researchers have developed a series of test programs and ultimate-load analyses to study safety in these shells. A summary of these results appears in Table 9-1 with load deflection curves shown in Fig. 9-5.[32]

The results generalize and confirm the conclusion of Sec. 6-8 that the shell has roughly three states: a linear-elastic one up to working loads, a more flexible one after initial cracking, and a very flexible one after yielding with the formation of plastic hinge lines. From Table 9-1 we see that shells for the most part designed on the basis of classical analysis (Chap. 6) have a substantial safety factor.

More refined analyses take into account cracking as it occurs by using finite elements. These studies depend upon complex computer programs and serve primarily for research. What the emphasis on overload does is to show that barrels,

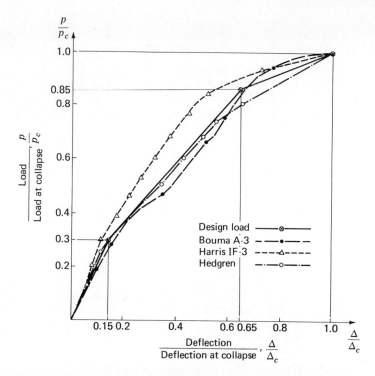

Fig. 9-5 Summary of barrel-shell tests. (From Darvall and Billington.)

designed by classical methods and built under close supervision, have a reserve of safety.

9-7 CREEP IN FLAT ROOFS

As Vandepitte's dome studies showed, flat concrete shells under permanent load have a lower buckling load because creep results in relatively large deflections which change the equilibrium. A striking example of this problem arose in 1970 with the collapse of a gabled hyperbolic-paraboloid roof in Henrico County, Virginia.[33] According to the investigators the failure occurred because of stresses which were "building up over a period of time." There was a flattening out of the roof, which was one of three similar designs. The other two were immediately studied, found to be flattening out also, and shored.

The form is closely patterned after the gabled hyperbolic paraboloid in Denver (described in Chap. 7) with three significant exceptions. First, the Virginia shell was flatter with $a = b = 50$ ft and a rise $c = 15$ ft, $k = 6 \times 10^{-3}$ ft compared to 7.6×10^{-3} ft at Denver; second, the Virginia shell was built of lightweight concrete; and third, the supports for the collapsed shell were flexible columns connected together at their tops by 14×22 in. prestressed ties spanning the 100 ft along each side of the building. At Denver the shell is held by heavy hinged buttresses tied together by a

Table 9-1 DIMENSIONS, FIALKOW LIMIT ANALYSIS, AND COLLAPSE LOADS FOR SIMPLY SUPPORTED THIN CYLINDRICAL CONCRETE SHELLS

Shell	Length (l)	Radius (r)	Thickness (t)	r/l	rt/l^2	Edge angle θ_0	Constant in Von Mises yield criterion $\sqrt{3}K = f'c$	Uniform pressure load at collapse from Fialkow chts — Upper bound	Lower bound	Actual collapse load	Collapse load factor
Example 1, ASCE Manual 31	62 ft (18.9 m)	31 ft (9.5 m)	3.75 in. (9.5 cm)	0.50	0.0025	40°	2640 psi (186 kgf/cm²)	1.25 psi (0.088 kgf/cm²)	0.88 psi (0.062 kgf/cm²)		
Parabolic shell model (Hedgren & Billington): Edge Thickening Ignored	84 in. (214 cm)	32 in.[a] (81 cm)	0.5 in. (1.3 cm)	0.38	0.0023	37°	4200 psi[b] (296 kgf/cm²)	2.56 psi (0.180 kgf/cm²)	1.90 psi (0.133 kgf/cm²)	2.3 psi (0.162 kgf/cm²)	4.4
Edge Thickening Included	84 in. (214 cm)	32 in.[a] (81 cm)	0.6 in. (1.5 cm)	0.38	0.0028	37°	4200 psi[b] (296 kgf/cm²)	2.88 psi (0.212 kgf/cm²)	2.16 psi (0.157 kgf/cm²)	2.3 psi (0.162 kgf/cm²)	4.4
Delft shell models (Bouma et al.): A-3	131 in. (333 cm)	39 in. (100 cm)	0.4 in.[c] (1 cm)	0.30	0.0009	41°[c]	3980 psi[d] (281 kgf/cm²)	1.16 psi (0.082 kgf/cm²)	0.85 psi (0.059 kgf/cm²)	0.97 psi (0.068 kgf/cm²)	2.7[e]

A-4	131 in. (333 cm)	39 in. (100 cm)	0.4 in.[c] (1 cm)	0.30	0.0009	41°[c]	3980 psi[d] (281 kgf/cm²)	1.16 psi (0.082 kgf/cm²)	0.85 psi (0.059 kgf/cm²)	1.04 psi (0.073 kgf/cm²)	2.9[e]
Harris and White models:										
IF-3	44 in. (112 cm)	13.1 in. (33.3 cm)	0.13 in.[c] (0.33 cm)	0.30	0.0009	41°[c]	4197 psi[b] (295 kgf/cm²)	1.23 psi (0.086 kgf/cm²)	0.89 psi (0.063 kgf/cm²)	1.22 psi[f] (0.085 kgf/cm²)	3.5[e]
IF-4	44 in. (112 cm)	13.1 in. (33.3 cm)	0.13 in.[c] (0.33 cm)	0.30	0.0009	41°[c]	4524 psi[b] (319 kgf/cm²)	1.33 psi (0.093 kgf/cm²)	0.97 psi (0.068 kgf/cm²)	1.00 psi (0.070 kgf/cm²)	2.8[e]
Mileykovsky model	236 in. (600 cm)	106 in. (270 cm)	0.8 in.[c] (2 cm)	0.45	0.0015	34°[c]	3410 psi[a] 240 kgf/cm²	1.58 psi (0.111 kgf/cm²)	1.11 psi (0.078 kgf/cm²)	1.41 psi (0.095 kgf/cm²)	4.5[e]

[a] Equivalent circular radius used.

[b] Cylinder strength.

[c] In using the Fialkow charts for shells with edge beams, the edge angle was increased to allow for the extra depth and the shell thickness increased to allow for the extra material.

[d] Cube strength × 0.80 to convert to equivalent cylinder strength.

[e] Dead load.

[f] "Corrected" result given by Harris and White.

[g] Design cube strength × 0.80 to convert to equivalent cylinder strength.

stiff floor and framing below. It is important to note that lightweight concrete not only has a lower E than regular concrete of comparable strength but it is also more prone to creep.

One further problem was beyond the state of the art in 1962 when the Virginia shells were designed; that was the inability of the membrane theory to predict shell and ridge beam stresses. Especially the shell stresses as shown in Chap. 7 can be over twice as great as assumed by the membrane theory and thus make the shell even more susceptible to creep, also in light of the fact that the ridge beams tend somewhat to load the shell rather than to take a significant portion of the load.

Tedesko had been worried about the flat central section of the Denver shell because "here on account of lack of curvature a critical point exists from the standpoint of buckling." He therefore put in 22 in.^{2} of steel in the center of the flat ridge beams compared to only 10 in.^{2} in the center of those in the Virginia shell.

The existence of large-scale computer programs makes it possible to study nonlinear behavior in such shells. Scordelis and his colleagues have made some such studies which include creep and cracking. Their results for hyperbolic paraboloids give some indication of how creep reduces safety.[34] In an 80-ft-square-plan gabled shell they show that the creep has no major influence; but this shell has a rise of 16 ft (double that of the shell studied by Schnobrich and by Shaaban and Ketchum) for a k = 0.01 compared to 0.0076 for Denver.

REFERENCES

1. Cross, H., "Dependability of the Theory of Concrete Arches," *Bulletin No. 203,* University of Illinois Engineering Experiment Station, 1929.
2. Popov, E. P., and S. J. Medwadowski, editors, *Concrete Shell Buckling,* ACI Publ. SP-67, 1981, 234 pages.
3. Timoshenko, S. P., and J. M. Gere, *Theory of Elastic Stability,* 2d ed., McGraw-Hill, New York, 1961, pp. 94–96.
4. Wang, Y-S, and D. P. Billington, "Buckling of Cylindrical Shells by Wind Pressure," *J. Engl. Mech. Div.,* ASCE, vol. 100, no. EM5, October 1974, pp. 1005–1023.
5. Kunderpi, P. S., G. Samevedam, and D. J. Johns, "Stability of Cantilever Shells under Wind Loads," *J. Eng. Mech. Div.,* ASCE, no. EM5, October 1975. For more complete details, see Godalacharyulu, S., and D. J. Johns, "Buckling of Thin Clamped-Free Circular Cylindrical Shells Subjected to Wind Load," no. TT 7113, Dept. Transp. Technol., Loughborough University of Technology, England, December 1971.
6. Abel, J. F., D. P. Billington, D. A. Nagy, and C. Wiita-Dworkin, "Buckling of Cooling Towers," Res. Rep. 79-SM-1, Dept. of Civ. Eng., Princeton University, N.J., January 1979, p. 25, and addendum to table 7, page 25.
7. Cole, P. P., J. F. Abel, and D. P. Billington, "Buckling of Cooling Tower Shells: Bifurcation Results," *J. Struct. Div.,* ASCE, vol. 101, no. ST-6, Proc. Paper 11365, June 1975, pp. 1205–1222.
8. "Specification for the Design and Construction of Reinforced Concrete Chimneys," ACI Committee 307-69, ACI, 1969.
9. Hayman, G., and A. H. Chilver, "The Effects of Structural Degeneracy on the Stability of Cooling Towers," *Rep. 71-77,* Dept. Eng., University of Leicester, England, June 1971.
10. Hayman, B., "Flexural Properties of Thin Reinforced Concrete Members," *Proceedings, Inst. Civ. Eng.,* pt. 2, vol. 65, June 1978, pp. 253–269.
11. Rish, R. F., "Forces in Cylindrical Shells Due to Wind," *Proceedings,* Inst. Civ. Eng., vol. 36, April 1967, pp. 791–803.

12. Gopalacharyulu, S., and D. J. Johns, "Cantilever Cylindrical Shells under Assumed Wind Pressure,"*J. Eng. Mech. Div.*, ASCE, vol. 100, no. EM5, Proc. Paper 10874, October 1974, pp. 1005–1023.

13. Mungan, I., "Buckling Stress States of Cylindrical Shells,"*J. Struct. Div.*, ASCE, vol. 100, no. ST11, Proc. Paper 10965, November 1974, pp. 2289–2306.

14. Mungan, I., "Buckling Stress States of Hyperboloidal Shells,"*J. Struct. Div.*, ASCE, vol. 102, no. ST10, Proc. Paper 12465, October 1976, pp. 2005–2020.

15. Veronda, D. R., and V. I. Weingarten, "Stability of Hyperboloidal Shells: An Experimental and Analytical Investigation," USCCE 009, School of Engineering, University of Southern California, Los Angeles, March 1973.

16. Der, T. J., and R. Fidler, "A Model Study of the Buckling Behavior of Hyperbolic Shells,"*Proceedings*, Inst. Civ. Eng., vol. 41, London, September 1968, pp. 105–118.

17. Cowdrey, C. F., and P. G. G. O'Neill, "Report of Tests on a Model Cooling Tower for the C.E.A. Pressure Measurements at High Reynolds Numbers," Natl. Phys. Lab., NPL Aero Rep. 316a, December 1956.

18. Cole, P. P., J. F. Abel, and D. P. Billington, "Buckling of Cooling-Tower Shells: State-of-the-Art,"*J. Struct. Div.*, ASCE, vol. 101, no. ST6, Proc. Paper 11364, June 1975, pp. 1185–1203.

19. "Reinforced Concrete Cooling Tower Shells: Practice and Commentary,"*J. ACI*, ACI-ASCE Committee 334, vol. 74, January 1977.

20. "Recommendations for the Design of Hyperbolic and Other Similarly Shaped Cooling Towers," IASS Working Group No. 3, Brussels, 1977.

21. Vandepitte, D., and J. Rathe, "An Experimental Investigation of the Buckling Load of Spherical Concrete Shells, Subjected to Uniform Radial Pressure,"*Proceedings*, RILEM, Buenos Aires, September 1971, pp. 427–442.

22. Weymeis, G., "Report on the Investigation of Instability of Concrete Domes Conducted from 1967–76 at the University of Ghent in Belgium" (in Dutch), Zwijnaarde, Belgium, 1977. See also D. Vandepitte, J. Rathe, and G. Weymeis, "Experimental Investigations into the Buckling and Creep Buckling of Shallow Spherical Caps Subjected to Uniform Radial Pressure,"*Proceedings*, IASS Congress, vol. 1, Madrid, 1979.

23. "Concrete Shell Structures, Practice and Commentary,"*J. ACI*, September 1964.

24. Billington, D. P., and H. G. Harris, "Test Methods for Concrete Shell Buckling,"*Proceedings*, ACI Symp. Concr. Shell Buckling, 1981.

25. Scordelis, A. C., "Stability of Reinforced Concrete Domes and Hyperbolic Paraboloid Shells,"*Concrete Shell Buckling*, ACI Publ. SP-67, 1981, pp. 63–110.

26. "Recommendations for Reinforced Concrete Shells and Folded Plates," IASS Working Group No. 5, Madrid, 1979, 66 pages.

27. "Design and Construction of Circular Prestressed Concrete Structures,"*J. ACI*, Rep. of ACI Committee 344, no. 9, Proc. V. 67, September 1970, p. 664.

28. "Natural Draught Cooling Towers—Ferrybridge and After,"*Proceedings*, Conf. held at Inst. Civ. Eng., London, June 12, 1967.

29. Report of the Committee of Inquiry into the Collapse of Cooling Towers at Ferrybridge, Monday 1 November 1965, Central Electricity Generating Board, London, August 1966.

30. Report of the Committee of Inquiry into the Collapse of the Cooling Tower at Ardeer Nylon Works, Ayrshire, on Thursday 27 September 1973, *Imperial Chemical Industries Ltd.*, London.

31. Kemp, K. D., and J. C. A. Croll, "Collapse of a Cooling Tower Due to Geometric Imperfections," presented at the Second Colloquium on Hyperbolic Cooling Towers, Brussels, June 1975.

32. Darvall, P. Le P., and D. P. Billington, "Correlation Study of the Behavior of Concrete Thin Shells to Collapse,"*Bull. Int. Assoc. Shell and Spat. Struct.*, no. 49, 1972, pp. 59–63.

33. "Students Clear Gym Moments Before Roof Fails,"*Eng. News-Rec.*, Sept. 24, 1970; see also *Eng. News-Rec.*, October 22, 1970.

34. Müller, G., A. F. Kabir, and A. C. Scordelis, "Nonlinear Analysis of Reinforced Concrete Hyperbolic Paraboloid Shells,"*Proceedings*, Conf. Nonlinear Behav. Reinf. Concr. Spat. Struct., vol. 1, Darmstadt, W. Germany, July 1978, pp. 191–203.

10
ROOF DESIGN

10-1 INTRODUCTORY IDEAS

This chapter is on structural design ideas rather than on analysis or detailed dimensioning as previous chapters were. The ideas discussed here spring from three observations. First, historical study of the past 75 years shows that there has arisen a new tradition in structures of which the thin shell roof is an important example; second, this new tradition is characterized by a group of designers who have practiced as structural engineers rather than as architects; and third, for these engineers, aesthetics has been a major motivation for their design which can be identified as a new art form, called structural art.

This chapter takes up three themes in the context of thin shell concrete roof design and deals specifically with the connection of engineering design to analysis as applied science, to construction as competitive contracting, and to appearance as architecture. The leading idea to emerge from these three themes is that proper engineering design of concrete thin shell roofs includes analysis, construction, and appearance and that each is an essential part of engineers' work. While engineers need to consider all three aspects in all of their structures, it is especially important in thin shell roofs of long span. To see why this is so we need only refer briefly to the half-century history of thin shell concrete roofs detailed in the first chapter. The principal designers identified there were engineers such as Maillart, Dischinger, Finsterwalder, Tedesko, Freyssinet, Nervi, Torroja, and Isler plus the architect Candela. Each was well trained in mathematical analysis, each either was himself a building contractor or worked closely with builders, and each considered appearance as a major goal for design. Even though Candela was trained as an architect, he had considerable mathematical skill and he worked as the building contractor himself for nearly all of his major works. He thus worked essentially as an engineer in most cases.

Although in many works of these designers there were collaborating architects, it can be demonstrated in most cases that the final visible form is the product not of collaboration but of the vision of one designer, the engineer. To be sure, there are many examples of thin shell roofs whose visible form was set by an architect. How then can we distinguish between works of engineering and those of architecture? To do this, we shall consider, with examples, some basic differences and then trace in more detail the works and ideas of one designer who demonstrates clearly the potential for engineering design in thin shell concrete roof structures.

10-2 ENGINEERING AND APPLIED SCIENCE

One view of engineering and architecture is that the architect sets the form and the engineer makes it stand up. In this view, the engineer becomes a kind of applied scientist. A fundamental distinction between engineering and science is that the engineer makes things which do not exist in nature, whereas the scientist discovers things that have always existed in nature. Thus engineers become applied scientists when they analyze forms given to them rather than when they set out to determine form themselves.

Many engineers have seen themselves as applied scientists and have thus focused their major attention on the means of analysis. This is an important activity but it is distinct from design. The designers mentioned earlier were aware of such analyses but their major objective was usually to decide on overall form with a design based on simple ideas backed up by simple analyses. A study of the works of some of them—Maillart, Nervi, Candela, and Isler, in particular—reveals several ideas on analysis that they have held in common.

The first idea was that the appropriate method of analysis depends upon the form chosen.[1] The various methods have developed from specific forms, not from general theories. The Geckeler approximation for domes was developed with one special form in mind, domes of rotation which are not too flat. Although the method can be derived from the general theory of shells, Bauersfeld did not do it that way but rather developed the idea based upon the specific dome form and the physical idea of rapid damping of edge effects.

An even more impressive demonstration of this idea is with hyperbolic paraboloids where early efforts to use a general membrane theory for all such forms failed, and it has since become clear that the membrane theory can only be used for certain forms such as groined vaults and not at all for others such as gabled roofs. The designers' study thus is devoted to setting form first and deciding on an appropriate simplest analysis to justify and confirm the design.

But designers would have little basis for setting form if they did not hold a second idea, namely that forms develop out of visual observations on full-scale thin shells supplemented by the use of small-scale models. Tedesko's observation of small cracks in the Denver hyperbolic paraboloid edge beams implied that these beams, rather than supporting the shell as assumed by the membrane theory analyses, were being supported by the shell and hence were more of a problem than a solution.[2] Isler, observing the same thing from model studies, grasped the design consequences and simply eliminated the edge beams on his 70-ft-span Camoletti house roof design of 1970. Later on computer analyses showed that those edge beams actually carried only a fraction of the load found from the membrane theory, but only Isler, stimulated by visual observations, seems to have converted that observation into a design which eliminated the edge beams entirely.[3]

The third idea common to the best-known designers grows out of the first two and is that their mature designs can be confirmed by very simple analyses. The most sophisticated designs, such as Maillart's Cement Hall, Nervi's little sports palace, Torroja's stadium cantilever roof, and Candela's groined vaults require only the simplest analyses. These forms all grew out of years of experience with full-scale structures and a sound background in mathematical analysis. [4]

Whereas the engineer as applied scientist develops new and more rigorous analyses, now relying on complex computer systems, the engineer as designer develops new and more subtle forms relying on very simple analyses and extensive experience observing full-scale behavior and model studies.

10-3 ENGINEERING AND COMPETITIVE CONTRACTING

Engineers and architects became fascinated with the visual possibilities for thin shell concrete roofs once the works of European engineers gained publicity in the 1950s. This fascination in many cases deemphasized the construction problems and highlighted only the final forms. Especially where architects wanted to express free curving form, construction problems did not directly influence design and as a consequence many structures ended up as very costly. The most publicized examples were the Sydney Opera House[5] and the Montreal Olympic Stadium.[6]

It was a curious paradox that the designers who heightened the interest in shells, such as Maillart, Nervi, and Candela, developed their styles out of a constant search for competitive solutions, whereas many of the works they stimulated produced the belief that thin shell roofs were naturally costly. Over against this result stands the work of a few American engineers such as Tedesko and Christiansen whose scores of thin shell designs were nearly all competitively built. The contrast between the Montreal stadium and Christiansen's Seattle Kingdome demonstrates how a nearly similar problem can produce radically different results depending upon the designers attitude toward construction. At Seattle, the cost per seat was under $1000, whereas at Montreal it was over $11,000 per seat. The Kingdome is fully covered and can be used for baseball as well as all other major sports, whereas the Montreal stadium is not covered and not so well suited for baseball. It was estimated in 1976 that when fitted with a cover, the total cost of the Montreal stadium will reach over $1 billion or about $17,000 per seat. These high costs were directly traceable to a design that did not carefully consider local construction practices. By contrast, Christiansen had already 20 years of experience at Seattle before the building of the Kingdome, and his design included detailed ideas on construction.

10-4 ENGINEERING AND AESTHETICS

It has often been held that the visual part of design belongs to the architect. Even for bridges, perhaps the purest example of engineering, some writers maintain that architects are needed to make them beautiful. Thin shell roofs are also thought to be architecture; Joedicke's influential book *Shell Architecture* announces by its title the belief that the shells of Maillart, Freyssinet, Nervi, Isler, and other engineers are works of architecture.[8]

This is more than a semantic question, although it does, in part, involve making clear and realistic use of the common words: engineering and architecture. Before the industrial revolution which began in the second half of the eighteenth century, the design of bridges and roofs was often in the hands of architects. However, with the new materials of industrialized metal and reinforced concrete there emerged a new group of designers, structural engineers, whose training was highly technical and whose practice was strongly conditioned by competitive contracting. A small group of these designers saw their work also in aesthetic terms. While most of their fellow engineers continued to think that aesthetics did not belong to engineering, this small group went ahead and designed works that gained significant attention for their beauty.

These engineers worked sometimes alone, sometimes in collaboration with architects, and sometimes merely as consultants. As consultants they had rarely any influence on form; the engineers on the Montreal stadium, for example, argued that it was not their business to influence the design, only to make it stand up.[9] On the other hand, the St. Louis Airport terminal roof was a collaboration between the architects Hellmuth, Yamasaki, and Leinweber and the engineer Tedesko. Yamasaki chose the groined-vault form but Tedesko insisted on the visually prominent diagonal arches.[10] One might say that this was a collaboration in which the architect chose the overall form but the engineer played a strong role in significantly modifying that form. In the Camoletti house, Isler as engineering collaborator played the same role as Tedesko at St. Louis. Here, however, Isler took the hyperbolic paraboloid form chosen by the architect Pierre Camoletti and modified it by eliminating ribs.

What is important to stress here are the many cases where the structural engineer sets the overall form and the collaborating architect either modifies it or has no influence directly on it. In the Berenplaat Filter Building the inverted umbrella hyperbolic paraboloid form was set by the engineer Adolph Bouma and the columns were shaped by the architect Quist (see Fig. 1-21).[11] In the case of Nervi's sports palaces in Rome, exhibition halls in Turin, and stadium roofs in Florence and Rome, it appears that the forms are entirely Nervi's with the collaborating architects playing no role in the visible structure. The same is certainly true of Maillart's Cement Hall and Torroja's cantilever stadium roof.

Structural art as distinguished from architectural art grows out of the design motives of controlling forces and of reducing costs. It arises most frequently where the scale is large and the use is simple. It requires the visualization of form as a means of controlling forces whereas for the architect, form is usually a means of controlling spaces.

The larger the scale the more significant become the engineer's ideas of design as opposed to the architect's. Ideas appropriate to row house roofs have little meaning for a 660-ft-span domed stadium. When an architect does not recognize this difference of scale, the results can be disastrous as the Montreal stadium demonstrates.

In addition to scale, the use helps determine whether a work is appropriate to engineering design or to architectural design. The less the use involves the intimate and changing activities of small groups of people the more it becomes an engineer's design. Put the other way around, the more the use is for machines (bridges), for materials (warehouses, treatment plants, piers), or for large groups of people in highly fixed conditions (roofs for covered stadiums, swimming pools, tennis courts), the more the form is an engineering design problem.

Finally and related closely to scale and use is the different way architects and engineers visualize form. For the architect the form serves primarily to define spaces used intensively by people and is therefore more a backdrop than a primary expressive element. For the engineer, the form serves primarily to define the carrying of loads and is therefore the main expressive element. Where the primary design problem is small-scale, intimate and complex human use, the design properly emphasizes the usable spaces and is prototypically architecture; however, where the major design problem is large-scale use of a simple type, the design focuses on the load-carrying form and is prototypically engineering. Clearly these distinctions overlap and blur at their edges, but they represent modern historical facts, not ideas based upon some theoretical premises.

The important conclusion to result from holding these distinctions between engineering design and architectural design is that both types of designers at their best are artists. It is well accepted to treat architecture as an art form and to speak of the best architects as artists. What the emerging tradition of engineering shows is that a series of structural engineers have arisen who, without being architects or sculptors, are also artists but of a new type which we might call structural artists.

Thus we can say that there are three types of professionals who create constructed forms: engineers, architects, and sculptors. Each at their best is an artist. To create such forms, each must consider three sorts of criteria: scientific—making the object stand; social—making it useful; and symbolic—making it visually expressive. For the engineer as structural artist, the primary discipline is scientific but the other two are essential; for the architect the primary discipline is social but again the others are crucial to the design; finally for the sculptor the primary discipline is symbolic or expressive, but the sculptor's objects must stand and be useful at least in the sense of appealing to some segment of society.

To help establish the idea of the structural artist, we shall close this chapter with a more detailed analysis of one such artist, Heinz Isler, and of one such American work, the Kingdome.

10-5 ROOF SHELLS AS STRUCTURAL ART: SCIENTIFIC IDEAS

For roof shells to be works of structural art they must visually express the aesthetic ideas of the designing engineer, and included in those ideas should be the drive for minimum materials and minimum costs. To see what these ideas are in the specific case of Heinz Isler, we shall analyze one of his representative works under the three headings already introduced: engineering and applied science, engineering and competitive contracting, and engineering and aesthetics.

Isler's basic scientific idea has been to evolve forms from physical analogies rather than from mathematical studies. Three examples are pneumatic forms, flowing forms, and the hanging reversed membrane. For the pneumatic forms, a membrane clamped along four straight edges is inflated under upward pressure creating a form in pure tension. This form will carry downward pressure by pure compression with the straight edges being in tension. Pneumatic forms are not ideal for gravity loads

Fig. 10-1 Eschmann factory, Switzerland, 1958. (Courtesy Heinz Isler.)

because pressure loads are normal to the curved surface whereas gravity loads are always vertical. However, where the surface is relatively flat the differences are small, and Isler has used this form since the mid-1950s with considerable success for industrial one-story structures (Fig. 10-1).[12]

The flowing form comes from allowing a foam to pass through a rectangular (or other shaped) opening. At the walls because of friction the velocity is zero, whereas it reaches a maximum at the center of the opening. A slowly expanding foam leaves a square tube and gives a dome shape. With this method, Isler found some unusual forms which could be built without any edge members. The example shown in Fig. 10-2 is on a 27-m-square (88.5-ft-square) plan and has a thickness of only 8 cm (3.15 in.). Its free edges have a slight tension, thus raising the buckling capacity. This bold new form has performed well in the harsh Swiss environment, but its deflections are greater than those of comparable pneumatic shells.[3] This means that the ratio of maximum dead-load vertical deflections (normally w at the crown) to span (diagonal distance in a square domelike roof, for example) was greater for the free-edge flowing forms than for the prestressed-edge-beam pneumatic forms.

The third form, the hanging reversed membrane, is in Isler's view the best of all because it can dispense with edge members while at the same time deflecting even less than comparable pneumatic forms. Here he suspends a cloth coated with a plastic material. The cloth thus loaded takes a shape which depends upon the sup-

Fig. 10-2 Bürgi garden center, Switzerland, 1971. (Courtesy Heinz Isler.)

Fig. 10-3 Heimberg tennis court roof, Switzerland, 1979.

ports and which puts the membrane into tension. Once the plastic hardens, the membrane is reversed and the resulting domelike shape is thus in compression under gravity loads. In this way Isler has created highly unusual forms both for regular boundaries (Fig. 10-3) and for irregular ones (Fig. 10-4).

In these various ways Isler applies scientific principles not in order to analyze forms but in order to design them. Of course, once created, a new form must be analyzed for its response, but that analysis need not be complex because the very design process has assured a form that is under low stress. The major analysis problem for such thin roofs is to estimate the buckling capacity, and for that Isler regularly tests precision small-scale models. There he measures deflections as well as strains and then he compares the deflections to those measured on the full-scale structure. In this way he controls his results and gains experience for his next designs. Thus Isler follows the three ideas noted under engineering and applied science: the forms are designed to make the internal forces small, the forms are controlled by model studies and measurements on full-scale structures, and finally the forms are set such that no complex mathematical analysis is necessary. As far as possible the analysis is done visually.

10-6 ROOF SHELLS AS STRUCTURAL ART: CONSTRUCTION IDEAS

The central idea about construction of roof shells is that of collaboration between the designer and the builder. In Isler's case he has trained a local contractor, the Bösiger company, who now builds nearly all his shell roofs. But more than that he has developed a system for building which is both simple and of wide application.

Figure 10-5 taken from a 1979 pneumatic form roof illustrates the system by showing first the carefully shaped laminated wooden arches which are reused many times and which give the accurate form that his designs require. These forming arches are set on light metal scaffolding and covered with quickly placed wooden slats. On top of the slats come wood fiberboard sheets to serve as a forming for the concrete and, left in place afterwards, as insulation for the roof (Fig. 10-4).

In this way, the roof is economically built because first the contractor is familiar with the process, second the costly forming arches are reused many times, and third the permanent forming insulation boards not only simplify construction but allow the concrete surface to be left without any roofing or insulation applied on top. This last remarkable saving arises because the shells are in almost pure compression and even in the harsh Swiss environment do not crack under gravity loads. Therefore, so long as there is sufficient insulation to prevent strong thermal gradients through the shell thickness, there will be no cracking and hence no leaking. After a quarter of a century even his earliest shells show no signs of deterioration. Sometimes, for aesthetic

Fig. 10-4   Sicli Building, Switzerland, 1969. (Courtesy Heinz Isler.)

Fig. 10-5   Lenzlenger Sons Company building, scaffolding and forming, Switzerland, 1979.

reasons an owner will cover the shell with a light roofing or a coating of paint but this is not done for reasons of utility.

10-7 ROOF SHELLS AS STRUCTURAL ART: AESTHETIC IDEAS

The strongest motivation for Isler is the aesthetic one. Like nearly all classical artists, Isler's major concern is to express through his own medium a personal vision of beauty.[13] If his scientific ideas emphasize visual analysis and his construction ones collaboration with builders, then his aesthetic ones center on the creation of a personal style. These latter ideas, perhaps unusual when applied to engineers, are nevertheless clearly apparent in the works of other structural artists such as Maillart, Nervi, Torroja, Candela, and the contemporary Swiss bridge designer Christian Menn.[14]

Thus to understand Isler's aesthetic ideas is to see his style as the search for the expression of form. At least three aspects of this expression are apparent: edge thinness, curved surfaces, and vivid contrast. Figure 10-3 emphasizes the thinness of the shell roofs, shows the strange curves of the roof exterior, and finally shows how the shell roof contrasts vividly with its natural setting. The shell forms provide a startling articulation amid the fields and mountains around.

But to see Isler's aesthetic ideas more fully, we need to compare his mature works with those of others who do not share his vision of structural art. Figure 10-6 shows part of a sports complex in Chamonix whose overall form was designed by the

Fig. 10-6 Sports center, Chamonix, France, 1971. (Courtesy Heinz Isler.)

Fig. 10-7 B. P. Gas Station roofs, Switzerland, 1968.

architect Roger Taillibert with Isler as collaborator. The shell forms are very flat spherical segments on three supports. Because these are geometrical rather than structural forms and because they are very flat, there had to be added to the shell form heavy edge ribs and heavy foundations. Compare these heavy forms to the light three-point supported roofs designed, without collaboration, by Isler using the hanging reversed membrane (Fig. 10-7). Here all edge beams are avoided and an extraordinarily light appearance results because the form came directly from structural ideas.

10-8 THE KINGDOME: STRUCTURAL ART AT LARGE SCALE

Isler's designs show what can be done at the scale of widely used industrial, recreational, and exhibition buildings. His work shows that properly designed thin shell concrete roofs have the potential for extensive use. The Kingdome, on the other hand, demonstrates that thin shell concrete roofs can compete with steel for any size of free span as yet designed. In fact, the cost per seat of the Kingdome was substantially less than comparable costs for either the Astrodome or the Super-dome.[15]

Fig. 10-8 International exhibition facility, New Orleans, Louisiana. (Courtesy J. V. Christiansen.)

Fig. 10-9 Kingdome, completed. (Courtesy J. V. Christiansen.)

At least one part of the reason for the Kingdome's relative economy is that its designer, J. V. Christiansen, had previously 20 years of experience in the design of thin shell roofs of which over seventy-five designs had been built by 1972 when the Kingdome design was made. Moreover, Christiansen had extensive experience not only with doubly curved roof shells but also with shells of unprecedented scale. His New Orleans Rivergate exhibition hall of 1968 used the longest span barrel shells in the world (Fig. 10-8),[16] and his 240-ft-span 1962 airport hangar in Seattle has the same arched hyperbolic paraboloidal shells that he was to use in the Kingdome (Fig. 10-9). In short, the immense concrete dome design followed years of experience in the design of similar forms which had been built and whose behavior had been observed at full scale.

Moreover, even though Christiansen's firm, like the firms of all modern structural engineers, is fully competent in advanced analysis, the forms of his shells did not come from mathematical analysis but from simple structural and constructional ideas. The Kingdome is a relatively simple form: pie-shaped arches for dead load (Fig. 10-10) and a dome for live load. The thin 5-in.-shell is stiffened locally by its hyperbolic paraboloidal form and overall by the 40 upstanding (about 2-ft-wide 6-ft-deep) ribs (Fig. 10-11) which fan outward from the central compression ring to the 24-ft-wide tension ring with a radius of about 325 ft. The ribs extend to the outer edge of the ring, while the shell transition is scalloped in plan because of its double curvature. The overall structural behavior is simple either as an arch or as a dome and only the buckling capacity presents a serious problem. [17]

A close look at the details reveals the subtlety of the form in the transition between arch rib and shell, between shell and tension ring, and between ribs and compressive ring. Close up these are not the polished details one would look for in the very small-scale (by contrast) church domes of the Renaissance, any more than are the bridge forms of Maillart, in detail, of the same cut granite smoothness of the most decorated city bridges built before the late nineteenth century.

The construction (Fig. 10-10) shows the isolated arch segments which had to be a major part of the design. The designer had to visualize the entire construction sequence and design for it. Contrast this result at Seattle with that at Montreal

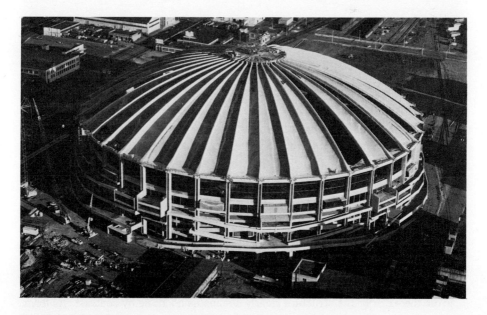

Fig. 10-10 Kingdome under construction. (Courtesy J. V. Christiansen.)

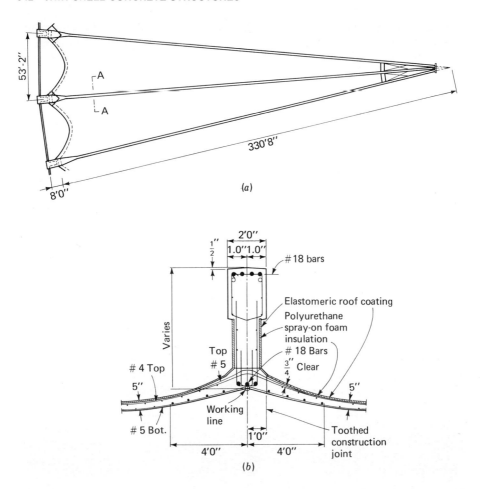

Fig. 10-11 Section of dome and cross section of stiffening rib. (Courtesy J. V. Christiansen.)

where the design provided no means for bracing the arching ribs and hence there they had to be built as individual cantilevers with the consequence of large scaffolding and huge dead-load bending moments, compared to the arch moments at Kingdome. Even the visual compression ring at Montreal is not structural but carries only utilities and no loads (see Hayward, Ref. 7).

We have given some idea of the efficiency of the Kingdome and of its utility as measured by relatively low cost per seat arising from the integration of construction thinking in the design stage. Finally comes the question of aesthetics. Such an immense structure can be evaluated in at least two entirely separate ways: outside and inside.

From the exterior the structure clearly dominates its locale by contrast. Its form, totally unlike anything around it, has three distinctive features: the dome ribbing, the transition between rib, ring, and columns, and the undulating ramps that surround the entire form.

The ribs are striking and give some visual intimation of the thinness. They are visually lighter than those at Montreal and they give a clearer image of the structure, unlike the smooth-topped Superdome and Astrodome.

The transition between ribs, ring, and columns is, on the other hand, visually the least striking part of the design. From the air one can see the elegant scallops of the shell edge, but that is missing in elevation. This design weakness arises from the fact that the ring is largely horizontal, and thus its structural stiffness is hidden in elevation. The engineer and probably the general public perhaps intuitively expect a stronger visual element for this transition. It is not easy to imagine how this could have been economically done.

Finally the ramps connected to the columns provide a strong visual image, perhaps somewhat disquieting at first, of both the central need for easy access and, less obviously, the equally important need for full bracing against the strong earthquake estimates upon which the design had to be based. Once these needs are recognized, the ramps seem less disturbing and become more of an integrating feature. Overall the exterior design is striking, gives some idea of the integration of form and the thinness of elements, and contrasts vividly with its surroundings. If the outside seems to fall short of its full visual potential, it can easily be noted that it is externally far superior aesthetically to such noted works as Nervi's large sports palace in Rome where all sense of Nervi's structural ideas are lost.

Fig. 10-12 Kingdome interior.

If the exterior of the Kingdome is not quite a visual masterpiece, the interior ranks visually with any work of concrete yet built anywhere. The space itself is overwhelming and the ribs show off more dramatically inside than outside because they extend visually right to the crown (Fig. 10-12). The transition at the outer ring is clearer and undisturbed by the vertical column lines seen only from without. The interior roof is certainly not as decoratively elegant as the roofs in Nervi's two Rome sports palaces, but the Kingdome interior does express clearly that the major behavior is archlike rather than domelike. Apart from the probably prohibitive cost of trying to create Nervi-like roof patterns, the stronger, less delicate Kingdome form is more appropriate to the much larger space in the American work. The space is not intimate and its use by spectators is inflexible in the literal sense of the fixed seating. It is thus a prototypical engineering design whose major significance for the future of thin shell concrete roofs is its demonstration that works of structural art can be built at the largest scale and with a relatively limited budget so long as engineering designers strive to achieve efficiency, utility, and aesthetic quality in all their works.[18]

REFERENCES

1. For a general discussion of this idea, see D. P. Billington, *Robert Maillart's Bridges: The Art of Engineering,* Chap. 9, Princeton University Press, N.J., 1979.
2. Tedesko, A., "Shell at Denver—Hyperbolic Paraboloid Structure of Wide Span," *J. ACI,* October 1960, pp. 403–413.
3. Isler, H., "New Shapes for Shells—Twenty Years Later," *Heinz Isler as Structural Artist,* The Art Museum, Princeton University, N.J., 1980.
4. For the designer's own description of these ideas, see D. P. Billington, *Robert Maillart's Bridges,* Chap. 10, Princeton University Press, N.J., 1979; P. L. Nervi, *Structures,* McGraw-Hill Book Company, New York, 1956; E. Torroja, *The Structures of Eduardo Torroja,* F. W. Dodge Corporation, New York, 1958; F. Candela, "The Shell as a Space Encloser," *Proceedings,* Conf. Thin Concr. Shells, Cambridge, Mass., 1954, pp. 5–11.
5. Arup, O. N. and C. J. Kunz, "Sydney Opera House," *Civ. Eng.,* December 1971, pp. 50–54. The article states that the building "is not really of this age and in concept is more appropriate to the product of autocratic rule of a former era." The original cost estimate was $7.8 million in 1957, and by completion in 1973, 9 years late, it had cost $142 million, had eliminated the planned parking garage, and had scaled down the main concert hall. See *Eng. News-Rec.,* Apr. 5, 1973, p. 13.
6. "Montreal's XXI Olympiad Sports a Lavish Design at Record Costs," *Eng. News-Rec.,* June 10, 1976, pp. 22–37.
7. Christiansen, John V., "The Kingdome," *Proceedings,* World Congr. Shell and Spat. Struct., Montreal, 1976. The total Kingdome cost reported here was $58.2 million. Later the original contractor lost a suit and was ordered to pay back $10.9 million. Thus, based upon 60,000 seats for baseball, the cost was slightly less than $800 per seat. The Montreal stadium costs by mid-1976 were estimated at $825 million for 60,000 seats (there is also provision for 14,000 standees). Thus the costs then were about $13,750 per seat, without a roof. Once that is added it is estimated that the total costs will exceed $1 billion, and hence the cost per seat will be about $17,000. David Hayward, "The Montreal Story," *New Civ. Eng.,* London, June 3, 1976, pp. 29–36.
8. Joedicke, J., *Shell Architecture,* New York, 1963.
9. Ostenfeld, K., "Discussion," *Civ. Eng.,* December 1976. In a debate on the appropriateness of the Montreal design, Ostenfeld stated, "I am entirely in agreement with the engineering criticisms made here. But if we agree that these structures are to be considered sculptures with a certain function, esthetics should be left to the architect. . . . If we agree that the esthetics should not be touched by the consulting engineer, I think another objection to these structures—cost—can not be discussed

either." This is a clear statement of the idea that the engineer is merely a technician hired to make the architect's vision stand up.

10. Tedesko, A., "The St. Louis Air Terminal Shells," *Proceedings,* 1962 World Conf. Shell Struct., San Francisco, 1964, pp. 469–474.
11. Aarnoudse, J., "Het Drinkwaterproduktiebedrijf Berenplaat" (The Water Treatment Plant on the Berenplaat Island near Rotterdam), *Cem.,* vol. 17, no. 2, 1965, pp. 75–82.
12. Isler, H., "Twelve Years of Applications of Pneumatically-Formed Shells," *Proceedings,* Int. Congr. Shells, Mexico City, 1967.
13. By classical here we mean to exclude certain modern artists and some from earlier times who have used art to shock, horrify, or parody their society.
14. *The Bridges of Christian Menn,* Cat. of an Art Museum Exhib., organized by the Art Museum and the School of Engineering and Applied Science, Princeton University, N.J., 1978.
15. In Ref. 6 the Superdome cost is given as $178 million, or $2382 for the 74,669-seat stadium. These are roughly comparable in time to costs for the Kingdome and the Montreal stadium since all three were completed between 1975 and 1976. For the $31.6 million Astrodome of 1966, a construction cost index increase of 2.28 is needed to bring the 50,000 seat stadium prices up to at least 1975, making the comparable cost per seat of $1440. See Carrigan, M. C., "Louisiana Superdome: World's Most Versatile Building," *Civ. Eng.,* November 1974, p. 53; and also *Eng. News-Rec.,* Mar. 23, 1978, p. 73 for the construction cost indexes.
16. Christiansen, J. V., "Shell Roof for International Exhibition Facility, New Orleans," *Concr. Thin Shells,* ACI, no. SP-28, Detroit, 1971, pp. 139–148.
17. Christiansen, J. V., "The Kingdome," ASCE Fall Conv., San Francisco, October 1977, Preprint 2946.
18. Christiansen, J. V., "Aesthetic Choices Leading to the Kingdome," ASCE Spring Conv., New York, May 1981, Preprint 81-072.

APPENDIX TABLES

The following tables are reprinted from "Design of Cylindrical Concrete Shell Roofs," Manual 31, American Society of Civil Engineers, New York, 1952.

The following notations used in these tables differ from those used in the text:

Tables	Text
T_x	N_x
S	$N_{xy} = N_{yx}$
T_ϕ	N_ϕ
t	h
l	L
I	$h^3/12$

Table 1A SYMMETRICAL EDGE LOADS ON SIMPLY SUPPORTED CYLINDRICAL SHELLS

(a) Basic Formulas and Loading Diagrams

	VERTICAL EDGE LOAD	HORIZONTAL EDGE LOAD	SHEAR EDGE LOAD	EDGE MOMENT LOAD
	Longitudinal Force T_x— $V_L\left[\left(\frac{l}{r}\right)^2\times\text{Col. (1)}\right]\sin\frac{\pi x}{l}$ Shearing Force S— $V_L\left[\frac{l}{r}\times\text{Col. (2)}\right]\cos\frac{\pi x}{l}$ Transverse Force T_ϕ— $V_L\times\text{Col. (3)}\times\sin\frac{\pi x}{l}$ Transverse Moment M_ϕ— $V_L\left[r\times\text{Col. (4)}\right]\sin\frac{\pi x}{l}$	Longitudinal Force T_x— $H_L\left[\left(\frac{l}{r}\right)^2\times\text{Col. (5)}\right]\sin\frac{\pi x}{l}$ Shearing Force S— $H_L\left[\frac{l}{r}\times\text{Col. (6)}\right]\cos\frac{\pi x}{l}$ Transverse Force T_ϕ— $H_L\times\text{Col. (7)}\times\sin\frac{\pi x}{l}$ Transverse Moment M_ϕ— $H_L\left[r\times\text{Col. (8)}\right]\sin\frac{\pi x}{l}$	Longitudinal Force T_x— $S_L\left[\left(\frac{l}{r}\right)^2\times\text{Col. (9)}\right]\sin\frac{\pi x}{l}$ Shearing Force S— $S_L\left[\frac{l}{r}\times\text{Col. (10)}\right]\cos\frac{\pi x}{l}$ Transverse Force T_ϕ— $S_L\times\text{Col. (11)}\times\sin\frac{\pi x}{l}$ Transverse Moment M_ϕ— $S_L\left[r\times\text{Col. (12)}\right]\sin\frac{\pi x}{l}$	Longitudinal Force T_x— $\frac{M_L}{r}\left[\left(\frac{l}{r}\right)^2\times\text{Col. (13)}\right]\sin\frac{\pi x}{l}$ Shearing Force S— $\frac{M_L}{r}\left[\frac{l}{r}\times\text{Col. (14)}\right]\cos\frac{\pi x}{l}$ Transverse Force T_ϕ— $\frac{M_L}{r}\times\text{Col. (15)}\times\sin\frac{\pi x}{l}$ Transverse Moment M_ϕ— $M_L\times\text{Col. (16)}\times\sin\frac{\pi x}{l}$

(b) $l/r = 100$ and $r/l = 0.1$

ϕ	T_x (1)	S (2)	T_ϕ (3)	M_ϕ (4)	T_x (5)	S (6)	T_ϕ (7)	M_ϕ (8)	T_x (9)	S (10)	T_ϕ (11)	M_ϕ (12)	T_x (13)	S (14)	T_ϕ (15)	M_ϕ (16)
$\phi_k=30$: 30	− 5.382	0	− 3.471	− 0.3278	− 0.0013	0	+ 0.9997	+ 0.1310	− 0.0904	0	− 0.0207	− 0.0010	− 0.0038	0	+ 0.0008	+ 0.9738
20	+ 3.575	− 2.619	− 2.656	− 0.2779	+ 0.0017	− 0.0001	+ 0.9848	+ 0.1162	− 0.0396	− 0.0403	− 0.0081	− 0.0007	+ 0.0172	+ 0.0021	+ 0.0022	+ 0.9767
10	+ 1.809	− 3.265	− .775	− 0.1512	+ 0.0034	+ 0.0017	+ 0.9399	+ 0.0722	+ 0.1118	− 0.0250	+ 0.0142	+ 0.0002	+ 0.0264	+ 0.0173	+ 0.0014	+ 0.9854
0	+ 10.69	0	+ .500	0	− 0.0151		+ 0.8660	0	+ 0.3600	+ 0.1000	0	0	− 0.1370	0	0	+ 1.000

347

Table 1A (Continued)

(b) (Continued)

	VERTICAL EDGE LOAD				HORIZONTAL EDGE LOAD				SHEAR EDGE LOAD				EDGE MOMENT LOAD			
ϕ	T_x (1)	S (2)	T_ϕ (3)	M_ϕ (4)	T_x (5)	S (6)	T_ϕ (7)	M_ϕ (8)	T_x (9)	S (10)	T_ϕ (11)	M_ϕ (12)	T_x (13)	S (14)	T_ϕ (15)	M_ϕ (16)
$\phi_k=35$:																
35	−3.430	0	−2.959	−0.3703	0.0061	0	+0.9986	+0.1754	−0.0774	0	−0.0243	−0.0015	0.0310	0	0.0051	+0.9642
30	−3.220	−0.921	−2.820	−0.3594	0.0043	−0.0068	+0.9951	+0.1717	−0.0694	−0.0205	−0.0213	−0.0014	0.0211	−0.0076	0.0037	+0.9650
20	−1.538	−2.302	−1.812	−0.2781	0.0060	−0.0136	+0.9663	+0.1425	−0.0054	−0.0439	−0.0015	−0.0008	0.0373	−0.0043	0.0040	+0.9710
10	1.822	−2.301	−0.330	−0.1444	0.0080	+0.0041	+0.9071	+0.0848	−0.1212	−0.0149	+0.0185	−0.0002	0.0469	+0.0251	0.0045	+0.9825
0	6.845	0	+0.574	0	0.0316		+0.8192	0	+0.3076	+0.1000		0	0.2051	0	0	+1.000
$\phi_k=40$:																
40	−2.210	0	−2.534	−0.4048	0.0165	0	+0.9954	+0.2247	−0.0675	0	−0.0279	−0.0022	0.0712	0	0.0184	+0.9530
30	−1.806	−1.138	−2.168	−0.3683	0.0049	−0.0068	+0.9836	+0.2041	−0.0461	−0.0331	−0.0179	−0.0018	0.0196	−0.0293	0.0074	+0.9570
20	−0.578	−1.830	−1.205	−0.2691	0.0176	−0.0032	+0.9420	+0.1672	−0.0181	−0.0423	+0.0054	−0.0009	0.0815	−0.0118	0.0112	+0.9653
10	1.516	−1.614	−0.045	−0.1352	0.0156	+0.0081	+0.8685	+0.1047	+0.1236	−0.0057	+0.0221	−0.0002	0.0724	+0.0998	0.0106	+0.9798
0	4.534	0	+0.643	0	0.0631		+0.7660	0	+0.2682	+0.1000		0	0.3099	0	0	+1.000
$\phi_k=45$:																
45	−1.602	0	−2.231	−0.4355	0.0359	0	+0.9873	+0.2779	−0.0596	0	−0.0315	−0.0031	0.1280	0	0.0437	0.9392
40	−1.546	−0.434	−2.163	−0.4274	0.0309	−0.0094	+0.9849	+0.2743	−0.0559	−0.0160	−0.0292	−0.0030	0.1102	−0.0325	0.0385	−0.9402
30	−1.098	−1.180	−1.646	−0.3654	0.0033	−0.0181	+0.9642	+0.2461	−0.0258	−0.0398	−0.0123	−0.0022	0.0128	−0.0642	0.0047	−0.9469
20	−0.166	−1.550	−0.763	−0.2562	0.0412	−0.0050	+0.9138	+0.1899	+0.0338	−0.0389	+0.0120	−0.0010	0.1523	−0.0158	0.0277	−0.9595
10	1.310	−1.263	+0.179	−0.1255	0.0260	+0.1175	+0.8251	+0.1072	+0.1222	+0.0025	+0.0250	−0.0002	0.0981	+0.0680	0.0212	+0.9771
0	3.407	0	+0.707	0	0.1185		+0.7071	0	+0.2378	+0.1000		0	0.4613	0	0	+1.000
$\phi_k=50$:																
50	−1.051	0	−1.931	−0.4541	0.0693	0	+0.9696	+0.3336	−0.0531	0	−0.0350	−0.0042	0.2048	0	0.0885	+0.9218
40	−0.948	−0.558	−1.741	−0.4265	0.0213	−0.0326	+0.9650	+0.3198	−0.0423	−0.0271	−0.0268	−0.0037	0.1192	−0.0964	0.0573	+0.9258
30	−0.619	−0.999	−1.208	−0.3492	0.0284	−0.0366	+0.9430	+0.2786	−0.0098	−0.0424	−0.0057	−0.0025	0.0849	−0.1086	0.0111	+0.9371
20	0.001	−1.184	−0.451	−0.2379	0.0820	−0.0039	+0.8851	+0.2104	+0.0440	−0.0340	+0.0181	−0.0010	0.2303	−0.0099	0.0576	+0.9537
10	0.998	−0.930	+0.317	−0.1141	0.0374	+0.0357	+0.7738	+0.1167	+0.1188	+0.0077	+0.0275	−0.0002	0.1176	+0.1119	0.0373	+0.9745
0	2.488	0	+0.766	0	0.2047		+0.6428	0	+0.2138	+0.1000		0	0.6695	0	0	+1.000

(c) $r/l = 100$ and $r/l = 0.2$

	VERTICAL EDGE LOAD				HORIZONTAL EDGE LOAD				SHEAR EDGE LOAD				EDGE MOMENT LOAD			
ϕ	T_x (1)	S (2)	T_ϕ (3)	M_ϕ (4)	T_x (5)	S (6)	T_ϕ (7)	M_ϕ (8)	T_x (9)	S (10)	T_ϕ (11)	M_ϕ (12)	T_x (13)	S (14)	T_ϕ (15)	M_ϕ (16)
$\phi_k=30$:																
30	−5.160	0	−3.420	−0.3028	0.0742	0	+0.9815	+0.1219	−0.1946	0	−0.0414	−0.0005	0.549	0	0.0813	+0.9009
20	−3.440	−2.531	−2.632	−0.2565	0.0123	−0.0292	+0.9757	+0.1084	−0.0755	−0.0776	−0.0145	0	0.073	−0.2142	0.0153	+0.9124
10	−1.763	−3.228	−0.787	−0.1395	0.0700	−0.0087	+0.9427	+0.0678	−0.0208	−0.0469	+0.0287	+0.0009	0.557	−0.0429	0.0568	+0.9456
0	10.92	0	+0.500	0	0.1140		+0.8660	0	+0.7119	+0.2000		0	1.067	0	0	+1.000

Table 1A (Continued)

ϕ	Vertical Edge Load				Horizontal Edge Load				Shear Edge Load				Edge Moment Load			
	T_x (1)	S (2)	T_ϕ (3)	M_ϕ (4)	T_x (5)	S (6)	T_ϕ (7)	M_ϕ (8)	T_x (9)	S (10)	T_ϕ (11)	M_ϕ (12)	T_x (13)	S (14)	T_ϕ (15)	M_ϕ (16)
$\phi_k=35$:																
35	-3.087	0	-2.871	-0.3347	-0.1552	0	+0.9603	+0.1591	-0.1487	0	-0.0458	-0.0024	0.912	0	-0.1914	+0.8637
30	-2.939	-0.833	-2.744	-0.3248	-0.1260	-0.0399	+0.9622	+0.1559	-0.1334	-0.0394	-0.0401	-0.0022	0.742	0.2343	-0.1585	+0.8669
20	-1.639	-2.156	-1.807	-0.2514	-0.0561	-0.0634	+0.9649	+0.1301	-0.0106	-0.0845	-0.0018	-0.0012	0.327	0.3747	+0.0315	+0.8914
10	+1.501	-2.296	-0.365	-0.1302	+0.1526	+0.0074	+0.9224	+0.0781	-0.2378	-0.0280	+0.0369	-0.0001	0.928	0.0486	+0.1212	+0.9362
0	+7.434		+0.574	0	0.3244	0	+0.8192	0	+0.6159	+0.2000	0	0	2.045	0	0	+1.000
$\phi_k=40$:																
40	-1.764	0	-2.398	-0.3556	-0.3096	0	+0.9087	+0.1962	-0.1273	-0.0626	-0.0525	-0.0035	1.421	0	-0.3885	+0.8156
30	-1.640	-0.947	-2.105	-0.3239	-0.1336	-0.1389	+0.9351	+0.1845	-0.0880	-0.0817	-0.0335	-0.0029	0.631	0.6461	-0.2007	+0.8306
20	-0.965	-1.697	-1.251	-0.2374	-0.2243	-0.1133	+0.9680	+0.1487	+0.0323	-0.0111	+0.0109	-0.0013	1.015	0.5396	-0.1574	+0.8707
10	+1.111	-1.753	-0.109	-0.1188	+0.2753	+0.0557	+0.9101	+0.0870	+0.2408		+0.0434	-0.0007	1.316	0.2523	+0.2259	+0.9271
0	+5.836		+0.643	0	0.7210	0	+0.7660	0	+0.5460	+0.2000	0	0	3.452	0	0	+1.000
$\phi_k=45$:																
45	-0.775	0	-1.934	-0.3619	-0.5679	0	+0.7941	+0.2293	-0.1067	-0.0287	-0.0577	-0.0050	2.074	0	-0.7365	+0.7509
40	-0.820	-0.217	-1.897	-0.3556	+0.4987	-0.1493	+0.8118	+0.2269	-0.0988	-0.0726	-0.0535	-0.0048	1.828	0.5460	-0.6582	+0.7558
30	-1.062	-0.731	-1.584	-0.3068	+0.0210	-0.3078	+0.9229	+0.2071	-0.0506	-0.0737	-0.0228	-0.0034	0.116	1.138	-0.1410	+0.7920
20	-1.244	-1.317	-0.910	-0.2175	+0.5381	-0.1552	+1.008	+0.1641	+0.0570	+0.0028	-0.0224	-0.0014	1.958	0.5987	+0.3936	+0.8510
10	+0.700	-1.478	-0.039	-0.1073	+0.4245	+0.1611	+0.9154	+0.0946	+0.2333		+0.0482	0	1.641	0.5811	+0.3778	+0.9196
0	+5.344		+0.707	0	1.392	0	+0.7071	0	+0.4995	+0.2000	0	0	5.299	0	0	+1.000
$\phi_k=50$:																
50	+0.052	0	-1.437	-0.3493	-0.9469	0	+0.5813	+0.2514	-0.0818	-0.0439	-0.0601	-0.0065	2.814	0	-1.245	+0.6613
40	+0.291	-0.028	-1.413	-0.3315	+0.5827	-0.4585	+0.7040	+0.2452	-0.0714	-0.0723	-0.0468	-0.0056	1.762	1.381	-0.8336	+0.6869
30	-1.016	-0.383	-1.246	-0.2786	+0.2885	-0.5526	+0.9656	+0.2228	-0.0258	-0.0639	-0.0113	-0.0035	0.797	1.690	-0.0859	+0.7523
20	-1.244	-1.059	-0.745	-0.1953	+0.1015	-0.1620	+1.112	+0.1764	+0.0664	+0.0125	+0.0312	-0.0013	3.038	0.536	+0.7607	+0.8338
10	+0.381	-1.427	+0.110	-0.0943	+0.5624	+0.3610	+0.9471	+0.1015	+0.2257		+0.0513	0	1.811	1.083	+0.5718	+0.9154
0	+5.670		+0.766	0	2.398	0	+0.6428	0	+0.4748	+0.2000	0	0	7.473	0	0	+1.000

(c) (Continued)

(d) $r/l = 100$ and $r/l = 0.3$

ϕ	T_x (1)	S (2)	T_ϕ (3)	M_ϕ (4)	T_x (5)	S (6)	T_ϕ (7)	M_ϕ (8)	T_x (9)	S (10)	T_ϕ (11)	M_ϕ (12)	T_x (13)	S (14)	T_ϕ (15)	M_ϕ (16)
$\phi_k=30$:																
30	-4.709	0	-3.362	-0.2769	-0.2662	0	+0.9571	+0.1099	-0.2618	0	-0.0596	-0.0024	2.101	0	-0.3444	+0.7859
20	-3.568	-2.381	-2.632	-0.2347	+0.0094	-0.0961	+0.9752	+0.0482	-0.1182	-0.1174	-0.0231	-0.0016	0.108	0.7669	-0.0802	+0.8119
10	+1.053	-3.290	-0.837	-0.1275	+0.2942	+0.0035	+0.9639	+0.0622	-0.3251	-0.0750	+0.0421	-0.0003	2.372	0.0141	+0.1945	+0.8845
0	+12.43		+0.500	0	0.6606	0	+0.8660	0	+1.102	+0.3000	0	0	5.456	0	0	+1.000

Table 1A (Continued)

φ	Vertical Edge Load T_x (1)	S (2)	T_ϕ (3)	M_ϕ (4)	Horizontal Edge Load T_x (5)	S (6)	T_ϕ (7)	M_ϕ (8)	Shear Edge Load T_z (9)	S (10)	T_ϕ (11)	M_ϕ (12)	Edge Moment Load T_z (13)	S (14)	T_ϕ (15)	M_ϕ (16)
								(d) (Continued)								
φk=35:																
35	-2.156	0	-2.682	-0.2903	-0.6039	0	+0.8706	+0.1362	-0.2140	0	-0.0685	-0.0036	3.493	0	-0.7613	+0.7084
30	-2.203	-0.595	-2.591	-0.2821	-0.4823	-0.1544	+0.8892	+0.1338	-0.1937	-0.0568	-0.0603	-0.0033	2.808	-0.8945	+0.6319	+0.7161
20	-2.106	-1.817	-1.830	-0.2201	-0.2698	-0.2300	+0.9859	+0.1137	-0.1244	-0.1248	-0.0046	-0.0018	1.506	-1.358	+0.7566	+0.7726
10	+0.513	-2.456	-0.482	-0.1148	+0.6242	+0.0755	+0.9800	+0.0698	+0.3456	-0.0471	+0.0535	-0.0002	3.764	+0.4236	+0.4372	+0.8689
0	+10.03		+0.574	0	+1.543	0	+0.8192	0	+0.9676	+0.3000	0	0	9.403	0	0	+1.000
φk=40:																
40	-0.216	0	-1.977	-0.2853	-1.176	0	+0.6733	+0.1553	-0.1668	0	-0.0736	-0.0048	5.193	0	-1.469	+0.6071
30	-1.020	-0.275	-1.892	-0.2631	-0.1865	-0.5149	+0.8208	+0.1488	-0.1238	-0.0837	-0.0485	-0.0039	2.222	-2.288	-0.7475	+0.6444
20	-2.210	-1.203	-1.430	-0.1982	+0.9141	-0.3958	+1.069	+0.1252	+0.0263	-0.1164	+0.0125	-0.0018	3.993	-1.809	+0.5716	+0.7382
10	-0.295	-2.141	-0.338	-0.1018	+1.073	+0.2676	+1.040	+0.0762	+0.3397	-0.0252	+0.0611	-0.0001	5.004	+1.180	+0.7566	+0.8572
0	+10.01		+0.643	0	+3.068	0	+0.7660	0	+0.8934	+0.3000	0	0	14.38	0	0	+1.000
φk=45:																
45	+1.400	0	-1.178	-0.2565	-1.970	0	+0.3055	+0.1591	-0.1142	0	+0.0719	-0.0059	6.814	0	-2.463	+0.4746
40	+1.083	+0.354	-1.224	-0.2537	-1.728	-0.5178	+0.3766	+0.1589	-0.1112	-0.0310	+0.0674	-0.0056	6.032	-1.802	-2.262	+0.4871
30	+1.040	+0.432	-1.454	-0.2296	+0.0472	-1.061	+0.8385	+0.1550	-0.0776	-0.0851	-0.0328	-0.0040	0.316	-3.743	+0.4779	+0.5768
20	-3.110	-0.788	-1.326	-0.1743	+1.941	-0.5051	+1.279	+0.1338	+0.0382	-0.1013	-0.0241	-0.0016	6.770	-1.867	+1.312	+0.7124
10	-0.946	-2.214	-0.330	-0.0899	+1.513	+0.6328	+1.158	+0.0821	+0.3209	-0.0128	+0.0645	-0.0001	5.696	+2.230	+1.229	+0.8510
0	+11.24		+0.707	0	+5.257	0	+0.7071	0	+0.8743	+0.3000	0	0	19.61	0	0	+1.000
φk=50:																
50	+2.605	0	-0.326	-0.2300	-2.821	0	-0.2407	+0.1409	-0.0565	0	-0.0611	-0.0063	7.871	0	-3.553	+0.3153
40	+1.276	+1.180	-0.676	-0.2032	-1.744	-1.346	-0.1577	+0.1474	-0.0651	-0.0327	-0.0514	-0.0056	4.994	-3.782	-2.394	+0.3754
30	+1.837	+1.062	-1.342	-0.1926	+0.8732	-1.624	+1.033	+0.1561	-0.0643	-0.0700	-0.0208	-0.0036	2.207	-4.666	+0.2204	+0.5257
20	-4.099	-0.690	-1.432	-0.1527	+3.129	-0.4486	+1.623	+0.1422	+0.0214	-0.0879	+0.0276	-0.0013	8.917	-1.422	+2.171	+0.6995
10	-1.187	-2.497	-0.379	-0.0801	+1.766	+1.155	+1.315	+0.0885	+0.3005	-0.0113	+0.0639	+0.0002	5.588	+3.317	+1.025	+0.8508
0	+12.66		+0.766	0	+7.773	0	+0.6428	0	+0.8992	+0.3000	0	0	23.76	0	0	+1.000
								(e) r/l = 100 and r/l = 0.4								
φk=30:																
30	-3.508	0	-3.175	-0.2404	-0.736	0	+0.881	+0.0947	-0.3318	0	-0.0772	-0.0028	5.33	0	-0.879	+0.6558
20	-3.500	-1.942	-2.592	-0.2048	-0.038	-0.2682	+0.958	+0.0856	-0.1568	-0.1503	-0.0304	-0.0019	0.14	-1.920	+0.196	+0.6995
10	-0.244	-3.281	-0.947	-0.1121	+0.804	-0.0008	+1.008	+0.0544	-0.4153	-0.0998	+0.0545	-0.0003	6.27	+0.187	+0.502	+0.8169
0	+15.23		+0.500	0	+1.772	0	+0.866	0	+1.508	+0.4000	0	0	15.27	0	0	+1.000

Table 1A (Continued)

ϕ	Vertical Edge Load				Horizontal Edge Load				Shear Edge Load				Edge Moment Load			
	T_x	S	T_ϕ	M_ϕ	T_x	S	T_ϕ	M_ϕ	T_x	S	T_ϕ	M_ϕ	T_x	S	T_ϕ	M_ϕ
	(1)	(2)	(3)	(4)	(5)	(6)	(7)	(8)	(9)	(10)	(11)	(12)	(13)	(14)	(15)	(16)
$\phi_k=35$:																
35	−0.217	0	−2.261	−0.2351	−1.518	0	+0.670	+0.1091	−0.2518	0	−0.0842	−0.0040	−8.48	0	−1.890	+0.5364
30	−0.656	−0.102	−2.241	−0.2293	−1.214	−0.3878	+0.723	+0.1078	−0.2316	−0.0672	−0.0746	−0.0037	−6.82	−2.173	−1.574	+0.5500
20	−2.982	−1.086	−1.911	−0.1833	−0.680	−0.5777	+1.014	+0.0952	−0.0480	−0.1540	−0.0073	−0.0019	+3.69	−3.300	+0.248	+0.6463
10	−1.512	−2.715	−0.712	−0.0982	+1.596	+0.1982	+1.092	+0.0612	−0.4268	−0.0671	+0.0673	−0.0001	+9.37	+1.106	+1.064	+0.8000
0	+15.11	0	+0	0	+3.990		+0.819	0	+1.373	−0.4000		0	+23.99	0	0	+1.000
$\phi_k=40$:																
40	+2.485	0	−1.221	−0.2044	−2.664	0	+0.250	+0.1088	−0.1542	0	−0.0818	−0.0049	−11.15	0	−3.211	+0.3896
30	+0.117	+0.913	−1.512	−0.1948	+1.123	−1.171	+0.606	+0.1093	−0.1315	−0.0851	−0.0567	−0.0039	−4.83	−4.928	−0.164	+0.4541
20	−4.296	−0.278	−1.730	−0.1580	+2.058	−0.9139	+1.236	+0.1013	+0.0023	−0.1351	+0.0099	−0.0017	+8.71	−3.904	+1.250	+0.6098
10	−2.814	−2.749	−0.749	−0.0862	+2.495	+0.6036	+1.272	+0.0665	−0.4228	−0.0471	+0.0728	0	+11.30	+2.733	+1.764	+0.7918
0	+17.20	0	+0.643	0	+7.122		+0.766	0	+1.368	−0.4000		0	+33.39	0	0	+1.000
$\phi_k=45$:																
45	+4.329	0	−0.119	−0.1513	+3.817	0	−0.372	+0.0899	−0.0653	0	−0.0657	−0.0051	−12.50	0	−4.647	+0.2210
40	+3.663	+1.126	−0.277	−0.1526	+3.361	−1.005	−0.230	+0.0923	−0.0723	−0.0186	−0.0630	−0.0049	−11.09	−3.299	−4.168	+0.2412
30	+0.918	+2.029	−1.249	−0.1564	+0.150	−2.079	−0.703	+0.1060	−0.1030	−0.0673	−0.0384	−0.0035	+0.92	−6.970	+0.971	+0.3846
20	−6.034	−0.008	−1.889	−0.1377	+3.796	−1.015	+1.638	+0.1083	−0.0344	−0.1142	+0.0158	−0.0013	+12.56	+3.637	+2.441	+0.5941
10	−3.355	−3.194	−0.854	−0.0780	+3.104	+1.260	+1.497	+0.0731	+0.3651	−0.0449	+0.0716	+0.0003	+11.51	+4.262	+2.363	+0.7936
0	+19.64	0	+0.707	0	−10.78		+0.707	0	+1.388	+0.4000		0	+39.59	0	0	+1.000
$\phi_k=50$:																
50	+4.973	0	+0.784	−0.0881	−4.477	0	−1.043	+0.0545	+0.0216	0	−0.0395	−0.0045	−11.45	0	−5.493	+0.0676
40	+2.829	+2.328	+0.079	−0.1028	+2.858	−2.155	−0.394	+0.0716	−0.0276	+0.0025	−0.0408	−0.0041	−7.65	−4.182	−3.773	+0.1539
30	+2.433	+2.516	−1.399	−0.1279	+1.235	−2.673	+1.065	+0.1057	−0.1165	−0.0382	−0.0313	−0.0029	+2.61	−3.745	+0.216	+0.3594
20	+6.881	−0.213	−2.118	−0.1244	+5.125	−0.8202	+2.120	+0.1184	−0.0779	−0.1025	+0.0109	−0.0010	+13.82	+1.157	+3.391	+0.5981
10	−3.049	−3.540	−0.904	−0.0724	+3.212	+1.909	+1.707	+0.0810	+0.3428	−0.0541	+0.0667	+0.0003	+10.19	+5.359	+2.668	+0.7998
0	+20.35	0	+0.766	0	−13.59		+0.643	0	+1.459	+0.4000		0	+41.80	0	0	+1.000

(e) (Continued)

ϕ	Vertical Edge Load				Horizontal Edge Load				Shear Edge Load				Edge Moment Load			
	T_x (1)	S (2)	T_ϕ (3)	M_ϕ (4)	T_x (5)	S (6)	T_ϕ (7)	M_ϕ (8)	T_x (9)	S (10)	T_ϕ (11)	M_ϕ (12)	T_x (13)	S (14)	T_ϕ (15)	M_ϕ (16)
$\phi_k=30$:																
30	−1.757	0	−2.879	−0.2019	−1.420	0	+0.767	+0.0789	−0.3849	0	−0.0912	−0.0030	−10.64	0	−1.779	+0.5185
20	−3.508	−1.323	−2.524	−0.1738	+0.037	−0.5108	+0.932	+0.0727	−0.1963	−0.1773	−0.0368	−0.0020	+0.48	−3.877	+0.421	+0.5825
10	−2.292	−3.378	−1.108	−0.0969	+1.616	+0.0381	+1.071	+0.0489	+0.4854	−0.1263	+0.0654	−0.0002	+12.53	+0.236	+0.998	+0.7480
0	+20.22	0	+0.500	0	+3.767		+0.866	0	+1.964	+0.5000		0	+30.35	0	0	+1.000

(f) $r/l = 100$ and $r/l = 0.5$

φ	VERTICAL EDGE LOAD				HORIZONTAL EDGE LOAD				SHEAR EDGE LOAD				EDGE MOMENT LOAD			
	T_x (1)	S (2)	T_ϕ (3)	M_ϕ (4)	T_x (5)	S (6)	T_ϕ (7)	M_ϕ (8)	T_x (9)	S (10)	T_ϕ (11)	M_ϕ (12)	T_x (13)	S (14)	T_ϕ (15)	M_ϕ (16)
φk=35:																
35	+2.471	0	−1.677	−0.1788	−2.776	0	+0.395	+0.0816	−0.2558	0	−0.0962	−0.0039	−14.99	0	−3.345	+0.8692
30	+1.493	+0.587	−1.758	−0.1758	−2.223	−0.7101	+0.495	+0.0816	−0.2428	−0.0837	−0.0875	−0.0037	−12.12	−4.668	−2.786	+0.3888
20	−4.209	−0.060	−2.002	−0.1480	−1.246	−1.062	+1.056	+0.0775	−0.0879	−0.1706	−0.0227	−0.0018	−6.40	−5.907	−0.453	+0.5266
10	−4.426	−3.090	−1.035	−0.0837	−2.986	+0.3737	+1.252	+0.0535	−0.4733	−0.0921	+0.0629	+0.0000	+17.12	+1.978	+1.931	+0.7345
0	+22.54	0	+0.574	0	+7.555	0	+0.819	0	+1.867	+0.5000		0	+44.56	0	0	+1.000
φk=40:																
40	+5.472	0	−0.358	−0.1324	−4.282	0	−0.220	+0.0682	−0.1143	0	−0.0775	−0.0034	−17.17	0	−5.060	+0.2033
30	+1.404	+2.232	−1.069	−0.1354	−1.839	−1.889	+0.363	+0.0754	−0.1408	−0.0683	−0.0585	−0.0034	−7.86	−7.678	−2.645	+0.2912
20	+6.661	+0.757	−2.062	−0.1254	−3.336	−1.488	+1.415	+0.0819	−0.0853	−0.1408	+0.0027	−0.0014	−13.14	−6.372	−1.904	+0.5001
10	−5.826	−3.503	−1.214	−0.0751	+4.196	+1.017	+1.528	+0.0593	+0.4218	−0.0802	+0.0780	+0.0003	+18.72	+4.129	+2.821	+0.7347
0	+26.12	0	+0.643	0	−12.17	0	+0.766	0	+1.908	+0.5000		0	+54.72	0	0	+1.000

(f) (Continued)

φ	T_x (1)	S (2)	T_ϕ (3)	M_ϕ (4)	T_x (5)	S (6)	T_ϕ (7)	M_ϕ (8)	T_x (9)	S (10)	T_ϕ (11)	M_ϕ (12)	T_x (13)	S (14)	T_ϕ (15)	M_ϕ (16)
φk=45:																
45	+6.655	0	+0.771	−0.0750	−4.823	0	−0.923	+0.0405	+0.0153	0	−0.0477	−0.0038	−15.73	0	−6.128	+0.0523
40	+5.749	+1.741	+0.524	−0.0793	−4.625	−1.375	−0.727	+0.0448	−0.0054	+0.0023	−0.0480	−0.0037	−14.17	−4.167	−5.525	+0.0765
30	+0.652	+3.364	−1.053	−0.1039	+0.365	−2.002	+0.573	+0.0709	−0.1266	−0.0318	−0.0417	−0.0028	−2.11	−9.122	−1.418	+0.2487
20	−8.443	−0.730	−2.362	−0.1125	−5.266	−1.497	+1.929	+0.0901	−0.1395	−0.1180	+0.0012	−0.0011	+16.12	−5.253	+3.200	+0.5015
10	−5.738	−4.033	−1.326	−0.0708	+4.705	+1.781	+1.798	+0.0670	+0.3807	−0.0882	+0.0726	+0.0003	+17.30	+5.658	+3.294	+0.7426
0	+27.76	0	+0.707	0	−16.12	0	+0.707	0	+2.002	+0.5000		0	+58.57	0	0	+1.000
φk=50:																
50	+6.074	0	+1.416	−0.0253	−5.122	0	−1.462	+0.0088	+0.1028	0	−0.0139	−0.0027	−11.38	0	−6.094	−0.0520
40	+3.738	+2.900	+0.542	−0.0469	−3.478	−2.507	−0.713	+0.0302	+0.0147	+0.0398	−0.0269	−0.0027	−8.59	−5.742	−4.363	+0.0339
30	+2.427	+3.394	−1.371	−0.0900	+1.103	−3.275	+1.025	+0.0759	−0.1664	−0.0015	−0.0410	−0.0022	+0.92	−8.235	−0.093	+0.2533
20	−8.536	+0.212	−2.512	−0.1065	−6.245	−1.164	+2.385	+0.1024	−0.1868	−0.1144	−0.0079	−0.0010	+15.24	−3.756	+3.776	+0.5154
10	−4.760	−4.221	−1.261	−0.0672	+4.596	+2.387	+1.966	+0.0752	+0.3719	−0.1018	+0.0671	+0.0003	+14.96	+6.079	+3.305	+0.7497
0	+26.68	0	+0.766	0	−18.51	0	+0.643	0	+2.065	+0.5000		0	+56.51	0	0	+1.000

(g) $r/l = 100$ and $r/l = 0.6$

φ	T_x (1)	S (2)	T_ϕ (3)	M_ϕ (4)	T_x (5)	S (6)	T_ϕ (7)	M_ϕ (8)	T_x (9)	S (10)	T_ϕ (11)	M_ϕ (12)	T_x (13)	S (14)	T_ϕ (15)	M_ϕ (16)
φk=30:																
30	+0.690	0	−2.477	−0.1646	−2.375	0	+0.609	+0.0636	−0.4130	0	−0.1015	−0.0031	−17.24	0	−2.909	+0.3912
20	+3.149	−0.438	−2.432	−0.1447	+0.079	−0.8580	+0.895	+0.0605	−0.2349	−0.1448	−0.0424	−0.0010	−0.99	−6.327	−0.707	+0.4746
10	−5.088	−3.429	−1.334	−0.0837	+2.720	+0.0560	+1.160	+0.0431	+0.5284	−0.1530	+0.0739	−0.0001	+16.89	+0.297	+1.631	+0.6844
0	+26.76	0	+0.500	0	+6.379	0	+0.866	0	+2.481	+0.6000		0	+50.48	0	0	+1.000

Table 1A (Continued)

φ	Vertical Edge Load				Horizontal Edge Load				Shear Edge Load				Edge Moment Load			
	T_x (1)	S (2)	T_ϕ (3)	M_ϕ (4)	T_x (5)	S (6)	T_ϕ (7)	M_ϕ (8)	T_x (9)	S (10)	T_ϕ (11)	M_ϕ (12)	T_x (13)	S (14)	T_ϕ (15)	M_ϕ (16)

(g) (Continued)

φ	(1)	(2)	(3)	(4)	(5)	(6)	(7)	(8)	(9)	(10)	(11)	(12)	(13)	(14)	(15)	(16)
φk=35:																
35	+5.530	0	+0.999	−0.1294	−4.187	0	+0.085	+0.0577	−0.2245	0	−0.0937	−0.0036	−21.54	0	−4.900	+0.2286
30	+3.964	+1.371	−1.194	−0.1291	+3.368	−1.072	+0.231	+0.0589	−0.2258	−0.0617	−0.0849	−0.0034	−17.56	−5.539	−4.097	+0.2534
20	+5.521	+1.142	−2.096	−0.1187	+1.835	−1.625	+1.098	+0.0628	−0.1434	−0.1738	−0.0163	−0.0014	−8.78	−8.709	+0.609	+0.4257
10	−7.802	−3.492	−1.452	−0.0731	+4.617	+0.5579	+1.438	+0.0477	−0.4825	−0.1215	+0.0821	+0.0005	+25.70	+2.778	+2.864	+0.6789
0	+31.29	0	+0.574	0	−11.74	0	+0.891	0	+2.453	+0.6000	0	0	+67.87	0	0	+1.000
φk=40:																
40	+8.029	0	+0.412	−0.0781	−5.611	0	+0.632	+0.0378	−0.0378	0	−0.0653	−0.0034	−20.91	0	−6.413	+0.0741
30	+2.640	+3.389	−0.660	−0.0907	+2.518	−2.498	+0.140	+0.0500	−0.1275	−0.0385	−0.0561	−0.0020	−10.34	−9.508	−3.449	+0.1748
20	+8.598	+1.754	−2.347	−0.1015	+4.328	−2.036	+1.563	+0.0673	−0.1799	−0.1354	−0.0078	−0.0012	+16.02	−8.383	+2.316	+0.4147
10	−8.783	−4.118	−1.651	−0.0674	+5.859	+1.341	+1.421	+0.0538	−0.4136	−0.1192	+0.0787	+0.0004	+25.53	+5.071	+3.686	+0.6842
0	+34.71	0	+0.643	0	−17.02	0	+0.766	0	+2.558	+0.6000	0	0	+75.81	0	0	+1.000
φk=45:																
45	+7.904	0	+1.346	−0.0293	−5.855	0	−1.258	+0.0115	+0.1009	0	−0.0267	−0.0026	−15.56	0	−6.655	−0.0364
40	+6.935	+2.078	+1.051	−0.0352	−5.259	−1.550	−1.038	+0.0166	+0.0669	+0.0245	−0.0302	−0.0025	−14.36	−4.156	−6.057	−0.0122
30	+0.196	+4.195	−0.884	−0.0712	+0.727	−3.373	+0.459	+0.0490	−0.1460	+0.0082	−0.0439	−0.0022	−3.96	−9.685	+1.817	−0.1633
20	−9.811	+1.352	−2.657	−0.0959	+6.025	−1.905	+2.094	+0.0775	−0.2510	−0.1174	−0.0148	−0.0010	+16.43	+6.537	+3.383	+0.4285
10	−7.893	−4.563	−1.682	−0.0659	+6.150	+2.089	+2.020	+0.0623	−0.3813	−0.1339	+0.0714	+0.0004	+22.53	+6.087	+3.873	+0.6947
0	+34.56	0	+0.707	0	−20.64	0	+0.707	0	+2.659	+0.6000	0	0	+74.68	0	0	+1.000
φk=50:																
50	+6.029	0	+1.647	+0.0038	−4.847	0	−1.571	−0.0114	+0.1678	0	+0.0073	−0.0015	−8.26	0	−5.666	−0.0950
40	+4.096	+2.955	+0.283	−0.0192	+3.656	−2.446	−0.851	+0.0100	+0.0523	+0.0699	−0.0151	−0.0017	−8.00	−4.483	−4.344	−0.0164
30	+1.803	+3.754	−1.264	−0.0627	+0.458	−3.469	+0.919	+0.0574	−0.2030	+0.0305	−0.0491	−0.0018	+1.93	−7.812	+0.623	+0.1835
20	−9.145	+0.659	−2.736	−0.0932	+6.530	−1.539	+2.469	+0.0897	−0.2862	−0.1225	−0.0245	−0.0010	+13.86	+5.044	+3.549	+0.4425
10	−6.445	−4.528	−1.501	−0.0624	+6.010	+2.570	+2.141	+0.0696	−0.3899	−0.1463	+0.0675	+0.0003	+20.05	+5.942	+3.658	+0.6986
0	+31.83	0	+0.766	0	−22.64	0	+0.643	0	+2.688	+0.6000	0	0	+70.37	0	0	+1.000

(h) r/l = 200 and r/l = 0.1

φ	(1)	(2)	(3)	(4)	(5)	(6)	(7)	(8)	(9)	(10)	(11)	(12)	(13)	(14)	(15)	(16)
φk=30:																
30	−5.388	0	−3.482	−0.3278	+0.0129	−0.0043	+0.9979	+0.1310	−0.0905	0	−0.0207	−0.0009	+0.1005	0	−0.0164	+0.9730
20	−3.600	−2.628	−2.666	−0.2779	+0.0014	+0.0020	+0.9845	+0.1162	−0.0397	−0.0403	−0.0080	−0.0006	+0.0123	−0.0332	−0.0024	+0.9761
10	+1.787	−3.290	−0.781	−0.1511	+0.0164		+0.9410	+0.0722	+0.1117	−0.0251	+0.0142	−0.0002	+0.1340	+0.0177	+0.0099	+0.9852
0	+10.83	0	+0.500	0	+0.0447		+0.8660	0	+0.3604	+0.1000	0	0	+0.3807	0	0	+1.000

353

Table 1A (Continued)

φ	Vertical Edge Load				Horizontal Edge Load				Shear Edge Load				Edge Moment Load			
	T_x (1)	S (2)	T_ϕ (3)	M_ϕ (4)	T_x (5)	S (6)	T_ϕ (7)	M_ϕ (8)	T_x (9)	S (10)	T_ϕ (11)	M_ϕ (12)	T_x (13)	S (14)	T_ϕ (15)	M_ϕ (16)
φ_k = 35:																
35	−3.373	0	−2.947	−0.3692	−0.0349	0	+0.9926	+0.1751	−0.0773	0	−0.0243	−0.0015	−0.2044	0	−0.0435	+0.9623
30	−3.175	−0.907	−2.810	−0.3584	−0.0272	−0.0089	+0.9901	+0.1715	−0.0693	−0.0205	−0.0212	−0.0014	−0.1601	−0.0520	−0.0336	+0.9632
20	−1.566	−2.282	−1.814	−0.2774	+0.0192	−0.0121	+0.9675	+0.1424	−0.0055	−0.0439	−0.0015	−0.0008	+0.1134	+0.0710	+0.0090	+0.9700
10	−1.763	−2.310	−1.635	−0.1441	+0.0378	+0.0075	+0.9107	+0.0832	+0.1211	−0.0150	+0.0185	−0.0002	+0.2305	+0.0464	+0.0256	+0.9823
0	+6.999		+0.574	0	−0.1053		+0.8192		+0.3082	+0.1000		0	−0.6605	0	0	+1.000
φ_k = 40:																
40	−2.193	0	−2.526	−0.4041	−0.0791	0	+0.9782	+0.2237	−0.0670	0	−0.0278	−0.0022	−0.3607	0	−0.1000	+0.9479
30	−1.834	−1.137	−2.168	−0.3677	−0.0303	−0.0341	+0.9742	+0.2095	−0.0458	−0.0329	−0.0178	−0.0018	−0.1407	−0.1563	−0.0487	+0.9521
20	−0.678	−1.866	−1.220	−0.2690	+0.0667	−0.0239	+0.9491	+0.1669	+0.0177	−0.0425	+0.0054	−0.0009	+0.3060	+0.1102	+0.0428	+0.9635
10	+1.483	−1.699	−0.048	−0.1354	+0.0718	+0.0226	+0.8777	+0.0965	+0.1231	−0.0058	+0.0220	−0.0002	+0.3419	+0.1068	+0.0540	+0.9795
0	+4.960		+0.643	0	−0.2243		+0.7660		+0.2696	+0.1000		0	−1.083	0	0	+1.000
φ_k = 45:																
45	−1.415	0	−2.167	−0.4305	−0.1591	0	+0.9446	+0.2747	−0.0585	0	−0.0311	−0.0031	−0.5815	0	−0.2039	+0.9270
40	−1.384	−0.385	−2.106	−0.4226	−0.1386	−0.0418	+0.9469	+0.2713	−0.0549	−0.0127	−0.0288	−0.0030	−0.5076	−0.1526	−0.1817	+0.9284
30	−1.097	−1.081	−1.635	−0.3622	+0.0019	−0.0838	+0.9561	+0.2441	−0.0259	−0.0392	−0.0122	−0.0021	+0.0014	+0.3085	+0.0344	+0.9392
20	−0.350	−1.505	−0.798	−0.2550	+0.1623	−0.0353	+0.9371	+0.1892	+0.0326	−0.0387	+0.0118	−0.0010	+0.5987	+0.1323	+0.1135	+0.9566
10	+1.175	−1.323	+0.148	−0.1254	+0.1167	+0.0570	+0.8460	+0.1071	+0.1213	+0.0021	+0.0248	−0.0002	+0.4491	+0.2141	+0.0996	+0.9768
0	+3.887		+0.707	0	−0.4371		+0.7071		+0.2407	+0.1000		0	−1.684	0	0	+1.000
φ_k = 50:																
50	−0.838	0	−1.831	−0.4457	−0.2911	0	+0.8743	+0.3250	−0.0506	0	−0.0340	−0.0041	−0.8725	0	−0.3803	+0.8948
40	−0.842	−0.461	−1.678	−0.4195	−0.1746	−0.1378	+0.9021	+0.3125	−0.0408	−0.0260	−0.0262	−0.0036	−0.5294	−0.4145	−0.2512	+0.9030
30	−0.759	−0.908	−1.225	−0.3455	+0.1012	−0.1615	+0.9516	+0.2745	−0.0107	−0.0411	−0.0176	−0.0024	+0.2947	+0.4901	+0.0344	+0.9238
20	−0.314	−1.227	−0.517	−0.2369	+0.3240	−0.0358	+0.9439	+0.2092	+0.0369	−0.0337	+0.0174	−0.0010	+0.9856	+0.1131	+0.2362	+0.9497
10	+0.907	−1.111	+0.281	−0.1142	+0.1645	+0.1245	+0.8204	+0.1167	+0.1175	+0.0087	+0.0271	−0.0001	+0.5268	+0.3830	+0.1664	+0.9741
0	+3.402		+0.766	0	−0.7874		+0.6428		+0.2199	+0.1000		0	−2.487	0	0	+1.000

(h) (Continued)

(i) r/l = 200 and r/l = 0.2

φ	T_x (1)	S (2)	T_ϕ (3)	M_ϕ (4)	T_x (5)	S (6)	T_ϕ (7)	M_ϕ (8)	T_x (9)	S (10)	T_ϕ (11)	M_ϕ (12)	T_x (13)	S (14)	T_ϕ (15)	M_ϕ (16)
φ_k = 30:																
30	−4.804	0	−3.394	−0.3054	−0.2444	0	+0.9630	+0.1216	−0.1770	0	−0.0407	−0.0017	−1.972	0	−0.3643	+0.8904
20	−3.563	−2.414	−2.649	−0.2592	−0.0144	−0.0894	+0.9776	+0.1084	−0.0792	−0.0792	−0.0160	−0.0012	−0.154	−0.7298	−0.0618	+0.9051
10	+1.170	−3.280	−0.836	−0.1412	+0.2619	−0.0028	+0.9629	+0.0679	+0.2192	−0.0502	+0.0279	−0.0003	+2.142	+0.0427	+0.1925	+0.9435
0	+12.12		+0.500	0	−0.5594		+0.8660		+0.7297	+0.2000		0	−4.639	0	0	+1.000

354

Table 1A (Continued)

φ	VERTICAL EDGE LOAD				HORIZONTAL EDGE LOAD				SHEAR EDGE LOAD				EDGE MOMENT LOAD			
	T_x (1)	S (2)	T_ϕ (3)	M_ϕ (4)	T_x (5)	S (6)	T_ϕ (7)	M_ϕ (8)	T_x (9)	S (10)	T_ϕ (11)	M_ϕ (12)	T_x (13)	S (14)	T_ϕ (15)	M_ϕ (16)
(i) (Continued)																
$\phi_k=35$:																
35	−2.269	0	−2.720	−0.3304	−0.5645	0	+0.8832	+0.1560	−0.1455	0	−0.0466	−0.0027	−3.344	0	−0.7035	+0.8373
30	−2.290	−0.624	−2.624	−0.3211	−0.4523	−0.1446	+0.9000	+0.1531	−0.1313	−0.0386	−0.0410	−0.0025	−2.696	−0.8582	−0.5812	+0.8425
20	−2.035	−1.854	−1.855	−0.2507	−0.2416	−0.2183	+0.9868	+0.1291	−0.0150	−0.0842	−0.0033	−0.0014	+1.373	−1.319	+0.1200	+0.8787
10	+0.648	−2.423	−0.474	−0.1313	+0.5770	+0.0604	+0.9767	+0.0783	+0.2331	−0.0309	+0.0358	−0.0002	+3.519	+0.3350	+0.4303	+0.9335
0	+9.620	0	+0.574	0	−1.373	0	+0.8191	0	+0.6376	+0.2000			+8.444	0		+1.000
$\phi_k=40$:																
40	−0.324	0	−2.021	−0.3363	−1.133	0	+0.6921	+0.1847	−0.1144	0	−0.0503	−0.0037	−5.147	0	−1.422	+0.7549
30	−1.057	−0.319	−1.919	−0.3098	−0.4724	−0.4965	+0.8324	+0.1760	−0.0838	−0.0572	−0.0331	−0.0030	−2.213	−2.270	−0.7127	+0.7838
20	−2.100	−1.229	−1.425	−0.2331	+0.8653	−0.3869	+1.067	+0.1458	+0.0202	−0.0785	+0.0084	−0.0015	−3.880	−1.816	+0.5816	+0.8519
10	+0.178	−2.090	−0.327	−0.1202	+1.021	+0.2424	+1.036	+0.0873	+0.2295	−0.0158	+0.0411	−0.0002	+4.849	+1.085	+0.8003	+0.9255
0	+9.537	0	+0.643	0	−2.851	0	+0.7660	0	+0.5858	+0.2000			+13.55	0		+1.000
$\phi_k=45$:																
45	+1.358	0	−1.210	−0.3140	−1.967	0	+0.3169	+0.1073	−0.0793	0	−0.0494	−0.0047	−7.076	0	−2.500	+0.6266
40	+1.046	+0.344	−1.254	−0.3103	−1.725	−0.5171	+0.3877	+0.1967	−0.0769	−0.0215	−0.0462	−0.0045	−6.231	−1.862	−2.231	+0.6378
30	−1.033	−0.270	−1.469	−0.2789	+0.050	−0.6113	+0.8459	+0.1884	−0.0516	−0.0642	−0.0223	−0.0032	+0.316	−2.536	+0.5827	+0.7167
20	−3.035	−0.788	−1.324	−0.2104	+1.916	−0.5110	+1.281	+0.1588	+0.0289	−0.0681	+0.0165	−0.0014	+6.897	−1.941	+1.367	+0.8267
10	−0.885	−2.169	−0.327	−0.1087	+1.486	+0.6094	+1.159	+0.0955	+0.2164	−0.0071	+0.0436	−0.0001	+5.708	+2.201	+1.273	+0.9200
0	+10.91	0	−0.707	0	−5.085	0	+0.7071	0	+0.5724	+0.2000			+19.32	0		+1.000
$\phi_k=50$:																
50	+2.686	0	−0.312	−0.2588	−2.918	0	−0.2665	+0.1807	−0.0395	0	−0.0419	−0.0052	−8.502	0	−3.770	+0.4461
40	+1.320	+1.218	−0.671	−0.2562	−1.798	−1.391	+0.1445	+0.1856	−0.0445	−0.0226	−0.0351	−0.0046	−5.350	−4.060	−2.523	+0.5106
30	+1.860	+1.105	−1.355	−0.2385	+0.904	−1.675	+1.046	+0.1889	−0.0420	−0.0476	−0.0140	−0.0031	−2.433	−4.983	+0.2750	+0.6671
20	−4.145	−0.669	−1.455	−0.1867	+3.198	−0.4671	+1.649	+0.1655	+0.0168	−0.0385	+0.0187	−0.0013	+9.459	−1.527	+2.336	+0.8314
10	−1.190	−2.490	−0.395	−0.0979	+1.779	+1.160	+1.332	+0.0992	+0.2016	−0.0062	+0.0428	−0.0001	+5.734	+3.392	+1.737	+0.9393
0	+12.58	0	+0.766	0	−7.777	0	+0.6428	0	+0.5914	+0.2000			+24.30	0		+1.000
(j) $r/l = 200$ and $r/l = 0.3$																
$\phi_k=30$:																
30	−2.875	0	−3.097	−0.2688	−1.018	0	+0.8433	+0.1065	−0.2481	0	−0.0585	−0.0023	−8.01	0	−1.251	+0.7583
20	−3.544	−1.727	−2.595	−0.2300	+0.025	−0.3675	+0.9558	+0.0961	−0.1187	−0.0759	−0.0236	−0.0016	−0.32	−2.906	−0.242	+0.7953
10	−1.006	−3.351	−1.021	−0.1270	+1.142	+0.0251	+1.038	+0.0619	+0.3098	−0.1126	+0.0402	−0.0003	+9.20	+0.158	+0.787	+0.8828
0	+17.18	0	+0.500	0	−2.618	0	+0.8660	0	+1.138	+0.3000			+21.66	0		+1.000

355

Table 1A (Continued)

(j) (Continued)

ϕ	Vertical Edge Load				Horizontal Edge Load				Shear Edge Load				Edge Moment Load			
	T_z (1)	S (2)	T_ϕ (3)	M_ϕ (4)	T_z (5)	S (6)	T_ϕ (7)	M_ϕ (8)	T_z (9)	S (10)	T_ϕ (11)	M_ϕ (12)	T_z (13)	S (14)	T_ϕ (15)	M_ϕ (16)
$\phi_k=35$:																
35	+1.157	0	−2.003	−0.2639	−3.540	0	+0.5377	+0.1233	−0.1777	0	−0.0616	−0.0033	−12.73	0	−2.708	+0.6320
30	+0.434	+0.250	−2.034	−0.2580	−1.763	−0.5643	+0.6157	+0.1220	−0.1650	−0.0475	−0.0548	−0.0031	−10.23	−3.260	−2.236	+0.6464
20	−3.630	−0.571	−1.989	−0.2098	−1.989	−0.8379	+1.051	+0.1087	−0.0418	−0.1115	−0.0066	−0.0017	+5.57	−4.934	+0.474	+0.7427
10	−2.937	−2.907	−0.903	−0.1149	+2.318	+0.2927	+1.184	+0.0698	+0.3091	−0.0522	+0.0487	−0.0002	+13.92	+1.650	+1.650	+0.8689
0	+18.66	0	+0.574	0	+5.780	0	+0.8192		+1.058	+0.3000		0	+35.32	0	0	+1.000
$\phi_k=40$:																
40	+4.491	0	−0.695	−0.2230	−4.689	0	−0.0553	+0.1198	−0.0958	0	−0.0550	−0.0039	−16.80	0	−4.723	+0.4470
30	+0.927	+1.787	−1.263	−0.2170	−1.599	−1.680	+0.4612	+0.1225	−0.0958	−0.0529	−0.0397	−0.0032	−7.30	−7.430	−2.330	+0.5253
20	−5.904	+0.374	−1.979	−0.1836	−2.992	−1.296	+1.378	+0.1168	−0.0298	−0.0936	−0.0040	−0.0016	+12.97	−5.929	+1.932	+0.6997
10	−4.643	−3.248	−1.057	−0.1038	+3.576	+0.8937	+1.451	+0.0774	+0.2786	−0.0416	+0.0506	−0.0001	+16.75	+3.932	+2.673	+0.8621
0	+22.55	0	+0.643	0	−10.29	0	+0.7660	0	+1.069	+0.3000	0	0	+48.57	0	0	+1.000
$\phi_k=45$:																
45	+6.463	0	+0.608	−0.1511	−5.443	0	−0.8614	+0.0899	−0.0114	0	−0.0370	−0.0038	−17.92	0	−6.551	+0.2280
40	+5.541	+1.688	+0.369	−0.1543	−4.617	−1.379	−0.6652	+0.0937	−0.0214	−0.0039	−0.0364	−0.0037	−15.90	−4.718	−5.866	+0.2542
30	+0.843	+3.190	−1.138	−0.1700	+0.203	−2.859	−0.6257	+0.1158	−0.0772	−0.0304	−0.0273	−0.0028	+1.34	−9.994	+1.292	+0.4354
20	−8.160	+0.555	−2.324	−0.1610	+5.220	−1.401	+1.929	+0.1250	−0.0627	−0.0765	+0.0042	+0.0013	+17.97	−5.233	+3.575	+0.6795
10	−5.055	−3.891	−1.245	−0.0955	+4.276	+1.728	+1.762	+0.0858	+0.2474	−0.0456	+0.0471	−0.0000	+16.41	+6.050	+3.437	+0.8640
0	+25.47	0	+0.707	0	−14.77	0	+0.7071	0	+1.134	+0.3000	0	0	+56.01	0	0	+1.000
$\phi_k=50$:																
50	+6.626	0	+1.510	−0.0742	−5.740	0	−1.603	+0.0441	−0.0544	0	−0.0135	−0.0031	−15.39	0	−7.343	+0.0368
40	+3.900	+3.127	+0.564	−0.0969	−3.696	−2.769	−0.7707	+0.0676	+0.0026	+0.0201	−0.0201	−0.0029	−10.44	−7.529	−5.048	+0.1442
30	−2.878	+3.516	−1.463	−0.1393	+1.552	−3.453	+1.111	+0.1158	−0.1000	−0.0070	−0.0254	−0.0023	+3.25	−9.817	+0.304	+0.4060
20	−8.823	+0.094	−2.598	−0.1467	+6.637	−1.072	+2.491	+0.1375	−0.1008	−0.0715	−0.0025	−0.0011	+18.60	+3.576	+4.612	+0.6832
10	−4.273	−4.274	−1.262	−0.0891	+4.214	+2.483	+1.987	+0.0953	+0.2354	−0.0563	+0.0424	−0.0000	+14.04	+7.094	+3.666	+0.8700
0	+27.68	0	+0.766	0	−17.75	0	+0.6428	0	+1.193	+0.3000	0	0	+56.30	0	0	+1.000

(k) $r/l = 200$ and $r/l = 0.4$

ϕ	T_z (1)	S (2)	T_ϕ (3)	M_ϕ (4)	T_z (5)	S (6)	T_ϕ (7)	M_ϕ (8)	T_z (9)	S (10)	T_ϕ (11)	M_ϕ (12)	T_z (13)	S (14)	T_ϕ (15)	M_ϕ (16)
$\phi_k=30$:																
30	+1.124	0	−2.479	−0.2211	−2.625	0	+0.5943	+0.0868	−0.2817	0	−0.0710	−0.0026	−20.19	0	−3.173	+0.5913
20	−3.463	−0.295	−2.481	−0.1937	−0.066	−0.9430	+0.9092	+0.0810	−0.1583	−0.1339	−0.0302	−0.0018	−0.87	−7.343	−0.619	+0.6619
10	−5.528	−3.468	−1.407	−0.1117	+2.974	+0.0700	+1.194	+0.0553	+0.3622	−0.1029	+0.0493	−0.0003	+23.57	+0.435	+1.959	+0.8155
0	+27.68	0	+0.500	0	+6.899	0	+0.8660	0	+1.634	+0.4000	0	0	+56.52	0	0	+1.000

Table 1A (Continued)

ϕ	VERTICAL EDGE LOAD				HORIZONTAL EDGE LOAD				SHEAR EDGE LOAD				EDGE MOMENT LOAD			
	T_x (1)	S (2)	T_ϕ (3)	M_ϕ (4)	T_x (5)	S (6)	T_ϕ (7)	M_ϕ (8)	T_x (9)	S (10)	T_ϕ (11)	M_ϕ (12)	T_x (13)	S (14)	T_ϕ (15)	M_ϕ (16)

(k) (Continued)

ϕ	(1)	(2)	(3)	(4)	(5)	(6)	(7)	(8)	(9)	(10)	(11)	(12)	(13)	(14)	(15)	(16)
$\phi_k = 35$:																
35	+ 6.873	0	−0.795	−0.1830	− 4.932	0	−0.0417	+0.0837	−0.1462	0	−0.0635	−0.0032	− 27.50	0	− 5.935	+0.3911
30	+ 5.012	+1.712	−1.039	−0.1820	− 3.951	−1.261	+0.1374	+0.0848	−0.1482	−0.0403	−0.0577	−0.0030	− 22.24	− 7.055	− 4.913	+0.4179
20	+ 6.203	+1.614	−2.206	−0.1647	− 2.208	−1.889	+1.153	+0.0862	−0.0986	−0.1152	−0.0124	−0.0017	− 11.73	−10.85	− 1.001	+0.5937
10	− 9.087	−3.682	−1.633	−0.1001	+ 5.303	+0.6597	+1.536	+0.0621	+0.3204	−0.0822	+0.0539	−0.0001	+ 31.27	+ 3.583	+ 3.651	+0.8049
0	+34.19	0	+0.574	0	−13.36	0	+0.8192	0	+1.639	+0.4000	+0.0539	0	− 80.78	0	0	+1.000
$\phi_k = 40$:																
40	+11.34	0	+0.768	−0.1216	− 7.637	0	−0.8602	+0.0620	+0.0319	0	−0.0448	−0.0032	− 32.54	0	− 8.301	+0.2027
30	+ 3.392	+4.247	−0.465	−0.1293	− 2.992	−3.037	+0.0122	+0.0723	−0.0781	−0.0171	−0.0365	−0.0026	− 13.32	−12.80	− 4.292	+0.2888
20	−10.20	+2.361	−2.620	−0.1409	+ 5.333	−2.419	+1.731	+0.0915	−0.1367	−0.0848	−0.0086	−0.0014	+ 21.73	−10.76	+ 3.315	+0.5642
10	−10.34	−4.580	−1.955	−0.0931	+ 6.841	+1.629	+1.957	+0.0704	+0.2620	−0.0845	+0.0492	−0.0000	+ 31.74	+ 6.881	+ 4.858	+0.8083
0	+39.09	0	+0.643	0	−19.78	0	+0.7660	0	+1.744	+0.4000	0	0	−92.98	0	0	+1.000
$\phi_k = 45$:																
45	+ 9.787	0	+1.934	−0.0436	− 7.229	0	−1.692	+0.0199	+0.0957	0	−0.0099	−0.0065	− 22.45	0	− 8.938	−0.0169
40	+ 8.573	+2.572	+1.567	−0.0513	− 6.456	−1.911	−1.419	+0.0264	+0.0691	+0.0238	−0.0133	−0.0043	− 20.42	− 5.970	− 8.074	+0.0165
30	− 0.227	+5.188	−0.825	−0.0982	+ 0.709	−4.101	−0.4128	+0.0678	−0.0987	+0.0199	−0.0292	−0.0020	+ 4.08	−13.43	− 2.116	+0.2506
20	−11.70	+1.769	−3.045	−0.1296	+ 7.382	−2.215	+2.378	+0.1027	−0.1953	−0.0728	−0.0161	−0.0012	+ 23.35	+ 8.260	+ 4.819	+0.5714
10	− 9.060	−5.153	−2.000	−0.0887	+ 7.017	+2.520	+2.255	+0.0800	−0.2396	−0.0981	+0.0429	−0.0000	+ 27.40	+ 8.293	+ 5.115	+0.8168
0	+38.68	0	+0.707	0	−23.69	0	+0.7071	0	+1.826	+0.4000	0	0	−91.16	0	0	+1.000
$\phi_k = 50$:																
50	+ 7.414	0	+2.207	+0.0042	− 6.022	0	−2.055	−0.0134	+0.1395	0	+0.0147	−0.0013	− 13.14	0	− 7.852	−0.1157
40	+ 4.947	+3.616	+1.124	−0.0267	− 4.389	−3.008	−1.161	+0.0151	+0.0505	+0.0598	+0.0042	−0.0015	− 11.30	− 6.932	− 5.801	−0.0061
30	+ 2.266	+4.540	−1.356	−0.0910	+ 0.8355	−4.146	−0.9927	+0.0776	−0.1446	+0.0347	−0.0352	−0.0017	+ 1.18	−10.95	+ 0.390	−0.2725
20	−10.69	+0.849	−3.073	−0.1245	+ 7.857	−1.711	+2.811	+0.1179	−0.2195	−0.0793	−0.0240	−0.0011	+ 19.84	− 6.154	+ 5.079	−0.5892
10	− 7.142	−5.081	−1.774	−0.0835	+ 6.637	+3.045	+2.378	+0.0892	+0.2513	−0.1076	+0.0402	−0.0001	+ 23.56	+ 7.944	+ 4.801	+0.8219
0	+35.00	0	+0.766	0	−25.44	0	+0.6428	0	+1.840	+0.4000	0	0	−84.42	0	0	+1.000

(l) $r/t = 200$ and $r/l = 0.5$

ϕ	(1)	(2)	(3)	(4)	(5)	(6)	(7)	(8)	(9)	(10)	(11)	(12)	(13)	(14)	(15)	(16)
$\phi_k = 30$:																
30	+ 6.908	0	−1.578	−0.1695	− 4.932	0	+0.2325	+0.0656	−0.2715	0	−0.0759	−0.0026	− 36.86	0	− 5.858	+0.4164
20	− 3.091	+1.791	−2.323	−0.1564	− 2.320	−1.780	+0.8452	+0.0654	−0.1974	−0.1375	−0.0355	−0.0018	− 2.00	−13.49	− 1.174	+0.5258
10	−12.18	−3.626	−1.972	−0.0957	+ 5.663	+0.128	+1.422	+0.0492	+0.3600	−0.1314	+0.05366	−0.0002	+ 43.95	+ 0.71	+ 3.711	+0.7505
0	+43.35	0	+0.500	0	−13.29	0	+0.8660	0	+2.253	+0.5000	0	0	−107.5	0	0	+1.000

357

Table 1A (Continued)

ϕ	VERTICAL EDGE LOAD				HORIZONTAL EDGE LOAD				SHEAR EDGE LOAD				EDGE MOMENT LOAD			
	T_x	S	T_ϕ	M_ϕ	T_x	S	T_ϕ	M_ϕ	T_x	S	T_ϕ	M_ϕ	T_x	S	T_ϕ	M_ϕ
	(1)	(2)	(3)	(4)	(5)	(6)	(7)	(8)	(9)	(10)	(11)	(12)	(13)	(14)	(15)	(16)
$\phi_k=35$:																
35	+12.92	0	+0.520	−0.1106	−7.760	0	−0.6687	+0.0486	+0.0566	0	−0.0528	−0.0027	−41.17	0	−9.181	+0.1851
30	+10.14	+3.193	+0.037	−0.1151	−6.371	−1.952	−0.3803	+0.0518	−0.0847	−0.0181	−0.0505	−0.0025	−33.79	−10.62	−7.611	+0.2222
20	−8.763	+4.022	−2.425	−0.1266	+3.367	−3.036	+1.253	+0.0670	−0.1818	−0.0966	−0.0204	−0.0015	+16.53	−16.90	+1.429	+0.4668
10	−16.09	−4.484	−2.447	−0.0897	+8.692	+1.030	+1.926	+0.0564	+0.2700	−0.1200	+0.0516	−0.0001	+50.25	+5.26	+5.747	+0.7504
0	+52.21	0	+0.574	0	−22.16	0	+0.8192	0	+2.374	+0.5000	0	0	−132.8	0	0	+1.000
$\phi_k=40$:																
40	+13.93	0	+1.963	−0.0443	−8.898	0	−1.529	+0.0185	−0.1093	0	−0.0175	−0.0020	−32.96	0	−10.30	−0.0072
30	+5.361	+5.888	+0.146	−0.0720	−4.097	−3.897	−0.3259	+0.0398	−0.0475	+0.0298	−0.0294	−0.0019	−17.70	−15.29	−5.586	+0.1389
20	−12.79	+3.938	−3.042	−0.1129	+6.622	−3.303	+1.951	+0.0737	−0.2613	−0.0652	−0.0242	−0.0012	+24.15	−14.33	+3.905	+0.4658
10	−15.31	−5.476	−2.659	−0.0866	+9.721	+2.099	+2.350	+0.0045	+0.2136	−0.1319	+0.0433	−0.0000	+44.95	+8.13	+6.422	+0.7611
0	+53.53	0	+0.643	0	−28.17	0	+0.7660	0	+2.509	+0.5000	0	0	−133.2	0	0	+1.000
$\phi_k=45$:																
45	+10.22	0	+2.435	+0.0045	−7.196	0	−1.959	−0.0106	+0.1830	0	+0.0132	+0.0011	−18.31	0	−8.981	−0.1089
40	+9.385	+2.771	+2.035	−0.0052	−6.757	−1.967	−1.675	−0.0034	+0.1446	+0.0466	+0.0066	−0.0012	−17.78	−4.98	−8.253	−0.0771
30	+0.957	+5.808	−0.565	−0.0616	+1.640	−4.438	+0.2286	+0.0431	−0.1100	−0.0637	−0.0303	−0.0014	+8.90	−13.02	−2.879	+0.1523
20	−12.61	+2.686	−3.298	−0.1102	+7.723	−2.850	+2.505	+0.0881	−0.3155	−0.0676	−0.0341	−0.0011	+20.73	−10.69	+4.696	+0.4876
10	−12.52	−5.653	−2.457	−0.0832	+9.498	+2.791	+2.554	+0.0750	+0.2211	−0.1467	+0.0389	−0.0001	+38.31	+8.33	+6.058	+0.7681
0	+48.96	0	+0.707	0	−30.94	0	−0.7071	0	+2.539	+0.5000	0	0	−123.1	0	0	+1.000
$\phi_k=50$:																
50	+5.720	0	+2.151	+0.0279	−4.333	0	−1.931	−0.0295	+0.1739	0	+0.0295	+0.0007	−4.66	0	−6.575	−0.1404
40	+5.105	+3.070	+1.265	−0.0015	−4.431	−2.438	−1.238	−0.0037	+0.0786	+0.0777	+0.0054	−0.0007	−9.77	−3.60	−5.629	−0.0518
30	−0.025	+4.652	−1.014	−0.0656	−1.252	−4.161	−0.6846	+0.0569	−0.1578	+0.0593	−0.0387	−0.0013	+10.31	−9.96	−1.682	+0.1945
20	−10.40	+1.897	−3.112	−0.1093	+7.442	−2.639	+2.813	+0.1043	−0.2949	−0.0794	−0.0364	−0.0011	+16.18	+9.65	+4.675	+0.5157
10	−12.65	−5.039	−2.643	−0.0852	+11.50	+3.115	+3.155	+0.0913	+0.2141	−0.1221	+0.0366	−0.0000	+41.78	+9.30	+7.575	+0.8127
0	+42.10	0	+0.766	0	−31.89	0	+0.6428	0	+2.441	+0.5000	0	0	−116.1	0	0	+1.000

(l) Continued

(m) $r/l = 200$ and $r/l = 0.6$

ϕ	T_x	S	T_ϕ	M_ϕ	T_x	S	T_ϕ	M_ϕ	T_x	S	T_ϕ	M_ϕ	T_x	S	T_ϕ	M_ϕ
	(1)	(2)	(3)	(4)	(5)	(6)	(7)	(8)	(9)	(10)	(11)	(12)	(13)	(14)	(15)	(16)
$\phi_k=30$:																
30	+13.41	0	−0.50	−0.1215	−7.494	0	−0.177	+0.0459	−0.2086	0	−0.0732	−0.0024	−53.95	0	−8.721	+0.2589
20	−2.95	+4.174	−2.106	−0.1221	+0.317	−2.725	+0.756	+0.0512	−0.2355	−0.1233	−0.0394	−0.0016	−4.01	−20.04	−1.828	+0.4042
10	−19.88	−3.751	−2.624	−0.0867	+8.756	+0.171	+1.682	+0.0443	+0.3019	−0.1613	+0.0532	−0.0001	+66.27	+0.71	+5.559	+0.6931
0	+61.86	0	+0.500	0	−20.84	0	+0.866	0	+3.003	+0.6000	0	0	−166.4	0	0	+1.000

Table 1A (Continued)

ϕ	VERTICAL EDGE LOAD				HORIZONTAL EDGE LOAD				SHEAR EDGE LOAD				EDGE MOMENT LOAD			
	T_x	S	T_ϕ	M_ϕ	T_x	S	T_ϕ	M_ϕ	T_x	S	T_ϕ	M_ϕ	T_x	S	T_ϕ	M_ϕ
	(1)	(2)	(3)	(4)	(5)	(6)	(7)	(8)	(9)	(10)	(11)	(12)	(13)	(14)	(15)	(16)

(m) (Continued)

ϕ	T_x (1)	S (2)	T_ϕ (3)	M_ϕ (4)	T_x (5)	S (6)	T_ϕ (7)	M_ϕ (8)	T_x (9)	S (10)	T_ϕ (11)	M_ϕ (12)	T_x (13)	S (14)	T_ϕ (15)	M_ϕ (16)
$\phi_k=35$:																
35	+17.48	0	+1.581	-0.0576	-9.794	0	-1.155	+0.0231	+0.0614	0	-0.0363	-0.0020	-48.68	0	-11.31	-0.0433
30	+13.76	+4.442	+0.942	-0.0648	-8.004	-2.520	-0.794	+0.0279	+0.0019	+0.0113	-0.0381	-0.0020	-40.77	-12.62	-9.488	-0.0867
20	-10.37	+6.000	-2.570	-0.0989	+4.010	-3.973	+1.313	+0.0531	-0.2759	-0.0663	-0.0294	-0.0012	+17.11	-21.22	+1.522	+0.3710
10	-22.33	-5.090	-3.136	-0.0827	+11.70	+1.270	+2.252	+0.0523	-0.1839	-0.1613	+0.0454	-0.0000	+66.22	+5.83	+7.338	+0.7029
0	+68.85	0	+0.574	0	-30.32		+0.819	0	+3.204	+0.6000			-180.9	0	0	+1.000
$\phi_k=40$:																
40	+14.46	0	+2.530	-0.0052	-8.900	0	-1.788	-0.0029	-0.2096	0	+0.0030	-0.0012	-29.22	0	-10.53	-0.0850
30	+6.70	+6.504	+0.519	-0.0397	-4.840	-4.143	-0.527	+0.0213	-0.0157	+0.0719	-0.0224	-0.0013	-20.21	-14.50	-6.191	+0.0578
20	-13.24	+4.992	-3.187	-0.0946	+6.608	-3.889	+2.003	+0.0630	-0.3733	-0.0458	-0.0384	-0.0010	+19.37	-16.37	+3.594	+0.3909
10	-19.32	-5.810	-3.114	-0.0812	+12.10	+2.231	+2.601	+0.0616	-0.1569	-0.1797	+0.0373	-0.0001	+56.09	+7.56	+7.285	+0.7139
0	+65.23	0	+0.643	0	-35.15		+0.766	0	+3.297	+0.6000			-169.2	0	0	+1.000
$q\ 0$																
$\phi_k=45$:																
45	+8.60	0	+2.397	+0.0217	-5.729	0	-1.858	-0.0205	-0.2347	0	+0.0286	-0.0004	+8.38	0	-7.554	-0.1241
40	+8.15	+2.319	+2.068	+0.0133	-5.622	-1.562	-1.636	-0.0140	+0.1924	+0.0605	+0.0199	-0.0005	-10.38	-2.49	-7.198	-0.0988
30	+2.68	+5.666	-0.311	-0.0418	-2.980	-4.165	+0.041	+0.0293	-0.1065	-0.0944	-0.0306	-0.0011	-15.36	-10.11	-3.686	-0.0939
20	-11.68	+3.500	-3.237	-0.0948	+6.764	-3.438	+2.416	+0.0759	-0.4030	-0.0896	-0.0476	-0.0001	+12.42	-13.03	+3.667	+0.4113
10	-16.02	-5.641	-2.752	-0.0772	+12.12	+2.709	+2.752	+0.0695	-0.1906	-0.1895	+0.0361		+50.54	+6.80	+6.668	+0.7160
0	+58.53	0	+0.707	0	-38.04		+0.707	0	+3.268	+0.6000			-158.3	0	0	+1.000
$\phi_k=50$:																
50	+3.09	0	+1.709	+0.0284	-1.816	0	-1.467	-0.0273	+0.1854	0	+0.0382	+0.0002	+7.18	0	-3.972	-0.1088
40	+4.42	+1.981	+1.179	+0.0063	-3.693	-1.374	-1.120	-0.0084	+0.1051	+0.0872	+0.0118	-0.0003	+5.77	+1.40	-4.634	-0.0572
30	+2.57	+4.287	-0.638	-0.0476	-3.650	-3.722	+0.325	+0.0406	-0.1481	+0.0826	-0.0419	-0.0012	+20.51	+6.83	-3.347	+0.1206
20	-9.53	+2.775	-2.920	-0.0908	+6.372	-3.425	+2.581	+0.0862	-0.3819	-0.0757	-0.0497	-0.0001	+7.78	-12.83	+3.041	+0.4149
10	-14.34	-5.142	-2.383	-0.0706	+13.11	+2.955	+2.877	+0.0757	+0.2199	-0.1883	+0.0384		+51.00	+5.85	+6.371	+0.7134
0	+53.46	0	+0.766	0	-41.80		+0.643	0	+3.218	+0.6000			-156.6	0	0	+1.000

Table 1B SYMMETRICAL EDGE LOADS ON SIMPLY SUPPORTED CYLINDRICAL SHELLS (Displacement of edge at $\phi = 0$)

Vertical Displacement ΔV—(Downward Positive)—

Horizontal Displacement ΔH—(Inward Positive)—

Rotation θ—

VERTICAL EDGE LOAD

ΔV: $V_L \dfrac{l^4}{r^4 t E} \times$ Col. (1) $\times \sin\dfrac{\pi x}{l}$

ΔH: $V_L \dfrac{l^4}{r^4 t E} \times$ Col. (2) $\times \sin\dfrac{\pi x}{l}$

Rotation θ: $V_L \dfrac{r^2}{EI} \times$ Col. (3) $\times \sin\dfrac{\pi x}{l}$

HORIZONTAL EDGE LOAD

ΔV: $H_L \dfrac{l^4}{r^4 t E} \times$ Col. (4) $\times \sin\dfrac{\pi x}{l}$

ΔH: $H_L \dfrac{l^4}{r^4 t E} \times$ Col. (5) $\times \sin\dfrac{\pi x}{l}$

Rotation θ: $H_L \dfrac{r^2}{EI} \times$ Col. (6) $\times \sin\dfrac{\pi x}{l}$

SHEAR EDGE LOAD

ΔV: $S_L \dfrac{l^4}{r^3 t E} \times$ Col. (7) $\times \sin\dfrac{\pi x}{l}$

ΔH: $S_L \dfrac{l^4}{r^3 t E} \times$ Col. (8) $\times \sin\dfrac{\pi x}{l}$

Rotation θ: $S_L \dfrac{r^2}{EI} \times$ Col. (9) $\times \sin\dfrac{\pi x}{l}$

MOMENT EDGE LOAD

ΔV: $M_L \dfrac{l^4}{r^4 t E} \times$ Col. (10) $\times \sin\dfrac{\pi x}{l}$

ΔH: $M_L \dfrac{l^4}{r^4 t E} \times$ Col. (11) $\times \sin\dfrac{\pi x}{l}$

Rotation θ: $M_L \dfrac{r}{EI} \times$ Col. (12) $\times \sin\dfrac{\pi x}{l}$

(a) $r/l = 100$

ϕ_k	ΔV (1)	ΔH (2)	θ (3)	ΔV (4)	ΔH (5)	θ (6)	ΔV (7)	ΔH (8)	θ (9)	ΔV (10)	ΔH (11)	θ (12)
$r/l = 0.1$:												
30	12.53	0.1445	0.1108	−0.1445	−0.0585	−0.04556	0.3401	0.000480	0.000363	−1.330	−0.5469	−0.5144
35	6.265	0.2508	0.1420	−0.2508	−0.1231	−0.07108	0.2178	0.001006	0.000544	−1.704	−0.8531	−0.5964
40	3.611	0.4009	0.1752	−0.4009	−0.2325	−0.1039	0.1443	0.002009	0.000822	−2.102	−1.247	−0.6749
45	2.591	0.5987	0.2088	−0.5981	−0.4040	−0.1444	0.1084	0.003771	0.001223	−2.505	−1.734	−0.7542
50	2.043	0.8434	0.2410	−0.8434	−0.6558	−0.1925	0.0792	0.006516	0.001776	−2.892	−2.312	−0.8290
$r/l = 0.2$:												
30	17.38	2.031	0.0982	−2.031	−0.8742	−0.04288	0.6950	0.00726	0.000353	−18.86	−8.233	−0.4893
35	12.87	3.538	0.1259	−3.538	−1.794	−0.06530	0.4733	0.02065	0.000678	−24.18	−12.54	−0.5564
40	12.51	5.544	0.1524	−5.544	−3.282	−0.09256	0.3715	0.04590	0.001145	−29.25	−17.77	−0.6159
45	13.47	7.956	0.1748	−7.956	−5.447	−0.1233	0.3402	0.08862	0.001757	−33.56	−23.67	−0.6651
50	14.54	10.50	0.1899	−10.50	−8.270	−0.1546	0.3610	0.1526	0.002478	−36.46	−29.70	−0.6986
$r/l = 0.3$:												
30	35.24	9.474	0.0891	−9.474	−4.010	−0.03866	1.187	0.06357	0.000536	−86.64	−37.58	−0.4499
35	37.31	15.65	0.1092	−15.65	−7.890	−0.05606	0.9583	0.1474	0.000923	−106.2	−54.50	−0.4965
40	42.38	22.99	0.1245	−22.99	−13.59	−0.07595	0.9561	0.2931	0.001413	−121.0	−73.83	−0.5288
45	46.06	30.12	0.1317	−30.12	−20.69	−0.09384	1.073	0.5020	0.001927	−127.9	−91.21	−0.5434
50	45.75	35.11	0.1285	−35.11	−27.93	−0.1070	1.209	0.7422	0.002334	−124.9	−104.0	−0.5402
$r/l = 0.4$:												
30	75.06	26.10	0.07760	−26.10	−11.06	−0.03483	1.940	0.2256	0.000632	−239.2	−104.0	−0.4049
35	87.57	40.75	0.09046	−40.75	−20.25	−0.04716	1.924	0.5080	0.000994	−277.7	−145.2	−0.4320
40	96.68	55.00	0.09596	−55.00	−32.68	−0.05921	2.189	0.9067	0.001384	−293.8	−181.4	−0.4417
45	96.10	64.69	0.09256	−64.69	−44.95	−0.06746	2.500	1.372	0.001641	−284.1	−207.1	−0.4348
50	85.15	67.32	0.08360	−67.32	−54.74	−0.07216	2.591	1.731	0.001732	−255.7	−220.8	−0.4228

r/l												
r/l = 0.5:												
30	142.2	54.14	0.06634	− 54.14	− 22.98	−0.02900	3.218	0.5995	0.000644	−497.6	−216.6	−0.3602
35	163.2	78.77	0.07239	− 78.77	− 39.97	−0.03811	3.588	1.202	0.000945	−542.9	−285.9	−0.3706
40	167.7	97.43	0.07124	− 97.43	− 58.47	−0.04484	4.141	1.938	0.001160	−535.2	−336.5	−0.3668
45	152.4	104.9	0.06505	−104.9	− 74.31	−0.04883	4.418	2.566	0.001242	−487.9	−366.2	−0.3568
50	127.5	103.9	0.05747	−103.9	− 87.21	−0.05166	4.246	2.945	0.001199	−431.0	−387.5	−0.3493
r/l = 0.6:												
30	238.5	88.47	0.05559	− 88.47	− 39.78	−0.02434	5.112	1.218	0.000620	−864.6	−378.6	−0.3247
35	257.7	126.7	0.05714	−126.7	− 64.74	−0.03049	5.976	2.242	0.000834	−888.7	−474.2	−0.3185
40	245.4	144.9	0.05366	−144.9	− 88.10	−0.03396	6.629	3.250	0.000932	−834.6	−528.2	−0.3171
45	212.3	149.5	0.04783	−149.5	−108.3	−0.03700	6.601	3.942	0.000918	−743.9	−575.4	−0.3042
50	175.8	143.7	0.04304	−147.9	−127.9	−0.03967	6.678	4.324	0.000865	−669.4	−618.7	−0.3086

(b) r/l = 200

r/l												
r/l = 0.1:												
30	13.59	0.5598	0.10563	− 0.5598	− 0.3112	−0.04555	0.3446	0.001423	0.000252	− 5.070	2.186	−0.5143
35	7.732	0.9838	0.13813	− 0.9838	− 0.4916	−0.07100	0.2228	0.003352	0.000438	− 6.630	3.408	−0.5958
40	5.711	1.576	0.1719	− 1.576	− 0.9269	−0.1036	0.1579	0.007139	0.000718	− 8.250	4.974	−0.6749
45	5.186	2.354	0.2050	− 2.354	− 1.462	−0.1434	0.1237	0.01391	0.001117	− 9.842	6.883	−0.7503
50	5.265	3.292	0.2352	− 3.292	− 2.576	−0.1895	0.1083	0.02506	0.001649	− 11.29	9.097	−0.8195
r/l = 0.2:												
30	32.02	8.267	0.0986	− 8.267	− 3.492	−0.04257	0.7716	0.03561	0.000385	− 75.70	32.70	−0.4866
35	33.96	14.07	0.1245	− 14.07	− 7.090	−0.06423	0.6124	0.08739	0.000700	− 95.68	49.33	−0.5497
40	39.51	21.38	0.1468	− 21.38	− 12.62	−0.08898	0.6071	0.1815	0.001123	−112.7	68.34	−0.5990
45	44.27	28.96	0.1600	− 28.96	− 19.84	−0.1132	0.6944	0.3237	0.001601	−122.9	86.91	−0.6273
50	45.16	34.63	0.1600	− 34.63	− 27.44	−0.1321	0.8010	0.4950	0.002014	−122.8	101.0	−0.6299
r/l = 0.3:												
30	110.9	37.12	0.08756	− 37.12	− 15.70	−0.03796	1.641	0.2500	0.000532	−340.4	147.6	−0.4443
35	123.0	58.70	0.10342	− 58.70	− 29.63	−0.05150	1.782	0.5520	0.000868	−402.2	200.2	−0.4792
40	136.4	78.62	0.10899	− 78.62	− 46.73	−0.06707	2.154	0.9824	0.001193	−423.7	260.8	−0.4884
45	131.5	89.18	0.10274	− 89.18	− 62.16	−0.07481	2.432	1.410	0.001376	−399.4	290.9	−0.4749
50	111.5	89.27	0.09013	− 89.27	− 73.18	−0.07807	2.437	1.695	0.001382	−350.4	303.5	−0.4550
r/l = 0.4:												
30	247.6	98.82	0.07423	− 98.82	− 41.89	−0.03237	3.525	0.9471	0.000586	−912.9	397.8	−0.3936
35	283.3	139.7	0.07945	−139.7	− 70.98	−0.04180	4.354	1.701	0.000837	−976.3	513.6	−0.4011
40	273.8	161.7	0.07471	−161.7	− 97.64	−0.04717	4.977	2.518	0.000964	−918.0	579.7	−0.3884
45	232.0	163.1	0.06551	−163.1	−117.2	−0.04975	4.925	3.016	0.000945	−805.0	611.3	−0.3732
50	187.9	157.7	0.05768	−157.7	−135.5	−0.05270	4.456	3.239	0.000875	−708.8	647.5	−0.3670
r/l = 0.5:												
30	473.4	194.4	0.06058	−194.4	− 82.77	−0.02666	6.899	2.116	0.000570	−1817.	−800.0	−0.3421
35	486.4	244.3	0.05905	−244.3	−125.5	−0.03167	8.309	3.527	0.000707	−1772.	−950.2	−0.3343
40	424.2	256.6	0.05240	−256.6	−158.6	−0.03414	8.519	4.483	0.000653	−1572.	−1024.	−0.3210
45	349.1	254.2	0.04653	−254.2	−189.1	−0.03671	7.792	4.924	0.000616	−1396.	−1101.	−0.3156
50	295.1	260.4	0.04398	−260.4	−234.5	−0.04173	6.701	5.075		−1319.	−1252.	−0.3226
r/l = 0.6:												
30	756.0	315.3	0.04838	−315.3	−135.1	−0.02155	11.82	3.979	0.000510	−3009.	−1341.	−0.2961
35	706.1	360.6	0.04430	−360.6	−188.1	−0.02429	13.15	5.790	0.000555	−2756.	−1511.	−0.2847
40	591.7	367.3	0.03930	−367.3	−232.8	−0.02637	12.46	6.714	0.000519	−2445.	−1640.	−0.2778
45	498.8	374.8	0.03615	−374.8	−287.2	−0.02931	11.18	7.265	0.000486	−2249.	−1823	−0.2782
50	431.9	389.3	0.03392	−389.3	−357.3	−0.03284	10.21	7.983	0.000481	−2110.	−2043.	−0.2807

APPENDIX DEFINITIONS

These definitions are intended to clarify the concepts basic to the standard analysis of thin shell concrete structures, especially as detailed in Chapters 2, 3, 4, and 5. A full presentation of a more general theory appeared in Chapter 1 of the first edition of this book.

A *thin shell* is a curved slab whose thickness h is relatively small compared with its other dimensions and especially compared with its principal radii of curvature, r_x and r_y ($h/r \ll 1$), where α_x and α_y are coordinates along lines of principal curvature such as θ and ϕ for domes (Fig. 3-1, page 107).

The surface that bisects the shell thickness is called the *middle surface*, and by specifying the form of this surface and the thickness at every point, we completely define the geometry of the shell.

Stress resultants, N_x, N_y, N_{xy}, Q_x, Q_y, and *stress couples*, M_x, M_y, M_{xy}, defined as the total forces and moments respectively acting per unit length of middle surface, are the integrals of stress over the shell thickness, such as Eqs. (1-1) and (1-2), page 3. Because h/r can be neglected when compared to unity, $N_{xy} = N_{yx}$ and $M_{xy} = M_{yx}$.

The *displacements* in a thin shell are defined by the *translations, u, v, w,* of every point on the middle surface in three directions, and by the *rotations* of each of two tangents to the middle surface, ϕ_x and ϕ_y. These rotations are in turn expressed by the translations as in (*d*) and (*e*) on page 193.

The *middle surface strains* are the changes in unit lengths, ϵ_{x0} and ϵ_{y0} (axial or *extensional strain*) of the two coordinate lines, and the change in angle between them, γ_{xy0} (angular or *shearing strain*).

The *changes of the initial curvature* are defined by changes in the rotations, χ_x, χ_y, and χ_{xy}, with the last called the change in twist of the surface.

Thin shell theories are formulated by equations of statics (equilibrium), geometry (strain-displacement), and material (stress-strain).

Equilibrium equations connect the shell loads to the stress resultants and stress couples as given by Eqs. (3-29) on page 127, for example.

The *membrane theory* is defined by those equilibrium equations which result from neglecting all stress couples, such as in Eqs. (3-1) on page 108. The resulting equilibrium equations are statically determinate and the stress resultants, N'_x, N'_y, and N'_{xy}, depend only upon the loads and the initial shell geometry (primed values indicate membrane theory).

Strain-displacement equations connect the middle surface strains and the changes of initial curvature to the displacements under load. Examples of these appear in Eqs. (5-8) and (5-9) on page 193 and (a), (b), and (c) on page 197.

Stress-strain equations connect the statics to the geometry through the material properties. The equations used throughout this text are found presented on page 71 and are based upon the assumptions that reinforced concrete is linearly elastic, isotropic, and homogeneous. These assumptions define the *modulus of elasticity E* and *Poisson's Ratio ν* as constants: with level of stress (linear elasticity), in every direction on the surface (isotropy), and at every point within the shell (homogeneity).

The stress-strain equations lead to *force-strain equations* such as

$$N_x = K(\epsilon_{x0} + \nu\epsilon_{y0})$$
$$N_{xy} = Gh\gamma_{xy0} = \tfrac{1}{2}K(1 - \nu)\gamma_{xy0}$$
$$M_x = -D(\chi_x + \nu\chi_y)$$
$$M_{xy} = \tfrac{1}{6}Gh^3\chi_{xy} = D(1 - \nu)\chi_{xy}$$

in which the *extensional rigidity* $K = Eh/(1 - \nu^2)$, the *modulus of rigidity* (shear modulus) $G = E/2(1 + \nu)$, and the *flexural rigidity* $D = Eh^3/12(1 - \nu^2)$. The K and D correspond to EA and EI for the analysis of columns and beams.

A *thin shell theory* is expressed by a number of differential equations equal to the number of unknowns, usually the stress resultants, stress couples, and displacements. A variety of theories have been developed for thin shells; they are distinguished by various simplifications ranging from the *membrane theory* (the simplest) through the *shallow-shell theory* to a general theory as found in the first chapter of the first edition of this book.

The *shallow-shell theory* for roofs (derived on pages 195−201) is based upon two assumptions: first, that the shell is so flat that the out-of-plane translations w are much greater than the in-plane ones, u and v; and, second, that the shell is steep enough so that the in-plane stress resultants, N_x, N_y, and N_{xy}, carry far more of the load than the out-of-plane ones, Q_x and Q_y.

INDEX

ABOUT THE AUTHOR

David P. Billington, Professor of Civil Engineering, Princeton University, has worked on the design of aircraft hangars, piers, thin shell tanks, and missile-launch facilities. He has been a Visiting Professor at the Technical University of Delft, The Netherlands, and a Visitor in the School of Historical Studies at Princeton's Institute for Advanced Study. He has written many articles for professional journals as well as a book, *Robert Maillart's Bridges: The Art of Engineering,* which won the 1979 Dexter Prize of the Society for the History of Technology.

Prof. Billington is a member of numerous professional associations, including the American Concrete Institute, Prestressed Concrete Institute, American Society of Civil Engineering, American Society for Engineering Education, and American Association for the Advancement of the Humanities. From 1973 to 1979 he served as chairman of the ACI-ASCE Joint Committee on Concrete Shell Design and Construction.